P9-DZP-039

# Frozen Ground Engineering

**Prentice-Hall International Series in Civil Engineering and Engineering Mechanics**

*William J. Hall, editor*

# Frozen Ground Engineering

**ARVIND PHUKAN**
*University of Alaska, Anchorage*

PRENTICE-HALL, INC., Englewood Cliffs, New Jersey 07632

*Library of Congress Cataloging in Publication Data*

Phukan, Arvind.
Frozen ground engineering.

Bibliography: p.
Includes index.
1. Engineering—Cold weather conditions. 2. Building—Cold weather conditions. 3. Frozen ground. 4. Soil mechanics. I. Title.
TA153.P48 1985 624.1'5'0911 84–6927
ISBN 0–13–330705–0

*Editorial/production supervision and interior design: Linda Mihatov*
*Cover design: Diane Saxe*
*Manufacturing buyer: Anthony Caruso*
*Art production: Peter J. Ticola, Jr.*



Printed in the United States of America

10 9 8 7 6 5 4 3 2 1

ISBN 0-13-330705-0 01

Prentice-Hall International, Inc., *London*
Prentice-Hall of Australia Pty. Limited, *Sydney*
Editoria Prentice-Hall do Brasil, Ltda., *Rio de Janeiro*
Prentice-Hall Canada Inc., *Toronto*
Prentice-Hall of India Private Limited, *New Delhi*
Prentice-Hall of Japan, Inc., *Tokyo*
Prentice-Hall of Southeast Asia Pte. Ltd., *Singapore*
Whitehall Books Limited, *Wellington, New Zealand*

*To my wife Dipali,*
*and to my mother and father, Hema and Hari*

# Contents

## Appendices

# Preface

Engineering activity in the polar regions of the world has increased significantly in recent years, due mainly to the search for and development of energy, petroleum, and mineral resources as well as the desire to improve the standard of living of northern residents. The need to make available the most recent technology in frozen ground engineering has been felt by both students and practicing engineers, especially construction engineers. The purpose of this text is to present the state of the art in frozen ground engineering suitable for instruction at the university graduate level and for those interested in specializing in arctic engineering or cold regions engineering. In addition, many engineering design and construction problems encountered in permafrost areas are described, with typical solutions of foundation problems. For easy references, illustrations and tables are included to provide the background information needed in design and construction related to foundation engineering in frozen ground. Also, an effort has been made to present derivations of important relations.

The emphasis throughout is on a practical knowledge of frozen ground engineering. Because of the many fully worked example problems, the book is almost "self-teaching." For easy reference and comprehension, many terms used in permafrost engineering are defined in Appendix A. Both SI and U.S. customary/English units are used in the text. Appendix B contains useful conversion factors for SI units. Chapter 1 defines frozen ground and introduces characteristics of frozen ground features, temperature profile, and engineering problems. Distribution of permafrost and its relationship to environmental factors are also re viewed. Chapters 2 and 3 describe the frozen-soil classification system as well as the physical, mechanical, and thermal properties of frozen soils. These two chapters serve as a background for the next three chapters (Chapters 4, 5, and 6), which deal with foundation design philosophy and with shallow foundations and pile foundations in frozen ground. Roadway and airport design and construction techniques are discussed in Chapter 7. Design and foundation problems related to utility systems in permafrost areas are presented in Chapter 8. Slope stability in frozen ground is the topic of Chapter 9. This chapter involves the classification of soil mass measurements in permafrost and slope analysis in frozen and thawing ground. Chapter 10 discusses the field investigation of subsoil conditions in frozen ground, including laboratory testing of frozen-soil samples, ground-temperature measurements, and geophysical methods for the detection of permafrost areas.

The author's introduction of the course "Frozen Ground Engineering" at the University of Alaska in 1976 generated the writing of this textbook. The book is prepared based on the author's lectures on the subject and is supplemented by 12 years of consulting experience in both the United States and Canada.

The author gratefully acknowledges Professor N. R. Morgenstern and Mr. B. Sharky for review of the manuscript and many constructive suggestions. I thank the School of Engineering, University of Alaska, Anchorage, for support, and sincerely acknowledge my engineering colleagues for suggestions and encouragements, especially Dr. R. Miller for his review of Chapter 8. Sincere thanks and appreciation must be recorded for the many hours of effort given by Tanya Dzubay in drafting the illustrations. Our faithful secretaries, Mrs. Carol Burd, Miss Wyndra Dixson, and Mrs. Renee Wilcox, deserve particular thanks for typing and correcting the several drafts and the final manuscript for the book. Last, but not least, the author wishes to acknowledge with gratitude the interest and encouragement of his wife, Dipali Phukan, who provided the support that is necessary for an effort of this type.

ARVIND PHUKAN
*Anchorage, Alaska*

# Frozen Ground Engineering

CHAPTER 1

# Introduction

## 1.1 DEFINITION OF FROZEN GROUND

A ground consisting of earth materials such as soil and rock is said to be in a *frozen* condition when the temperature of the ground is below the freezing temperature 0°C (32°F). Depending on the impurities present, the surface tension, and the confining pressure, the water in the earth materials may not freeze completely at 0°C. In some cases, dry frozen ground may also be encountered, even when the earth material does not have any moisture in the soil particle voids.

Frozen ground may remain in a frozen condition throughout the year or seasonally. Ground that is frozen all year is called *Permafrost*. Permafrost has been recognized as a phenomenon encountered typically in the polar regions of the world. "Perennially frozen ground" and "permafrost" are both used to describe the thermal condition of soil or rock when the temperature of the ground remains below 0°C or 32°F continuously for more than 2 years. Seasonally frozen ground undergoes a cycle of frozen and unfrozen conditions whereby the ground may thaw to a certain depth during the warmer months and then refreeze during the colder months. In this textbook the emphasis is on perennially frozen ground and the term "permafrost" refers to the perennial condition only.

Frozen ground is the product of many factors, including climate, relief, vegetation, hydrology, snow cover, soil and rock type, and the growth and regime of glaciers and ice caps. Transient factors such as fire, which are discussed later in the chapter, may contribute to the degradation of frozen ground. Basically, it may be said that the formation of frozen ground is the result of a complex transfer or exchange of heat between the ground and the atmosphere. A frozen-ground condition may develop when the amount of heat the ground absorbs by radiation, conduction, and convection is less than the amount of heat dissipated.

## 1.2 THE SCOPE OF THIS BOOK

Rather than attempt to cover all aspects of arctic engineering, the primary emphasis in this text is on frozen-ground engineering. Major subjects to be introduced include (1) the engineering behavior of frozen soil materials subjected to loads and environmental conditions; (2) design concepts and analyses of foundations for different structures; (3) design considerations and problems associated with roads, airports, and utility systems; (4) slope stability; and (5) field investigation techniques in frozen ground.

The text is designed for graduate students as well as practicing engineers. The approach taken is to emphasize fundamentals with an eye toward their practical application. It is hoped that the book will help graduate engineering students, researchers, and practicing engineers, especially construction engineers, to gain an appreciation for and a working knowledge of the field of frozen-ground engineering.

In this introductory chapter, the basic characteristics of frozen ground are presented first. Following the basic discussion, systems for classifying frozen soils are presented. Just as is true of unfrozen soils, the classification of frozen soils is fundamental to understanding soil properties. Frozen-soil classification is very important because it is the "language" engineers use to communicate certain knowledge about the engineering behavior of frozen ground at particular sites. The discussion of classification systems is followed by a review of the physical, mechanical, and thermal engineering properties of frozen soils. An understanding of these properties is necessary for the design of foundations and other structures in frozen ground. The remaining chapters of the book present the principles of foundation design in frozen soils that can be applied to the design of various structures, such as roads, airports, and utilities. In addition, slope-stability problems and field investigation techniques in frozen ground are covered.

Some of the related topics also covered include how ice and negative temperatures affect the behavior of frozen soils, the long-term strength theory of frozen soil (which is important for the design of foundations), and frost heave and thaw settlement (which are important engineering phenomena that need to be understood when working in arctic and subarctic regions).

## 1.3 PERMAFROST DISTRIBUTION AND ITS RELATIONSHIP TO ENVIRONMENTAL FACTORS

The distribution of permafrost in the world has been reported by many authors (Black, 1954; Baranov and Kudryavtsev, 1966; Péwé, 1966; Stearns, 1966; Corte, 1969; Brown, 1970; Brown and Péwé, 1973; Research Institute of Glaciology,

1975). About 20% of the world, including approximately 85% of Alaska, 50% of Canada, 48% of the USSR, and 20% of China is underlaid by permafrost (Fig. 1–1). The permafrost region of the world is divided into two zones: a continuous

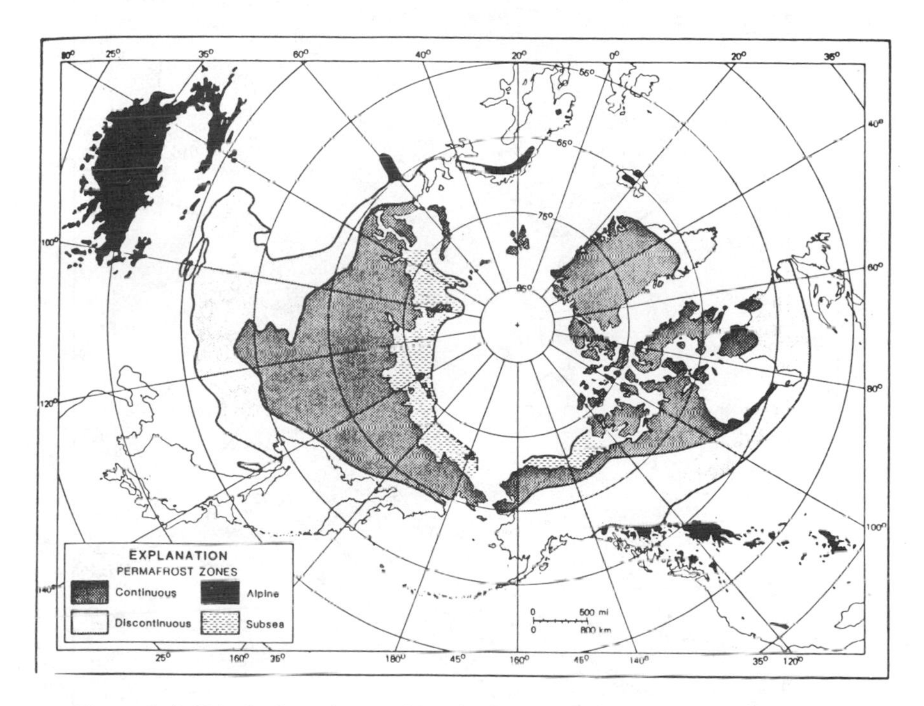


**Figure 1–1** Distribution of permafrost in the northern hemisphere. (After Péwé, 1982.)

zone in the north, followed by a discontinuous zone in the south. In the *continuous zone,* permafrost exists everywhere beneath the land surface except in newly deposited, unconsolidated sediments where the climate has just begun to impose its influence on the ground surface. In the *discontinuous zone,* some areas have permafrost beneath the land surfaces and other areas are free of permafrost. Distributions of permafrost zones in Alaska and Canada are presented in Figs. 1–2 and 1–3, respectively.

Bird (1967), Brown (1970), and Ives (1973) have published extensive information on permafrost in relation to climate. On the basis of various studies, it became known that the thickness of permafrost is about 60 to 90 m at the southern limit of the continuous zone, possibly increasing steadily to 1000 m in the northern part of the zone in Alaska and in the Canadian Arctic Archipelago. The temperature of the permafrost in this zone at a depth of zero or negligible temperature fluctuation (called the *level of zero annual amplitude*) ranges from −5°C (23°F) in the south to about −15°C (5°F) in the extreme north. The temperature of the

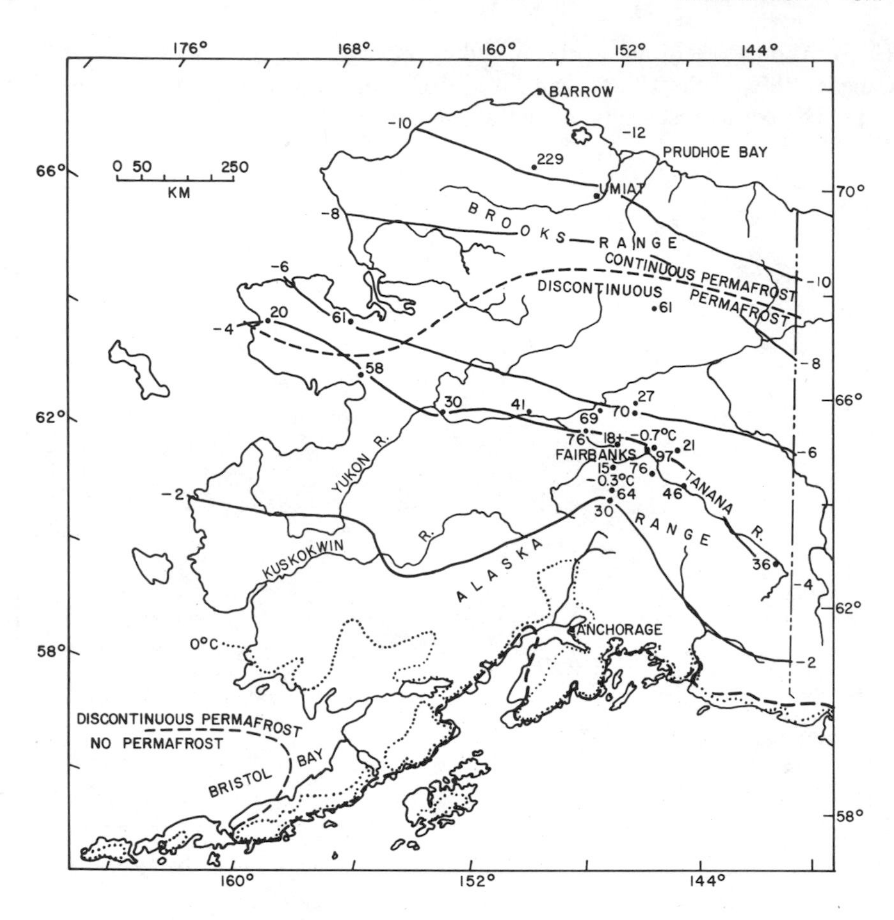


**Figure 1–2** Permafrost in Alaska. (After Brown and Péwé, 1973.)

discontinuous zone at the level of zero annual amplitude ranges from a few tenths of a degree below 0°C (32°F) at the southern limit to −5°C (23°F) at the boundary of the continuous zone (Howitt, 1971; Johnston and Brown, 1970; Williams, 1966).

Thus the permafrost area in the northern hemisphere covers considerable portions of the North American and Eurasian continents, the large islands of Greenland and Iceland, and the islands in the Arctic Ocean. Outside this area,

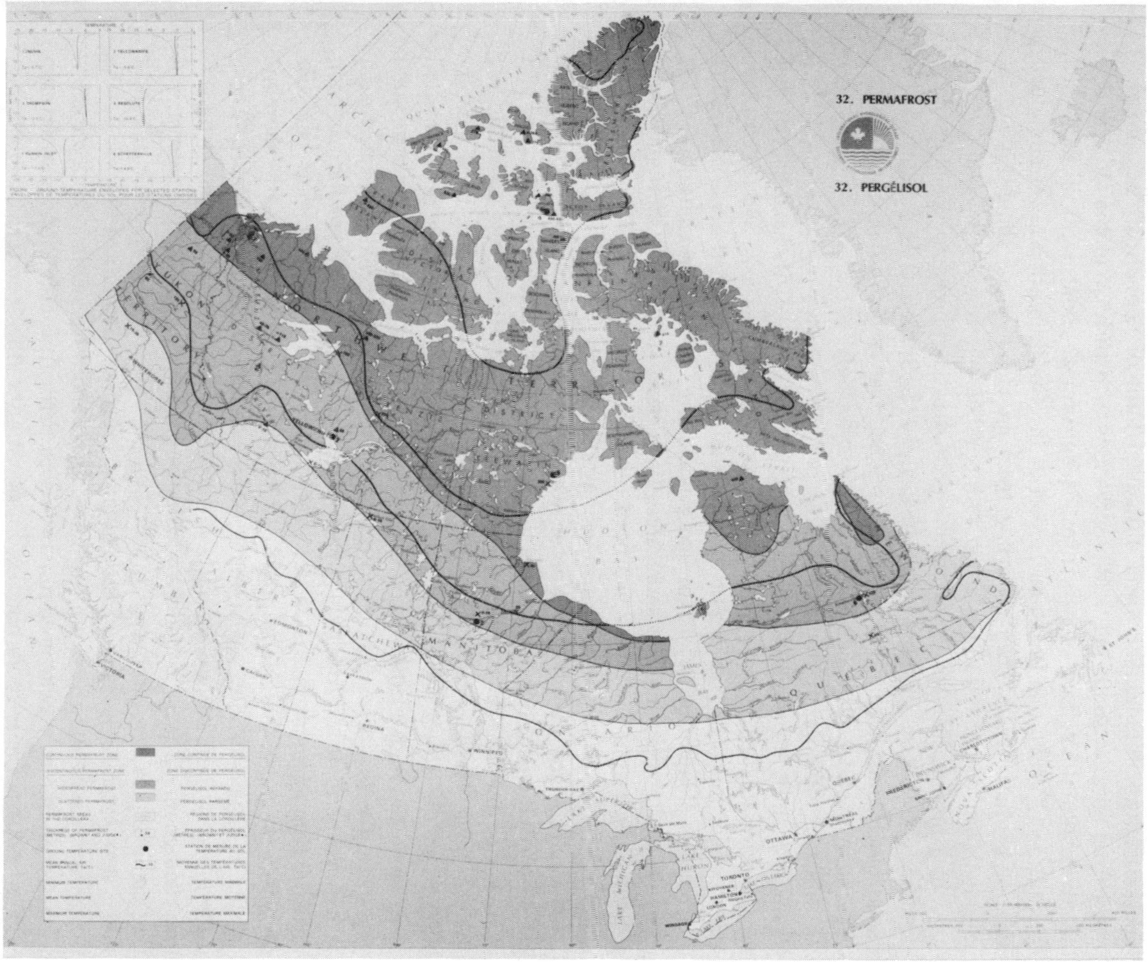

**Figure 1–3** Permafrost regions of Canada. (After National Research Council of Canada, 1975.)

permafrost also occurs in high mountain systems such as the Canadian and American Rockies. The characteristics of midlatitude and high-altitude (alpine) permafrost have been reported by Ives (1973).

In the Cordillera of Canada, the distribution of permafrost varies with latitude as well as altitude. The lower altitudinal limit of permafrost rises progressively from north to south and field observations in British Columbia indicate that the lower altitudinal limit of permafrost is uniform at about 1200 m above sea level (Brown, 1967b). In the southern part of the Cordillera, the lower altitudinal limit of permafrost has been estimated from an elevation of about 1200 m at latitude 54°30′ to about 2100 m at the 49th parallel.

In a totally different environment, permafrost is known to occur in many northern arctic coasts of the world (Mackay, 1972; Shearer, 1972; Judge, 1974; Lachenbruch and Marshall, 1969; Osterkamp and Harrison, 1976a). The thickness of permafrost under subsea conditions may vary widely.

Recent offshore probing and drilling projects reported by Lewellen (1973), Osterkamp and Harrison (1976b), and Sellmann et al. (1976), have proven that subsea permafrost exists near Barrow and Prudhoe Bay in Alaska and suggest that it exists over much of Alaska's continental shelf in the Beaufort Sea. The results of Canadian (Hunter et al., 1976) and Soviet (Are, 1973; Molochushkin, 1973) investigators in their arctic offshore areas have been similar. Offshore or subsea permafrost left after the ocean had been raised and submerged land areas, or it may have formed under subzero temperature conditions.

Muller (1966) reported that ocean water levels have fluctuated by 100 m or more on a geological time scale. This fluctuation has caused alternate flooding and exposing of the Beaufort Sea shelf and much of the Bering and Chukchi Sea shelves along Alaska's coast (Hopkins, 1967). These shelves were most recently exposed to the 100-m isobath about 18,000 years ago, allowing permafrost to form in them. With increasing ocean levels since then, and until 3000 to 4000 years ago, the shorelines continued to recede by thermal and hydrological erosion of ice-rich permafrost. Along the Beaufort Sea coast, shoreline erosion has been rapid, averaging more than 1 m per year, with rates of 10 m or more per year observed (Hopkins and Hartz, 1978). Along the Chukchi Sea coast, shoreline erosion is generally slow, with many areas considered relatively stable. These differences in shoreline erosion rates affect the timing of new thermal (temperature) and chemical (salt) boundary conditions applied to on-shore permafrost. Future complexities are introduced by variations in soil conditions, geothermal heat flow, sea water temperatures, topography, on-shore permafrost thickness, the presence of lakes and rivers prior to inundation, and other factors. With variation in these conditions, a wide spectrum of subsea permafrost conditions may develop with two distinct near-shore regimes. In one regime, the shoreline has been relatively stable for 3000 to 4000 years, with the subsea permafrost approaching equilibrium with the new thermal and chemical boundary conditions. In the other regime, the subsea permafrost has been inundated recently and is just beginning to respond to the new boundary conditions.

Numerous studies by Soviet scientists have documented the existence of subsea permafrost in the near-shore areas of the Arctic Ocean. It has been established that the mean annual seabed temperatures are negative and that the permafrost is very warm. In the Beaufort Sea along Alaska's coast, there is a thawed layer near the seabed where the temperatures are negative but where there is no ice in the interstitial pore spaces. Ice-bonded permafrost exists beneath this thawed layer. In some cases a transition zone, ice-bearing but not necessarily ice-bonded, exists between the thawed layer and the ice-bonded permafrost. The thawed layer must be produced by salt transport from the seabed after the permafrost has been inundated since the seabed temperatures are negative and the surface sediments usually contain fresh pore water. Methods of investigating the problems of salt transport and measurement of salt concentration profiles are addressed by Harrison and Osterkamp (1982).

In the Antarctic, permafrost is limited to areas where the earth's surface is covered by seasonal or perennial snow cover and by a "cold" ice sheet having a temperature below 0°C at the base (Baranov, 1959).

The thickness of permafrost is influenced by such factors as soil and rock type, surface characteristics (such as snow and vegetative cover) and proximity to bodies of water. Various investigators determined the thickness values shown in Fig. 1–4 by different methods such as direct observation in boreholes, mine shafts, and wells; ground temperature measurements using thermocouples and thermistor cables; and geophysical methods including seismic refraction resistivity.

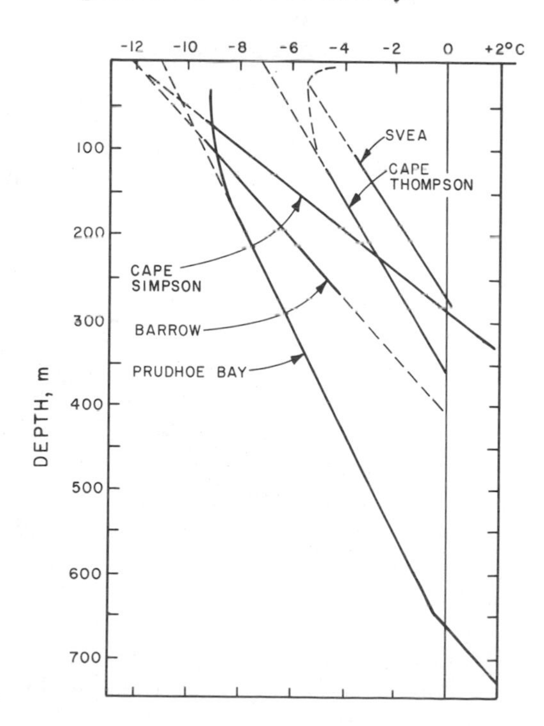


**Figure 1–4** Permafrost thickness profiles.

Of various climatic factors such as precipitation, wind, and energy exchange dynamics, air temperature is the fundamental factor initiating the formation and existence of permafrost. Observations indicate that a broad relationship exists between mean annual air and ground temperature in permafrost regions (Gold and Lachenbruch, 1973). Based on the regional studies in Alaska and Canada, correlations have been obtained between mean annual air temperature and the geographical distribution of the permafrost zones. It is found that the southern limit of permafrost generally coincides with the −1°C (30°F) mean annual air isotherm in Canada between James Bay and the Western Cordillera (Brown, 1967b). North of the southern unit between −1°C (30°F) and −4°C (25°F) the mean annual air isotherm is restricted mainly to the drier portions of the muskeg and organic soils, and some north-facing slopes and local shaded areas. North of the −4°C (25°F) mean annual air isotherm, permafrost becomes increasingly widespread and thicker. Scattered bodies of permafrost also occur on some north-facing slopes and in some heavily shaded areas. The boundary between the continuous and the discontinuous permafrost zones appears to be near the −8°C (17°F) mean annual air isotherm.

Terrain analysis factors such as relief, vegetation, hydrology, snow cover, glacier ice, soil and rock type, and fire also influence permafrost distribution. Relief determines the amount of solar radiation received by the ground surface and the accumulation of snow on it. The influence of orientation and degree of slope is particularly evident in mountainous regions (Ives, 1973). Variation in snow cover and surficial vegetative mat thickness may affect the distribution of permafrost on north- and south-facing slopes. Similar differences can occur even in areas of intensive microrelief, such as peat mounds and peat plateaus (Zoltai, 1971).

The most important effect that vegetation has in permafrost formations is its role in shielding the permafrost from solar heat during the summer period and thereby preventing thermal degradation or thawing of frozen ground. Disturbances of the vegetative mat may cause a deeper active layer. The degradation of permafrost may continue until a steady state of temperature is reached. The effect of vegetation removal is illustrated in Fig. 1–5. Surface vegetation plays a more important role in the discontinuous permafrost zone than in the continuous zone. Although the influence of the ground vegetation on permafrost is dominant, trees are also important. They shade the ground from solar radiation and intercept some of the snowfall in winter. The density and height of trees may influence the microclimatic effects of ground surface.

The effects of permafrost are mostly detrimental to plant development because of its cold temperatures and impermeability to water. Soil movement in the active layer may also influence the vegetation. Detrimental influences on the vegetation are produced by frost action, causing unevenness of the ground, by solifluction and downslope mass movements of earth materials, and by thermokarst, which changes the surface configuration of the ground.

The snow cover and the length of time that snow lies on the ground are

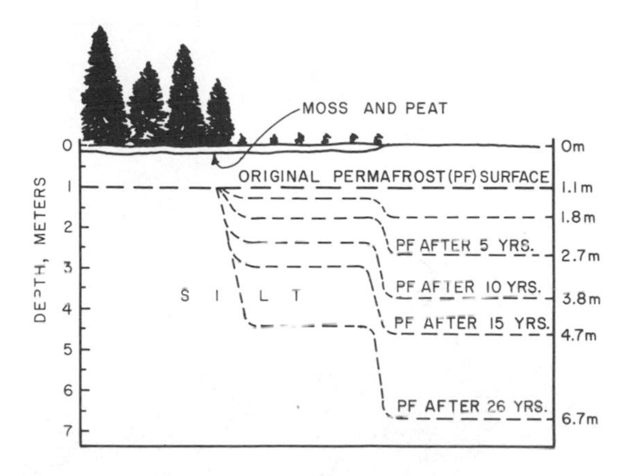


**Figure 1–5** Vegetation removal effects. (After Linell, 1973.)

important factors that influence the heat transfer between the air and the ground and hence affect the distribution of permafrost. A heavy fall of snow in the late autumn may prevent or delay frost penetration. A thick snow cover that remains until late in the spring may delay thawing of the frozen ground. The relation between these two conditions generally determines the net effect of snow cover on the ground thermal regime. Nicholson and Granberg (1973) reported that the permafrost is normally developed when the average snow depth for the area of influence is less than 65 to 75 cm.

Surface and subsurface moving water greatly influence the distribution and thermal regime of permafrost. Especially in areas of ice-rich ground, thawing and erosion are caused by moving water. The extent of the thawed zone varies with such factors as area and depth of the water body, water temperature, thickness of winter ice and snow cover, and general hydrology (Feulner and Williams, 1967). Shifting of stream channels changes the permafrost distribution and thermal regime below and adjacent to the channel. The impermeable nature of permafrost is responsible for the existence of many small, shallow lakes and ponds in the continuous permafrost regions. As shown in Fig. 1–6, the sources of water are: free water above permafrost (called *suprapermafrost water*), unfrozen zones (taliks) within permafrost (called *intrapermafrost water*), and the presence of water below the permafrost base (called *subpermafrost water*). Groundwater through unfrozen zones under artesian pressure may discharge through permafrost, causing complex ground thermal regimes, particularly near streams and lakes.

Bare soil and rock have considerable influence on the temperature of the ground. Thus the permafrost condition is controlled by the albedo and thermal properties of the materials, a topic to be discussed more fully in Chapter 3. Variations in the thickness of permafrost and the active layer are directly related to variations in soil and rock type when the ground is free of vegetation, water, or ice.

Glaciers and ice caps forming a layer on the ground surface affect the heat

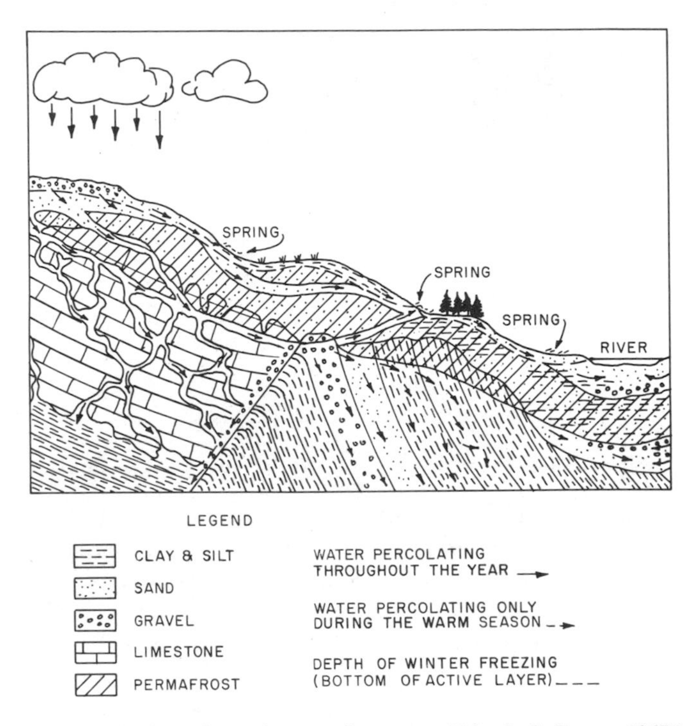


**Figure 1–6** Groundwater in permafrost areas. (After I. C. Brown, 1967.)

exchange between the atmosphere and permafrost. It has been postulated by many investigators that the temperature beneath much of a continental ice sheet is colder than 0°C (32°F) and that continental glaciation affected permafrost conditions in the past (Langway, 1967; Robin, 1972; Judge, 1973).

The degree of influence of a fire, which is a transient factor on permafrost, depends on the age and species composition of the vegetation, the rate of burning, and how quickly the fire progresses through an area. A fire may move rapidly through an area, burning trees, charring only the top surface of the ground vegetation, and thereby, the energy balance at the surface remains unchanged. If dry conditions have prevailed in an area for some time prior to a fire, considerable change in the permafrost may occur as the fire burns more intensely, resulting in a more thorough loss of vegetative cover (Viereck, 1973; Sykes, 1971).

## 1.4 FROZEN-GROUND TEMPERATURE PROFILE

The thickness of permafrost may be obtained by plotting the annual temperature variation in the ground, which fluctuates with air temperatures as shown in Fig. 1–7. The temperature profile in the ground is often referred to as the "whiplash" curve. The depth of zero annual amplitude indicated in Fig. 1–7 is commonly

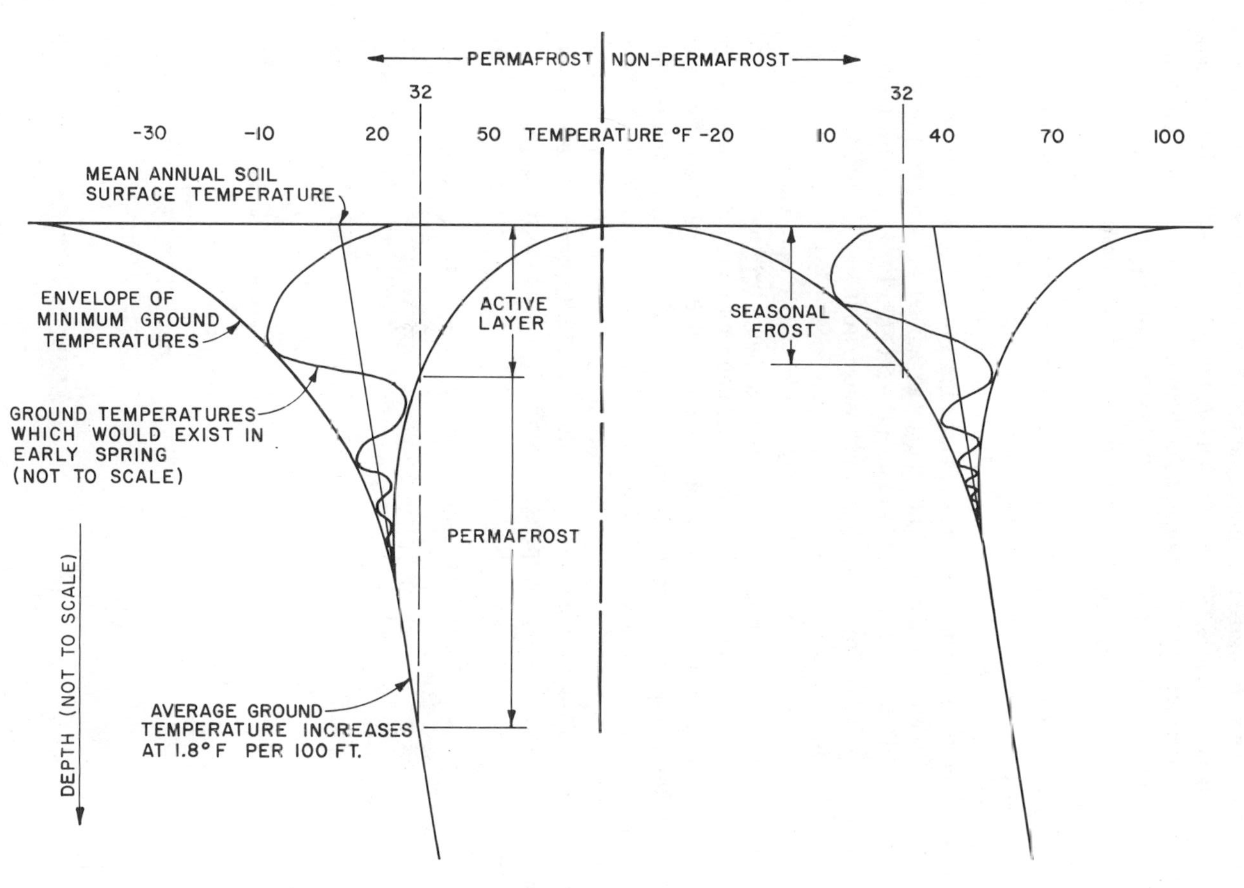


**Figure 1–7** Whiplash curve.

between 10 and 15 m (30 and 50 ft). Below the depth of zero annual amplitude, the temperature increases gradually under the influence of the geothermal gradient. Many investigators (Judge, 1973; Taylor and Judge, 1974, 1975, 1976; Lachenbruch et al., 1966) reported variations of geothermal gradient ranging from about 1°C per 22 m to 1°C per 160 m (1°F per 40 ft to 1°F per 300 ft) depending on the climatic, geological, and geographical factors. In general, an average value of 1°C per 30 m or 1.0°F in 100 ft may be used.

Many of the environmental factors discussed in previous sections also influence the surface and subsurface temperatures in the ground. As shown in Fig. 1–7, the magnitude of temperature fluctuation is reduced with increasing depth by the effect of the thermal characteristics of particular soil and rock types as well as surface cover. In the continuous zone, the mean annual air temperature varies from −8°C (17°F) to −19°C (−3°F) to about −5°C (23°F). Permafrost temperatures at the depth of zero annual amplitude range from −1°C (30°F) to about −8°C (17°F) (Brown, 1966).

The temperature regime of frozen ground depends on present and past climate, terrain factors, and the complex energy exchange between the ground and atmosphere. The factors that together affect the temperature regime determine the ground-temperature profile (Gold and Lachenbruch, 1973). Some typical ground-temperature profiles at various locations are shown in Figs. 1–8 to 1–10. For most civil engineering projects, a knowledge of the ground thermal regime is of prime importance, particularly for depths ranging from 3 to 20 m (10 to 70 ft). In some special cases involving drilling and the production of oil, gas, and water supply wells, the requirements for temperature profile data are necessary to greater depths.

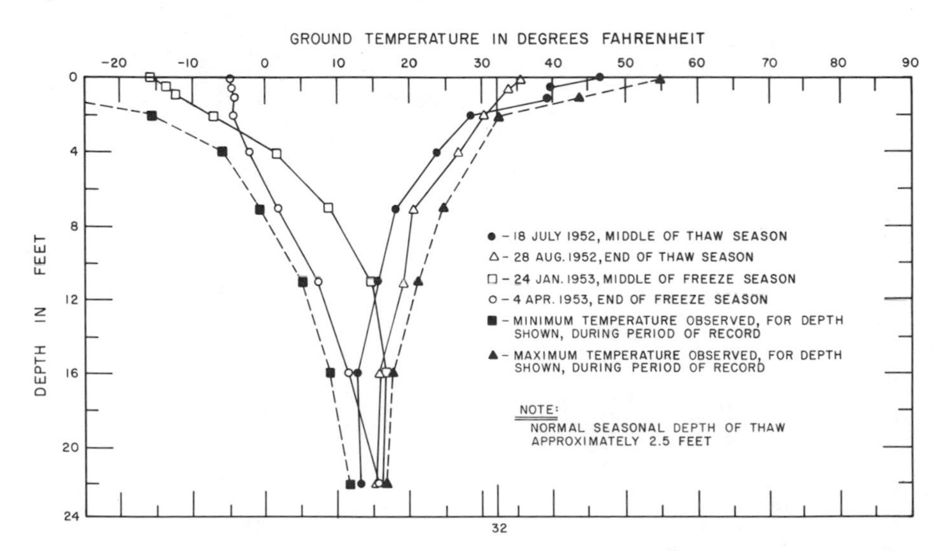


**Figure 1–8** Ground-temperature gradients and maximum/minimum curves, Barrow, Alaska. (After Aitken, 1966.)

An appreciation of the ground thermal regime and the changes that occur in it is of utmost importance to frozen-ground engineering. The ground thermal regime may be determined by the energy balance at the ground surface. The energy balance is calculated by the net flux of heat arising from absorbed solar and thermal radiation and from the ground and the overlying air. This net heat flux is given by

$$G = R + LW + SH + L \tag{1-1}$$

where $G$ = heat flux at the surface and is given by the temperature gradient
$R$ = net flux of solar radiation where the value depends on surface reflectivity (albedo) and orientation, cloud conditions, atmospheric conditions, latitude, season, and time of day
$LW$ = long-wave radiation, which depends on surface and air temperatures, cloud conditions, temperature, and humidity gradients in the atmosphere
$SH$ = sensible heat flux associated with air flow over the surface
$L$ = latent heat flux associated with evaporation of moisture from the surface; the plants from the vegetated surface may contribute to the latent heat flux through the process of evapotranspiration

The complex energy or heat-transfer mechanism exchange at the ground surface is influenced by surface cover, ground type, moisture conditions, geological structure, or geothermal gradient.

Given the complexity of the energy balance at the ground surface and the difficulty of estimating or measuring the individual heat-flux components in Eq. (1–1), one must make the ground temperature calculations using assumed surface

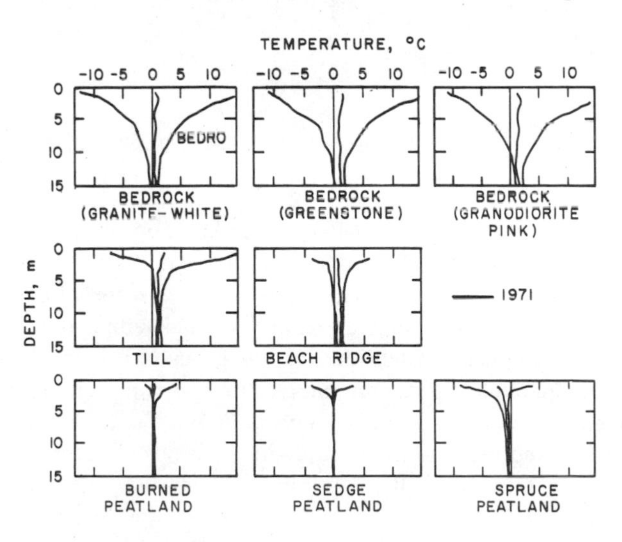


**Figure 1–9** Ground-temperature envelopes, Yellowknife, N.W.T. (After Brown, 1973.)

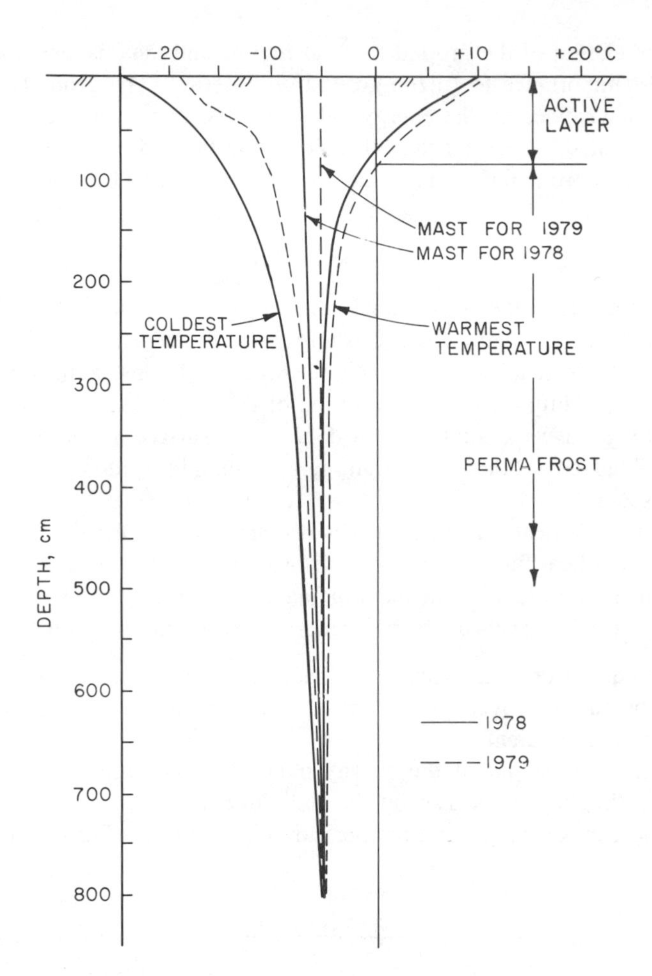


**Figure 1–10** Ground-temperature profile at Svea, Spitzbergen.

temperatures. As a result of the heat-balance process on the ground surface, the ground temperature roughly follows the seasonal variations in air temperature. Since air-temperature data are readily available from most meteorological stations, a correlation between the air temperature and ground temperature is required to define the thermal boundary conditions at the ground surface. For design purposes, surface temperatures are commonly estimated using an empirically determined coefficient called the *n factor,* which is defined as the ratio of the surface thawing or the freezing index to the air thawing or freezing index. The application of these concepts in determining the depth of thaw or frost penetration is discussed in Section 4.3.1.

The magnitude of thawing and freezing *n* factors depends on cloud cover, atmospheric conditions, latitude, wind speed, surface characteristics, and subsurface thermal properties. Lunardini (1978) presented a theoretical analysis for estimating $n$ factors. Table 4–2 gives approximate values of the $n$ factors for several different surfaces.

## 1.5 FROZEN-GROUND FEATURES

Often, frozen-ground conditions can be understood through visual analysis of such physical features as polygons, circles, nets, pingos, palsas, and thermokarst. The presence of such surface features gives an indication of different forms of ground ice. *Ground ice* refers to ice in unconsolidated surficial materials or in bedrock under permafrost conditions. From the engineering point of view, it is important to know how such surface features are formed in arctic and subarctic environments, as their presence suggests the physical and mechanical properties of the ground. The characteristic of ground ice and its distribution are of prime importance to the design and construction of structures in frozen ground.

The main factors that govern the amount, form, and distribution of ground ice are (1) climate, especially at the time of the onset of permafrost conditions; (2) the physical and mechanical properties of soil or rock materials, such as grain size, porosity, mineralogy (clay size material), permeability, and conductivity; (3) topographic form and vegetative cover; (4) hydrologic conditions at the onset of permafrost conditions; and (5) geologic origin and history of the materials. Mackay (1972) classified ground ice into different forms illustrated in Fig. 1–11. The engineering description of ice for the classification of frozen soils is discussed in Chapter 2.

Ground ice occurs in different forms within a wide range of geologic materials. Nearly all perennially frozen soil or rock having significant porosity contains pore ice that completely occupies voids (exceptions are "dry permafrost," which is a rare occurrence, and unfrozen water in clay). Due to moisture migration under different thermal gradients, ice in excess of that required to fill porosity is also formed in fine-grained soils, called *segregated ice*. Segregated ice occurs in forms ranging from veins to lenses to massive. Figure 1–12 shows massive ice encountered at a roadway cut in frozen ground. Most large masses of segregated ice exist at depths greater than 3 m, and they consist of generally horizontal layers (Mackay and Stager, 1966). The maximum depth to which such segregated ice can lie in permafrost is unknown, but Mackay (1976) has reported that it could grow to depths of at least 60 m (200 ft). All soils with segregated ice yield excess water when thawed, irrespective of the moisture conditions of soil layers between ice lenses.

*Reticulated vein ice* consists of a variety of segregated ice distinguished by a three-dimensional reticulated ice-lens pattern (Mackay, 1974). The ice is often found in fine-grained glacial silt, lake, and marine clay. Reticulate vein ice may become aggradational ice or horizontal ice lenses, depending on their upward or downward expansion. This type of ice is usually pure and sediment-free.

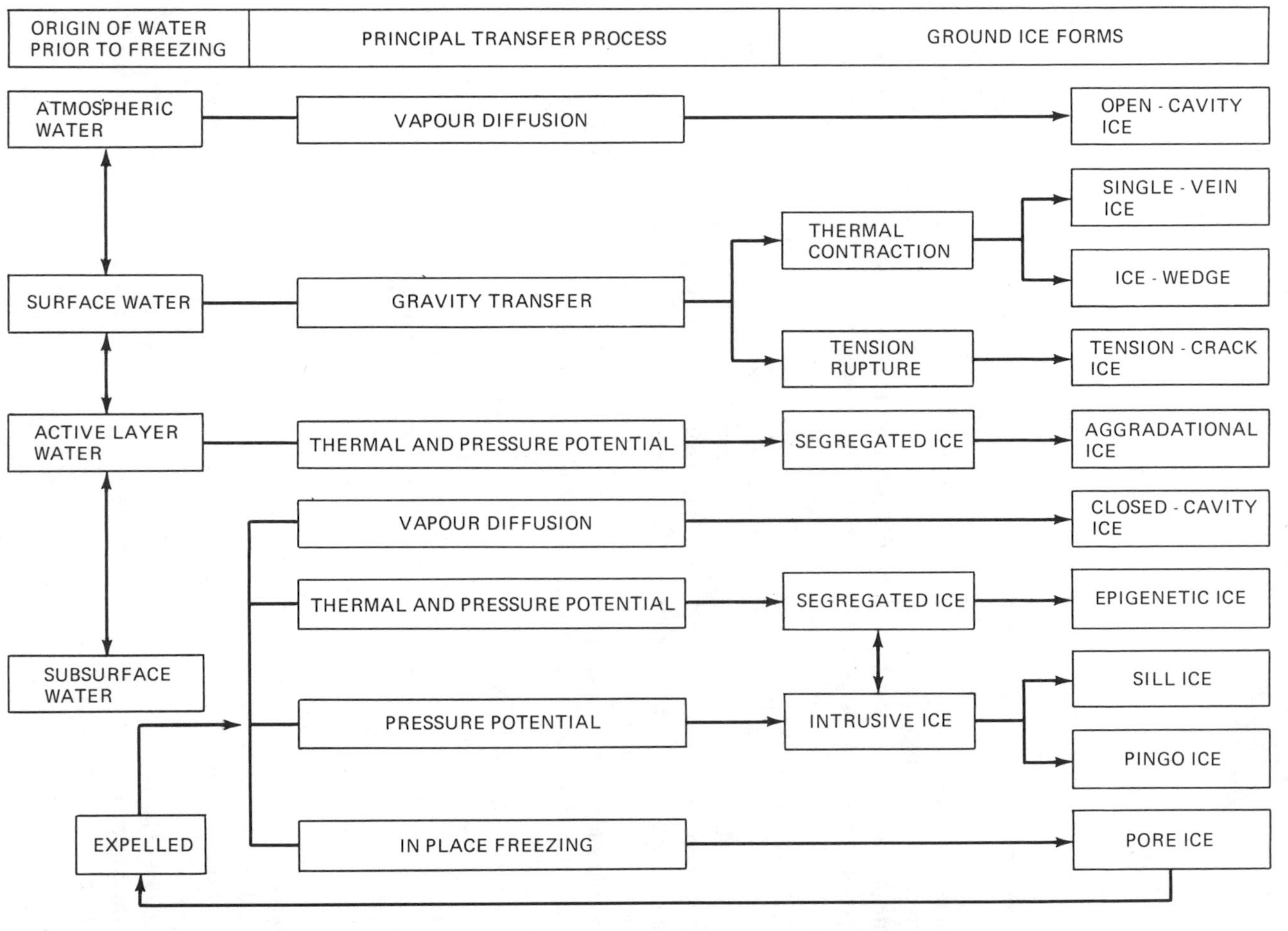


**Figure 1–11** Classification of underground ice. (After Mackay, 1972.)

**Figure 1–12** Massive ice exposed during road cut. (Courtesy of D. Esch.)

*Aggradational ice* grows in response to a rise of the permafrost table. A rise in the permafrost table may occur with climatic change, deposition of material on the existing ground surface (by soil creep, sedimentation, and organic accumulation), and other factors (Section 1.3). Aggradation ice consists of relatively thin ice lenses frozen at the bottom of the active layer and later preserved in permafrost. The lenses are roughly parallel to the undulations of the permafrost and no massive ice sheets will be formed. Ice lenses are generally found within the upper 1.5 to 3 m (5 to 10 ft) of the ground in many polar regions with fine-grained soils. The process and response of permafrost aggradation are further illustrated in Fig. 1–13.

The *ice-wedge polygon* is one of the most frequently described of the pattern ground forms. The formation of ice-wedge polygons is illustrated in Fig. 1–14. Initially, shrinkage cracks develop in the frozen ground due to changes in temperatures or thermal stress. Shrinkage cracks are generally filled with water during the summer, then freeze in the winter. Repeated cycles of melting and freezing will lead to the formation of ice wedges, which may create lateral compression forces. These forces cause the central area of the polygons to raise. Following the theoretical studies by Lachenbruch (1962) of the mechanics of ice-wedge cracking, attempts have been made to monitor the rate of ice-wedge cracking in the field.

At Point Barrow, Alaska, ground contraction was measured between vertical iron rods driven into permafrost in five sites of different ages: old, high-centered polygons in sandy-clayey silt with ice wedges up to 6 m wide; young, flat-cen-

PERMAFROST AGGRADATION

| Types of change | Processes | Results | Responses | Ground Ice Change |
|---|---|---|---|---|
| GEOMORPHIC | ACTIVE LAYER BURIAL FROM SEDIMENTATION, SOIL CREEP, SLUMPING, MUDFLOW, ETC. | UNFROZEN MATERIAL ADDS TO TOP OF ACTIVE LAYER | ACTIVE LAYER THICKENING INDUCES RISE OF PERMAFROST TABLE BY BOTH DOWN-FREEZING AND UP-FREEZING; RETURN TO THERMAL EQUILIBRIUM | |
| | ACCUMULATION OF ORGANIC MATTER | | | |
| VEGETATIONAL | | THERMAL INEQUILIBRIUM | GROUND UPLIFT EQUAL AMOUNT OF ICE TRAPPED | ICE AT TOP OF PERMAFROST IS TRAPPED ACTIVE LAYER ICE |
| | VEGETATION CHANGE LEADING TO GROUND COOLING | GROUND SURFACE TEMPERATURE DECREASED | IN PERMAFROST COOLING INDUCES RISE OF PERMAFROST TABLE AND RETURN TO THERMAL EQUILIBRIUM | |
| CLIMATIC | COOLING TREND, LONG- OR SHORT-TERM | | | |

**Figure 1–13** Permafrost aggradation. (After Mackay, 1971.)

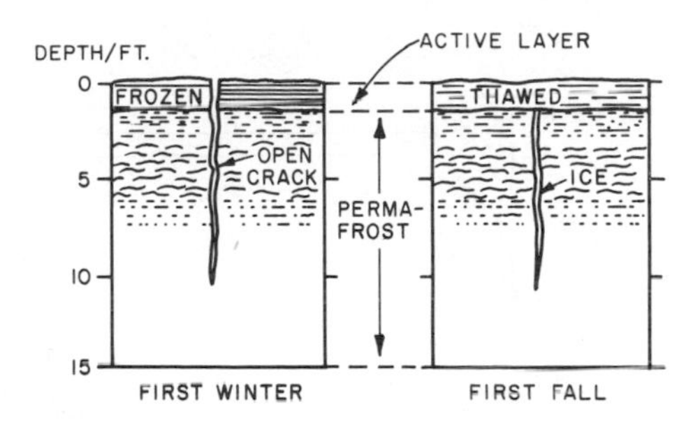


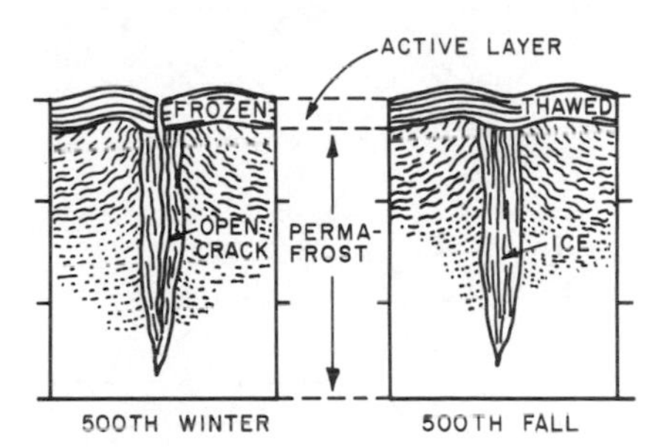


**Figure 1–14** Growth of an ice wedge. (After Lachenbruch, 1962.)

tered polygons in sandy-clayey silt with ice wedges up to 2 m wide; young, flat-centered polygons in sandy gravel with ice wedges up to 1 m wide; young, low-centered polygons in sandy-clayey silt with ice wedges up to 3 m wide; and old, high-centered polygons in sandy gravel ice wedges up to 3 m wide. In those five sites, 37, 64, 70 (est.), 59, and 46%, respectively, of the ice wedges cracked. Summer measurements showed that no contraction cracks closed completely. All cracks retained an increment of clear ice that ranged from 0.04 to 6.0 mm. A useful average increment was 2 mm; on the average, 50% of the wedges cracked in an area. Hence a round-figure growth rate of 1 mm per year was obtained. This suggested an age for most surfaces of 1000 to 4000 years in the vicinity of Point Barrow; others were as much as 10,000 years old (Mackay and Black, 1973). Circles and nets are also ground patterns, which usually occur as sorted patterns. These patterns are formed in a manner similar to polygons.

*Pingos* are conical or oval-based hills. They are ice-cored with an organic surface and may rise more than 60 m (90 ft). Pingos are found in Alaska, Baffin Island, MacKenzie Delta, Yukon, East Greenland, and scattered throughout Siberia (Mackay, 1963a; Hamelin and Cook, 1967). The growth of a pingo is illustrated in Fig. 1–15. A photographic view of a pingo is shown in Fig. 1–16.

*Palsas* are mounds of peat and ice and are dome-shaped structures 1 to 7 m (3 to 21 ft) high and 10 to 50 m (30 to 150 ft) in diameter. They occur in bogs and protrude prominently above the level of the bogs.

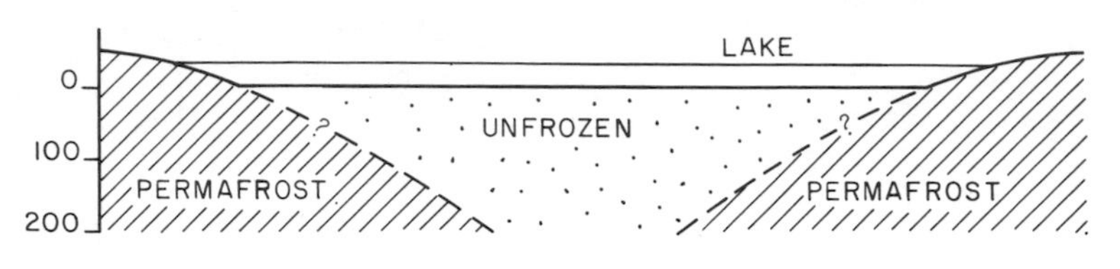


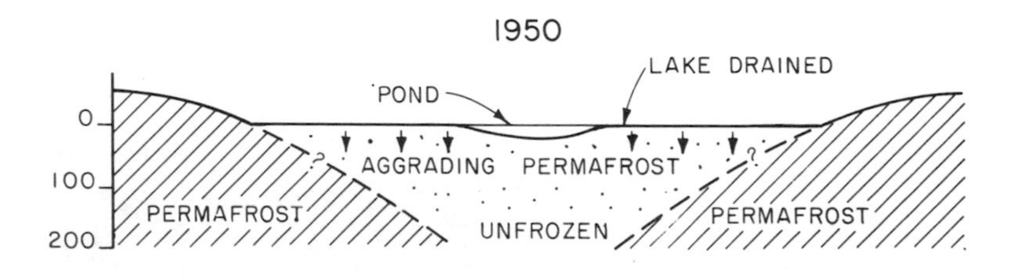


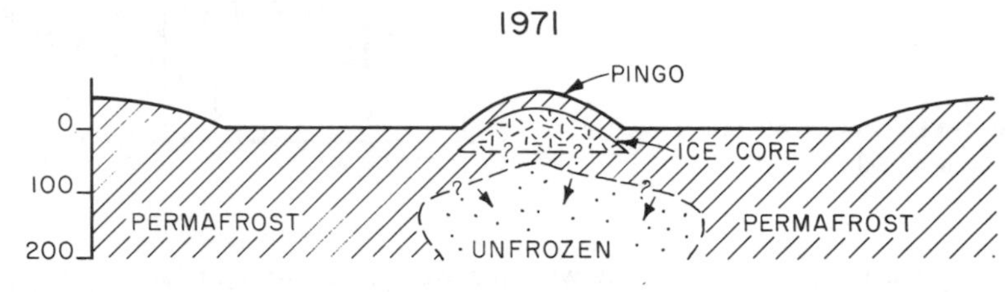


**Figure 1–15** Schematic diagram of growth of pingo. (After Mackay, 1972.)

**Figure 1–16** Pingo north of Brooks Range, Alaska.

Ground ice as reported by Mackay (1972) is discussed further under the following headings:

1. Soil ice (formed by in situ freezing of a local concentration of soil moisture)
   a. Needle ice (pipkrake, mushfrost, etc.): on surface of the ground
   b. Segregated (Taber) ice: supersaturated frozen soil
      (1) Epigenetic (formed later than enclosing material, migration of water; need source of water)
         (a) In active layer
         (b) At base of active layer
         (c) In permafrost
      (2) Penecontemporaneous (formed contemporaneously with surface aggradation)
         (a) Aggradation from sedimentation
         (b) Aggradation from organic accumulation
         (c) Aggradation from burial by soil movement
   c. Pore (cement, bonding) ice: usually saturated frozen soil
2. Intrusive ice (formed by freezing of "injected" water)
   a. Pingo ice
      (1) Open system (freezing of ground water under hydraulic gradient)
      (2) Closed system (freezing of expelled soil moisture)
   b. Sheet ice [freezing of intruded water larger area (thousands of square feet)]
3. Vein (foliated) ice (freezing of water and vapor in thermal contraction cracks)
   a. Vein ice (usually representing a single episode of freezing)
   b. Repeated vein ice (ice wedges)
      (1) Epigenetic (formed later than enclosing material)
      (2) Penecontemporaneous (formed contemporaneously with surface aggradation)
4. Extrusive ice (formed subaerially, melts usually)
   a. River icings (overflow from a constricted river channel)
   b. Ice fans (freezing from a spring)
5. Sublimation ice
   a. Crystals in open cavities and caverns
   b. Crystals in closed cavities
6. Buried ice
   a. Snowbank
   b. Sea, lake, river ice
   c. Glacier ice

*Thermokarsts* are formed as a result of thawing of ice-rich permafrost, which creates an uneven topography. Thermokarst features result from the disturbance or removal of the vegetation or from a warming climate. Thermokarst topography is most commonly formed where massive ice exists in the ground, such as ice wedges or thick layers of segregated ice. Thaw lake or cave-in, depression or thaw depression, and sinkholes are some of the thermokarst features found in arctic and subarctic North America. Mackay (1963b) has reported on the most outstanding oriented thermoskarst lakes on the North Slope of Alaska and east of the MacKenzie Delta on the Tuktoyaktuk Peninsula east to Cape Bathurst.

The most common form of mass wasting in permafrost environments is *solifluction,* the slow flowing of saturated mass underlain by permafrost or long-existing seasonal frost. Areas in which solifluction is active lie almost entirely above or beyond the forest limit. Solifluction areas in North America are widespread (Sigafoos and Hopkins, 1952; Taber, 1943; Kerfoot and Mackay, 1972). The engineering significance of solifluction is discussed in Chapter 9.

## 1.6 ENGINEERING PROBLEMS

Many engineering problems have been encountered in the construction of buildings, roads, airports, pipelines, and other facilities in frozen ground (Figs. 1–17 to 1–19). In the USSR, the problem of frozen ground has been a formidable obstacle in the development of Siberia and other areas. Intensive study of the permafrost has led to many engineering solutions. Similarly, in North America the construction of buildings, roads, utilities, warm pipelines, and other projects in permafrost regions have led to innovative design and construction procedures.

**Figure 1–17** Warm oil pipeline crossing fault zone.

The engineering problems in frozen ground may be categorized into the following groups:

1. Those involving subsidence under heated structures by heat input into the ground (discussed in Chapter 4)
2. Those involving subsidence under unheated structures such as roads and airports by natural freeze–thaw cycles (discussed in Chapter 7)
3. Those resulting from frost-heave forces (discussed in Chapter 4)
4. Those involving the negative or freezing temperature of frozen ground, causing buried water and sewer lines to freeze (discussed in Chapter 8)
5. Those involving construction difficulties such as excavation, drilling, and sampling and the disposal of sewage systems (discussed in Chapters 8, 9, and 10)
6. Those involving long-term deformation, such as structural loads on ice-rich frozen ground (discussed in Chapters 3 and 6)

**Figure 1–18** Water tank on pile foundations in continuous permafrost. (Courtesy of Alaska Area Native Health Services.)

**Figure 1–19** Roadways over ice-rich permafrost. (Courtesy of B. Connor.)

Frozen ground is not necessarily troublesome. Sometimes, it can be used for access to compressible organic terrain or off-road operations where it would be impossible to move during the thawed state. The bearing capacity of frozen soils is high. Construction problems in unfrozen soils due to seepage of water and weak soils can also be eliminated by the artificial freezing of soils. Artificial freezing techniques do not fall within the scope of this book. Maintaining frozen conditions or thermal equilibrium are some of the design strategies used for solving many engineering problems in frozen ground. Thus, ''keep frozen'' is one of the most useful tools used in the design and construction of structures in frozen ground.

The most common engineering problem in frozen ground is the thermal degradation of existing ground conditions. Special precautions must be taken in areas where extensive ground-ice conditions prevail. Often, undetermined ground-ice conditions have caused many thaw-settlement problems, causing significant maintenance cost or even major remedial design of structures. Another serious problem is frost action under natural conditions where thawing and freezing of the active or thawed layer caused serious damage to structures such as roads and airports. Cognizance of frozen-ground conditions and instituting appropriate design measures are also required for water and sewer lines subject to freezing in frozen ground. Frozen ground is generally considered as an impermeable medium and the disposal of the liquid effluent from sewage collection system into impervious ground is not practical. These aspects of frozen-ground problems are discussed further in Chapter 8.

The matrix of different soil constituents under frozen conditions must be considered for excavations. It is very difficult to excavate frozen soils at low temperatures. The compressive and shear strength of frozen soils are significantly increased at low temperatures. As such, standard construction equipment may prove ineffective in the excavation of frozen gravelly soils.

## 1.7 SYMBOLS AND UNITS

Pertinent symbols used in each of the succeeding chapters are presented in a table at the beginning of the chapter. The symbols conform to those adopted by the International Society for Soil Mechanics and Foundation Engineering, ''Terms and Symbols Relating to Soil Mechanics.'' In a few cases where there is no standard, the author has introduced an appropriate symbol.

Most of the world has officially adopted the SI system (based on the metric system), based on the meter of length and the kilogram of mass. However, many engineers continue to think in terms of the older units of pounds or kilograms of force. Throughout the book, both the older system (U.S. customary/English units) and the SI system have been used. Whenever possible, the equivalent conversion has been given in parentheses. Some useful conversion factors for both units are presented in Appendix A.

CHAPTER 2

# Frozen-Soil Classification

## 2.1 INTRODUCTION

The primary objective of this chapter is to establish a vocabulary for use in a frozen-soil classification system. A well-established engineering classification system for unfrozen soils is introduced. The importance of including ice description, terrain, and topographic features in the classification system is discussed. The available techniques (site investigations and laboratory measurements) to establish relevant engineering properties of frozen soil are presented in Chapter 10.

The notation used in this chapter is given in Table 2–1.

**TABLE 2–1 Notation**

| Symbol | Dimension[a] | Unit | Definition |
|---|---|---|---|
| $C_c$ | — | — | Coefficient of curvature (Eq. 2–2) |
| $C_u$ | — | — | Coefficient of uniformity (Eq. 2–1) |
| $D_{10}$ | $L$ | mm | Diameter for 10% finer by weight |
| $D_{30}$ | $L$ | mm | Diameter for 30% finer by weight |
| $D_{60}$ | $L$ | mm | Diameter for 60% finer by weight |
| LL | — | — | Liquid limit |
| PL | — | — | Plastic limit |
| PI | — | — | Plasticity index |
| $N$ | — | — | Nonvisible ice |
| $V$ | — | — | Visible ice |
| $n$ | — | % | Porosity |

[a]$L$, length.

A soil classification is generally used to arrange different soils into groups having similar properties. Having a classification system provides a language of communication between engineers. Various classification systems, such as the Unified Soil Classification System (USCS) and those of the American Association of Highway and Transportation Officials (AASHTO) and the Federal Aviation Administration (FAA), have typically been used by geotechnical, highway, and transportation engineers. The USCS is the most widely accepted system for the classification of soils and has been extended to include frozen soils (Linell and Kaplar, 1966). The definition of "frozen soil" discussed in Chapter 1 gave a general idea of how soils are found in the frozen state. Both the description of soils in the unfrozen state and the characteristics of frozen soils are utilized in the classification of frozen soils. The following steps are introduced to classify frozen soils:

STEP 1. The unfrozen-soil phase independent of the frozen state is described by the Unified Soil Classification System (USCS), shown in Table 2–2.

STEP 2. Ice intrusion in frozen soils is examined and the details, shown in Table 2–4, are added to the classification given in step 1.

STEP 3. The basic descriptions of terrain characteristics of a site where frozen soils are recovered at different depths are to be noted during the field investigation, and such data, in terms of information as shown in Table 2–5, are added to the description obtained by steps 1 and 2. The first two steps have been used commonly (Linell and Kaplar, 1966) for the classification of frozen soils. The introduction of step 3 by the author will give more basic background information on frozen soils which will be useful in engineering applications.

**TABLE 2–2 Unified Soil Classification System (ASTM Designation D-2487)**

| Major Division | Group Symbol | Typical Names | Classification Criteria |
|---|---|---|---|
| Coarse-grained soils (more than 50% retained on No. 200 sieve) Gravels 50% or more coarse fraction | GW | Well-graded gravels and gravel-sand mixtures, little or no fines | $C_u = D_{60}/D_{10}$ greater than 4; $C_2 = (D_{30})^2/(D_{10} \times D_{60})$ between 1 and 3 |
| | GP | Poorly graded gravels and gravel-sand mixtures, little or no fines | Not meeting both criteria for GW |
| | GM | Silty gravels, gravel-sand-silt mixtures | Atterberg limits plot below A line or plasticity index less than 4 |

TABLE 2–2 *(Cont.)*

| Major Division | Group Symbol | Typical Names | Classification Criteria |
|---|---|---|---|
| | GC | Clayey gravels, gravel-sand-clay mixtures | Atterberg limits plot below A line or plasticity index greater than 7 |
| Sands more than 50% coarse fraction | SW | Well-graded sands and gravelly sands, little or no fines | $C_u = D_{60}/D_{10}$ greater than 6; $C_2 = (D_{30})^2/(D_{10} \times D_{60})$ |
| | SP | Poorly graded sands and gravelly sands, little or no fines | Not meeting both criteria for SW |
| | SM | Silty sands, sand-silt mixtures | Atterberg limits plot below A line or plasticity index less than 4 |
| | SC | Clayey sands, sand-clay mixtures | Atterberg limits plot below A line or plasticity index greater than 7 |
| Fine-grained soils (50% or more passes No. 200 sieve) Silts and clays LL 50% or less | ML | Inorganic silts, very fine sands, rock flour, silty or clayey fine sands | — |
| | CL | Inorganic clays of low to medium plasticity, gravelly clays, sandy clays, silty clays, lean clays | See Fig. 2–2 |
| | OL | Organic silts and organic silty clays of low plasticity | — |
| Silts and clays LL > 50% | MH | Inorganic silts, micaceous or diatomaceous fine sands or silts, elastic silts plasticity, fat clays | — |
| | CH | Inorganic clays of high plasticity, fat clays | — |
| | OH | Organic clays of medium to high plasticity | — |
| Highly organic soils | Pt | Peat, mulch, and other highly organic soils | Fibrous organic matter; will char, burn, or glow |

## 2.2 THE UNIFIED SOIL CLASSIFICATION SYSTEM

The Unified Soil Classification System (USCS) was adopted by the American Society for Testing and Materials (ASTM) as a standard method for classification of unfrozen soils for engineering purposes (ASTM D-2487) in 1969. The system was originally developed by A. Casagrande for the U.S. Army Corps of Engineers for use in airfield construction during World War II. Later, it was modified by the Corps and the U.S. Bureau of Reclamation as the Unified System. The system is based on the grain size distribution and the plasticity of soils. According to the system, the soils are divided into three main groups (Table 2–2): coarse-grained, fine-grained, and peat or high organic soils. The coarse-grained soils are those having 50% or more material retained on a No. 200 sieve (0.075 mm diameter) and are divided into two subgroups:

1. Gravels and gravelly soils; denoted by the symbol G
2. Sands and sandy soils; denoted by the symbol S

The gravels are those having the greater percentage of coarse fraction retained on a No. 4 sieve (4.75 mm in diameter) and the sands are those having the greater coarse fraction passing a No. 4 sieve. Depending on the grain size distribution and nature of fines, both the gravels and sands are subdivided into four subgroups.

1. Clean material (less than 5% passing a No. 200 sieve), well graded, has a wide representation of all particle sizes, and is denoted by the symbol W (GW, SW)
2. Clear material and gap graded or poorly graded material are denoted by the symbol P (GP, SP)
3. Coarse materials containing silt binder are denoted by GM, SM, and M and represent silt
4. Coarse materials containing clay binder are denoted by GC, SC, and C and represent clay

The coefficient of uniformity $C_u$ and the coefficient of curvature $C_c$ are determined from the grain size distribution to designate gravelly or sandy soils, whether they are well graded or not. These coefficients are defined as

$$C_u = \frac{D_{60}}{D_{10}} \tag{2–1}$$

where $D_{60}$ = grain diameter in millimeters corresponding to 60% passing by weight (Fig. 2–1)

$D_{10}$ = grain diameter in millimeters corresponding to 10% passing by weight (Fig. 2–1)

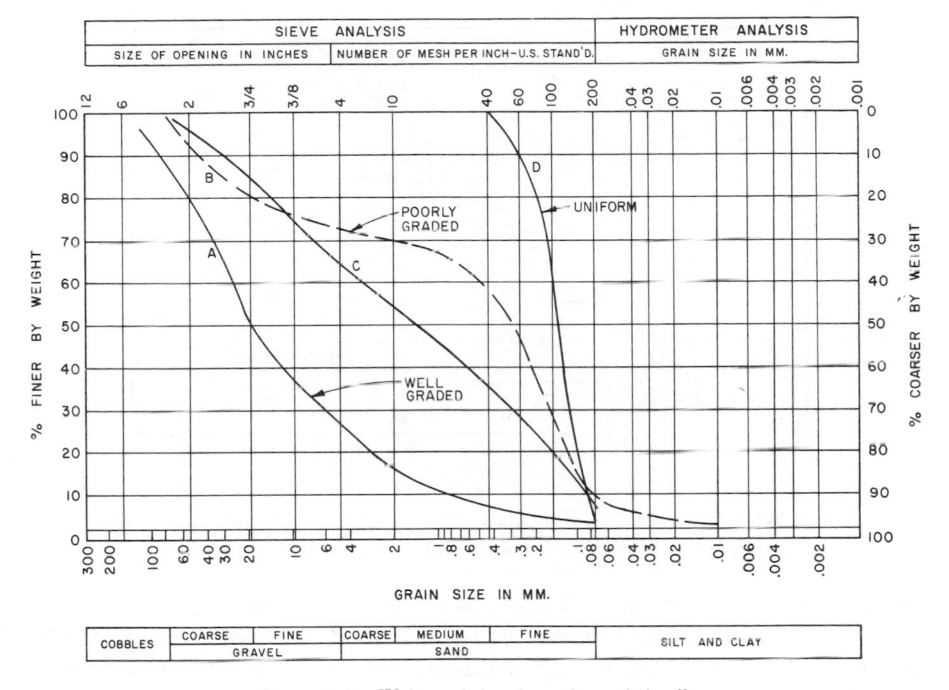


**Figure 2–1** Well-graded and poorly graded soils.

$$C_c = \frac{(D_{30})^2}{(D_{60}) \times (D_{10})} \tag{2–2}$$

where $D_{30}$ is the grain diameter in millimeters corresponding to 30% passing by weight (Fig. 2–1).

A gravelly or sandy soil with a coefficient of curvature between 1 and 3 is considered to be well graded as long as the coefficient of uniformity is also greater than 4 for gravels and greater than 6 for sands. When both criteria are not satisfied, the poorly graded designation is used.

The fine-grained soils (more than 50% passing a No. 200 sieve) are divided into the following three subgroups:

1. Inorganic silt; denoted by the symbol M
2. Inorganic clay; denoted by the symbol C
3. Organic silts and clays; denoted by the symbol O

The A line on Casagrande's (1948) plasticity relationship (Fig. 2–2) between liquid limit (LL) and plasticity index (PI) is used to determine whether a fine-grained soil is silt or clay. Fine-grained soils are considered silts (M) if their liquid limits and plasticity indices plot below the A line (Fig. 2–2). If their plots

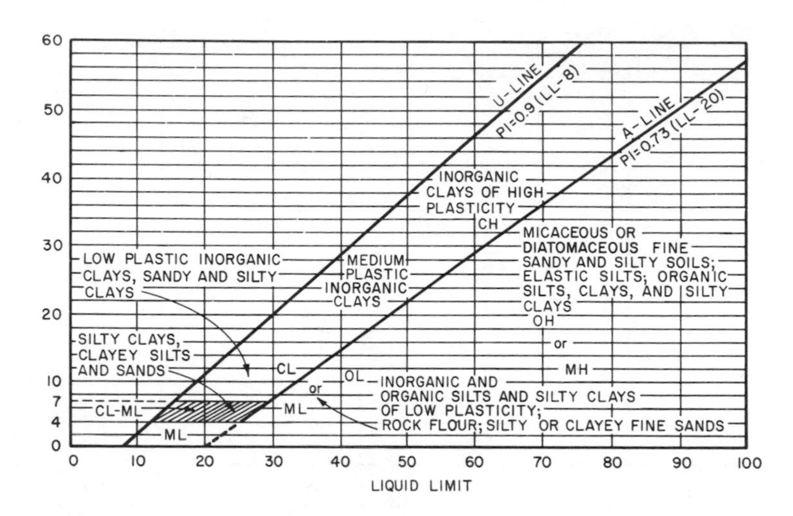


**Figure 2–2** Casagrande's plasticity chart, showing several representative soil types. (Developed from Casagrande, 1948, and Howard, 1977.)

are above the A line, the fine-grained soils are considered clays (C). As shown in Table 2–2, fine-grained soils having liquid limits of 50 or less are designated by the symbol L; for those having liquid limits greater than 50, the H symbol is used.

Generally, peat or highly organic soils ($P_t$) are classified based on visual description and identification. They are usually fibrous, dark black in color, and may have an unpleasant strong odor when moist.

A dual symbol designation is used for coarse-grained soils having between 5 and 12% fines passing a No. 200 sieve. The first symbol indicates whether the soil is well graded or poorly graded and the second part describes the nature of fines. For example, a soil classified as GW-GM means that it is well-graded gravel with between 5 and 12% silty fines. Fine-grained soils can also have dual symbols when the limits plotted fall within the shaded zone on Fig. 2–2.

A step-by-step procedure for USCS classification of unfrozen soils is presented in Fig. 2–3. The following steps, adapted from the U.S. Army Corps of Engineers, are useful in this process (U.S. Army, Corps of Engineers, 1960); classification should be done in conjunction with Table 2–2 and Fig. 2–3:

STEP 1. Determine if the soil is coarse grained, fine grained, or highly organic. This is done by visual inspection and/or by determining the amount of soil passing a No. 200 sieve.

STEP 2. If coarse-grained:

(a) Perform a sieve analysis and plot the grain size distribution curve. Determine the percentage passing a No. 4 sieve and classify the soil as gravel (greater percentage retained on a No. 4 sieve) or sand (greater percentage passing a No. 4 sieve).

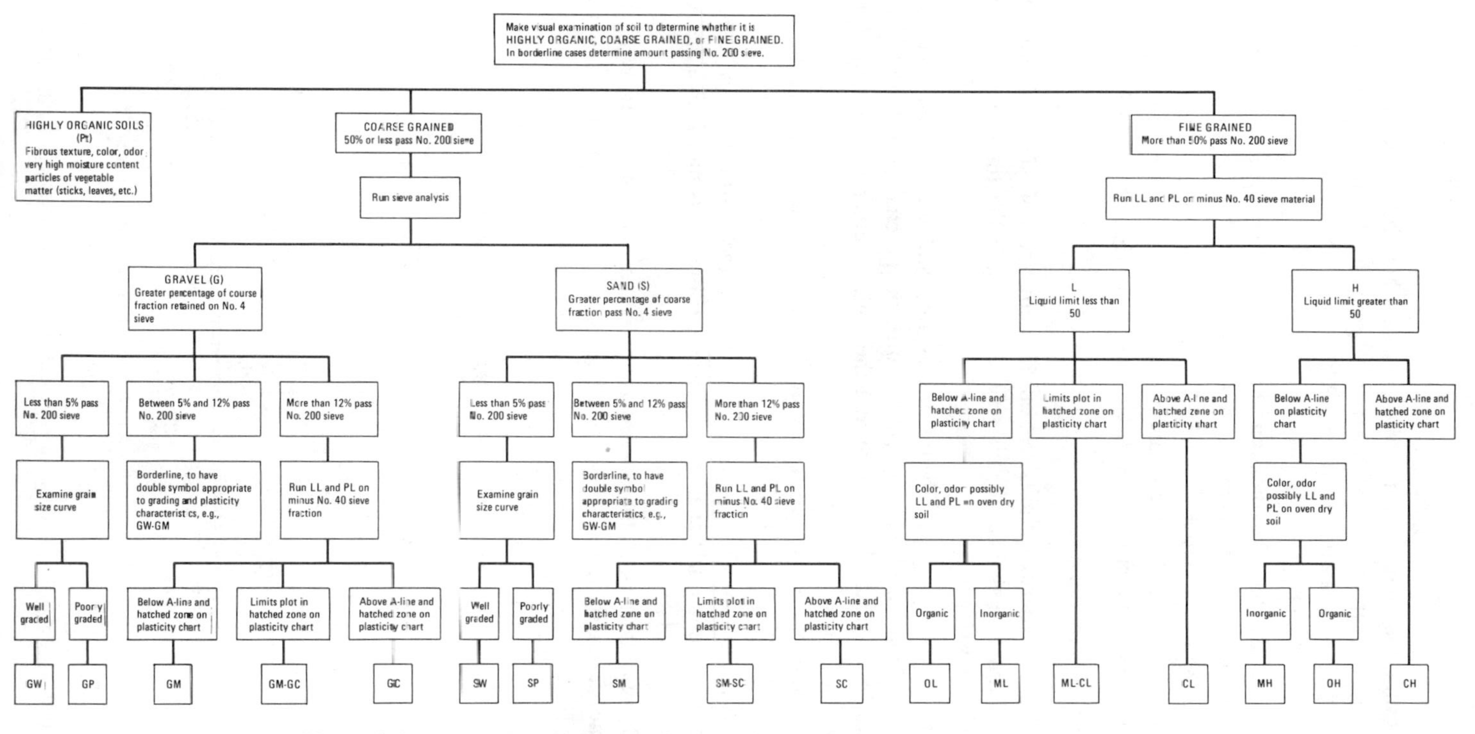


**Figure 2–3** Laboratory soil identification procedure. (After U.S. Army, Corps of Engineers, 1960.)

(b) Determine the amount of material passing a No. 200 sieve and examine the shape of the grain size curve; if well graded, classify as GW or SW; if poorly graded, classify as GP or SP.

(c) If between 5 and 12% of the material passes a No. 200 sieve, it is a borderline case, and the classification should have dual symbols appropriate to grading and plasticity characteristics (GW-GM, SW-SM, etc.).

(d) If more than 12% passes a No. 200 sieve, perform the Atterberg limits test on the minus No. 40 sieve fraction. Use the plasticity chart to determine the correct classification (GM, SM, GC, SC, GM-GC, or SM-SC).

STEP 3. If fine-grained:

(a) Perform the Atterberg limits test on the minus No. 40 sieve material. If the liquid limit is less than 50, classify as L, and if the liquid limit is greater than 50, classify as H.

(b) For L: If the limits plot below the A line and the hatched zone on the plasticity chart, determine by color, odor, or the change in liquid limit and plastic limit caused by oven-drying the soil, whether it is organic (OL) or inorganic (ML). If the limits plot in the hatched zone, classify as CL-ML. If the limits plot above the A line and the shaded zone on the plasticity chart (Fig. 2–2), classify as CL.

(c) For H: If the limits plot below the A line on the plasticity chart, determine whether organic (OH) or inorganic (MH). If the limits plot above the A line, classify as CH.

(d) For limits that plot in the shaded zone on the plasticity chart, close to the A line or around LL = 50, use dual (borderline) symbols, as shown in Fig. 2–4.

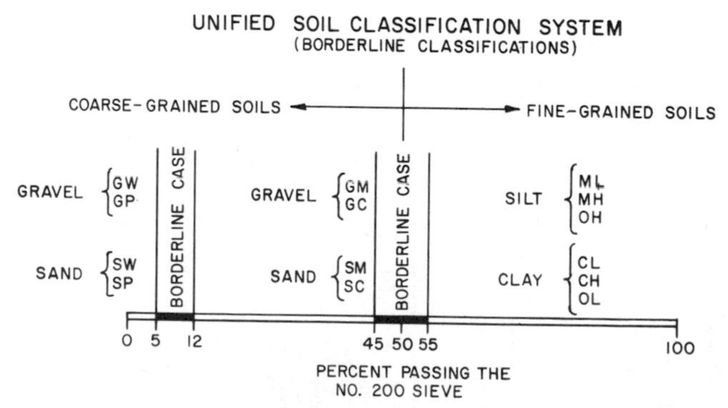


**Figure 2–4** Guide for borderline cases of soil classification. (After Howard, 1977.)

**TABLE 2–3 Information Required for Describing Soils[a]**

| Coarse-Grained Soils | Fine-Grained Soils |
|---|---|
| For undisturbed soils add information on stratification, degrees of compactness and cementation, moisture conditions, and drainage characteristics | Give typical name; indicate degree and character of plasticity, amount and maximum size of coarse grains, color in wet condition, odor (if any), local or geologic name, other pertinent descriptive information, and symbol in parentheses |
| Give typical name; indicate approximate percentages of sand and gravel, maximum size, angularity, surface condition and hardness of the coarse grains, local or geologic name, other pertinent descriptive information, and symbol in parentheses | For undisturbed soils add information on structure, stratification, consistency in undisturbed and remolded states, moisture and drainage conditions |
| *Example:* silty sand, gravelly; about 20% hard, angular gravel particles 12 mm maximum size, rounded and subangular sand grains coarse to fine, about 15% nonplastic fines with low dry strength, well compacted and moist in place, alluvial sand (SM) | *Example:* clayey silt, brown, slightly plastic, small percentage of fine sand, numerous vertical root holes, firm and dry in place, loess (ML) |

[a]Be prepared for wide variations in soil description among organizations and testing laboratories. They all have their own ways of doing things.

*Source:* After U.S. Army, Corps of Engineers, (1960).

Although the letter symbols in the USCS are convenient, they do not completely describe a soil or soil deposit. For this reason, descriptive terms should be used together with the letter symbols for a complete soil classification. Table 2–3 provides some useful information for describing soils.

## 2.3 ICE DESCRIPTION

Ice, which is defined as one of the solid states of water, is one of the main constituents in frozen soil. As temperatures of soils drop below 0°C or 32°F, a point is reached at which nucleation occurs and ice crystals begin to form within the soil pore spaces. Ice may be present in frozen soils in different forms (Fig. 2–5) which sometimes are not visible to the naked eye. A photographic view of an ice lens in a thin section is shown in Fig. 2–6. Table 2–4 gives guidance for the description of ice. This description should be included in the classification system. The basic knowledge of ground ice formation discussed in Chapter 1 is helpful in describing the ice in a soil mass. Knowledge of the geomorphology of the region one is working in generally provides useful information regarding the characteristics of ice inclusion in the frozen soils

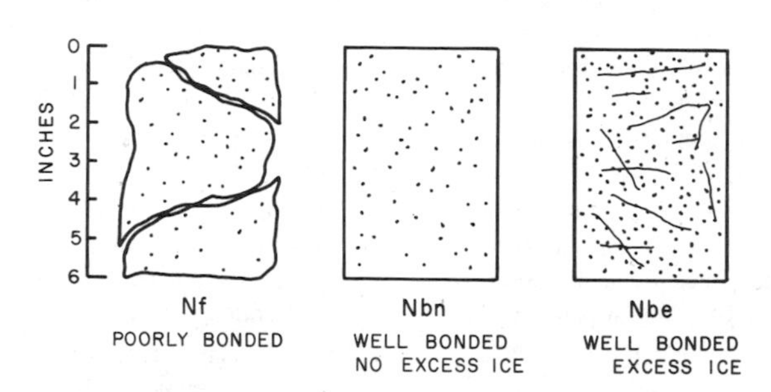


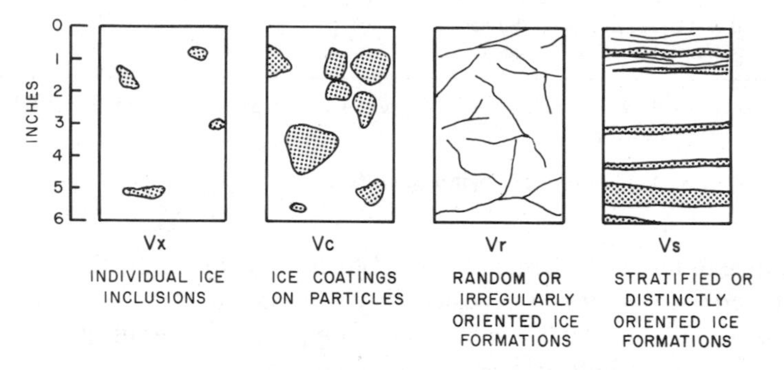


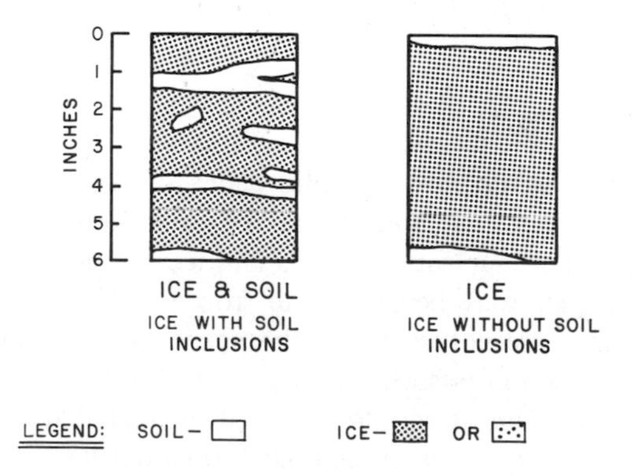


**Figure 2–5** Schematic description of ice inclusion.

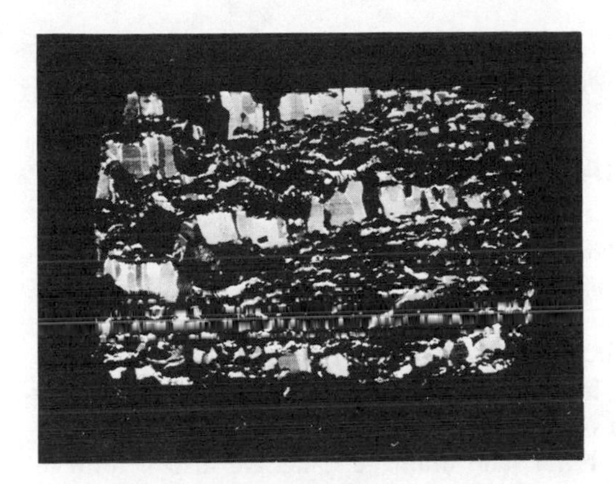

**Figure 2–6** Ice lenses under thin section. (Courtesy of T. E. Osterkamp.)

**TABLE 2–4 Ice Description**

| Major Group | | Subgroup | | | |
|---|---|---|---|---|---|
| Description | Designation | Description | | Designation | |
| Segregated ice not visible by eye | N | Poorly bonded or friable | | Nf | |
| | | Well bonded | No excess ice | Nb | n |
| | | | Excess ice | | e |
| Segregated ice visible by eye (ice 25 mm or less thick) | V | Individual ice crystals of inclusions | | Vx | |
| | | Ice coatings on particles | | Vc | |
| | | Random or irregularly oriented ice formations | | Vr | |
| | | Stratified or distinctly oriented ice formations | | Vs | |
| Ice greater than 25 mm thick | Ice | Ice with soil inclusions | | Ice + soil type | |
| | | Ice without soil inclusions | | Ice | |

*Source:* After Linell and Kaplar, 1966.

Only the descriptive aspects of ice are of concern in this classification system. The physical and mechanical aspects of ice are covered in Chapter 3.

## 2.4 SURFACE TERRAIN CHARACTERISTICS

A field description of the terrain features with respect to surface characteristics and topographic details should be included in the classification system. Surface vegetation is also an important factor to be considered. Table 2–5 lists the vegetation characteristics and topographic features that are included after the soil and ice description. The main intention in including these features is to provide guidance in the design and construction of structures in frozen ground. Such information, accompanied by snow cover, relief, and drainage data, are also helpful in formulating a thermal model for the prediction of temperature profile in the frozen-soil deposit under alternative design conditions.

**TABLE 2–5 Terrain Characteristics**

| Designation | Description |
|---|---|
| | Surface Characteristics or Details |
| A | Wooded area—spruce, larch, birch, pine trees, etc., of height greater than 5 ft |
| B | Nonwooded area—spruce, larch, birch, pine trees, etc., of height less than 5 ft |
| C | Nonwooded area—grass, sledges, and the like, up to 2 ft high |
| D | Nonwooded area—continuous vegetative mat up to 4 in. high |
| | Topographic Features |
| a | Mound, ridge, and hummock |
| b | Sloping area |
| c | Plain area (extensive exposed area) |
| d | Pond or lake margin within 50 ft |
| e | Visible ground ice characteristics; surface like patterned ground |

## 2.5 OTHER CLASSIFICATION SYSTEMS

Tsytovich (1975) classified frozen soils with the description of ice content and physical state categories. According to this system, frozen soils are classified into the following three main groups:

1. Hard-frozen (low-temperature) soils rendered firm by cohesive ice; these soils are practically incompressible.
2. Plastic-frozen (warm-temperature) soils with a high unfrozen water content and relatively low compressibility in the frozen state.
3. Friable frozen soils (coarse-grained soils) which are not cemented by ice, due to a negligible moisture content.

The iciness of frozen soils is described as:

- High ice; ice content above 50% by volume
- Low ice; ice content below 25% by volume
- Icy; ice content between 25% and 50% by volume

**TABLE 2–6 Frost-Susceptibility Soil Classification System**

| Frost Susceptibility[a] | Frost Group | Type of Soil | Amount Finer Than 0.02 mm (wt %) | Typical Soil Type under Unified Soil Classification System[b] |
|---|---|---|---|---|
| NFS[c,d] | None | Gravels | 0–1.5 | GW, GP |
| | | Sands | 0–3 | SW, SP |
| Possible | ? | Gravels | 1.5–3 | GW, GP |
| | | Sand | 3–10 | SW, SP |
| Very low to high | F1 | Gravels | 3–10 | GW, GP, GW-GM, GP-GM |
| Medium to high | F2 | Gravels | 10–20 | GM, GM-GC, GW-GM, GP-GM |
| Negligible to high | | Sands | 10–15 | SW, SP, SM, SW-SM, SP-SM |
| Medium to high | F3 | Gravels | >20 | GM, GC |
| Low to high | | Sands, except very fine silty sands | >15 | SM, SC |
| Very low to very high | | Clays, PI > 12 | — | CL, CH |
| Low to very high | F4 | All silts | — | ML, MH |
| Very low to high | | Very fine silty sands | >15 | SM |
| Low to very high | | Clays, PI < 12 | — | CL, CL-ML |
| Very low to very high | | Varved clays and other fine-grained, banded sediments | — | CL and ML; CL, ML, and SM; CL, CH, and ML; CL, CH, ML, and SM |

[a]Based on laboratory frost-heave tests.

[b]G, gravel; S, sand; M, silt; C, clay; W, well graded; P, poorly graded; H, high plasticity; L, low plasticity.

[c]Non-frost-susceptible.

[d]Requires laboratory frost-heave test to determine frost susceptibility.

*Source:* After U.S. Army, Corps of Engineers (1965).

A soil classification system that is used for the determination of the frost susceptibility of soils is also very useful in frozen-ground engineering. Chamberlain (1981) reviewed the various criteria used by several countries in the world for his frost-susceptibility-of-soil classification system. The system most commonly used in the United States is that developed by the U.S. Army Corps of Engineers (1965). As shown in Table 2–6, the frost susceptibility of soils is divided into six main categories and related to the USCS classification system. The original system was developed by A. Casagrande (1932), who formulated the system based on the presence of fines (size 0.02 mm) which inhabit ice formations given the availability of moisture and the freezing temperature gradient. Frost heave and resulting forces that arise from the freezing of frost-susceptible soils are discussed in Chapter 4. The frost-susceptibility classification system may also be obtained after the frozen-soil classification is determined by the use of Tables 2–2 and 2–6. Other frost susceptibility criteria are presented in Table 2–7.

**Example 2–1**

Determine the frozen-soil classification of all samples that were recovered from a soil boring made for a proposed structure. The data, obtained by mechanical analysis and plasticity tests and by visual descriptions of frozen-soil samples, are given below. During the field investigation, it was noted that near the surface, the boring area was covered with spruce trees of height generally greater than 5 ft and that the topography was plain.

| | | Percent Passing Sieve: | | | | | |
|---|---|---|---|---|---|---|---|
| Sample | Depth (ft) | No. 4 | No. 10 | No. 40 | No. 200 | Liquid Limit (LL) | Plastic Limit (PL) |
| 1 | 5 | — | — | 100 | 95 | 52 | 30 |
| 2 | 10 | 85 | 75 | 40 | 30 | 44 | 22 |
| 3 | 15 | 55 | 40 | 20 | 15 | 21 | 12 |
| 4 | 20 | 95 | 80 | 30 | 5 | — | — |

Visual descriptions:

- Sample 1: Ice nonvisible, well-bonded excess ice.
- Sample 2: Visible ice, less than 1 in. thick, ice coatings on particles.
- Sample 3: Same as sample 2 but ice lenses are randomly oriented.
- Sample 4: Same as sample 2 but ice distinctly oriented.

**Solution:**

1. For soil sample 1, more than 50% of the sample passed a No. 200 sieve; thus the sample is a fine-grained soil and the Atterberg limits are required to classify the sample further. With LL = 52 and plasticity index PI = LL − PL = 52 − 30 = 22, the sample plots below the A line on the plasticity chart (Fig. 2–2). As LL is

**TABLE 2–7 Frost-Susceptibility Criteria**

| CRREL[a] | | TRRL[b] | | West German[c] | |
|---|---|---|---|---|---|
| Frost Suscepti-bility | Average Rate of Heave (mm/day) | Frost Suscepti-bility | Frost Heave in 10 Days (in.) | Frost Suscepti-bility | $CBR_f$ |
| Negligible | 0–0.5 | None | <0.5 | None | >20 |
| Very low | 0.5–1.0 | Marginal | 0.5–0.7 | Low to medium | 4–20 |
| Low | 1.0–2.0 | High | >0.7 | High | <4 |
| Medium | 2.0–4.0 | | | | |
| High | 4.0–8.0 | | | | |
| Very high | >8.0 | | | | |

[a,b,c]Chamberlain, 1981.

greater than 50, the sample is MH. The given ice description indicates that the sample is $Nb_e$.

2. For soil sample 2, the sample is a coarse-grained soil as the percent passing a No. 200 sieve is less than 50%. Out of 70% (100 − 30) coarse fraction, sands are 55% (85% passing a No. 4 sieve − 30% passing a No. 200 sieve). Therefore, the sample is a sand rather than a gravel. As the percent of the sample passing a No. 200 sieve is greater than 12%, the sample is either SM or SC. With LL = 44 and PI = 20 (44 − 22), the sample plots above the A line on the plasticity chart (Fig. 2–2). Therefore, the sample is SC. The ice description gives the sample as Vc.
3. A quick look at the characteristics for sample 3 indicates that the sample is also coarse-grained (15% passing a No. 200 sieve). Since more than 50% coarse fraction is retained on a No. 4 sieve, the sample is a gravel. The percent of the sample passing a No. 200 sieve is greater than 12%, so the sample is either GM or GC. With LL = 21 and PI = 9, the sample is GC. The ice description gives the sample as Vr.
4. Sample 4 is immediately seen to be coarse-grained soil since only 5% passes a No. 200 sieve. Since 95% passes a No. 4 sieve (more than 50% coarse fraction passing a No. 4 sieve), the sample is a sand. Next, note the amount of material passing a No. 200 sieve (5%). From Table 2–2 and Fig. 2–3, the sample is borderline (5 to 12% passing a No. 200 sieve), and therefore has a dual symbol such as SP-SM or SW-SM, depending on the values of $C_u$ and $C_c$. From the grain size distribution curve (Fig. 2–7) we find that $D_{60} = 0.8$ mm, $D_{30} = 0.45$ mm, and $D_{10} = 0.1$ mm. The coefficient of uniformity $C_u$ is

$$C_u = \frac{D_{60}}{D_{10}} = \frac{0.8}{0.1} = 8 > 6$$

and the coefficient of curvature $C_c$ is

$$C_c = \frac{(D_{30})^2}{D_{10} \times D_{60}} = \frac{(0.45)^2}{0.1 \times 0.8} = 2.53 \quad \text{(between 1 and 3)}$$

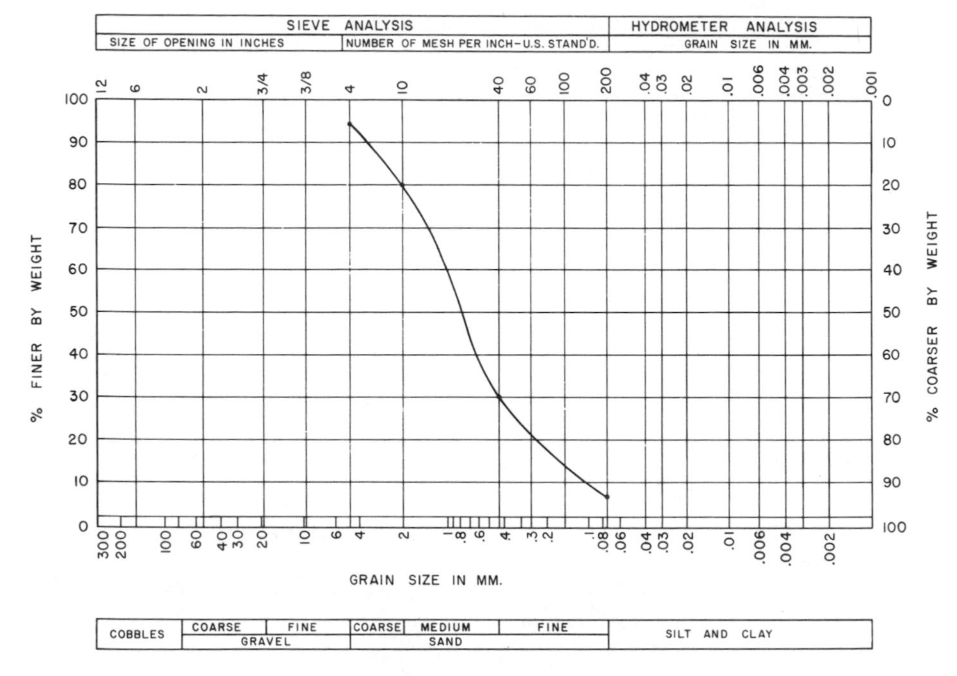


**Figure 2–7** Soil sample 4.

The sample meets the well-graded criteria as shown in Table 2–2 and its classification is SW-SM. The sample is SM because the fines are silty (nonplastic). From the ice description, classify the sample as Vs.

Answer:

- Sample 1: High-plasticity silt (MH, Nbe, A, c). From Table 2–6, the surface characteristics is category A, as spruce trees of height greater than 5 ft exist near the location and the topographic feature is category c due to its plain surface.
- Sample 2: Gravelly clayey sand (SC, Vc, A, c).
- Sample 3: Sandy clayey gravel (GC, Vr, A, c).
- Sample 4: Well-graded silty sand (SW-SM, Vs, A, c).

**Example 2–2**

Classify the four soil samples in Example 2–1 according to the frost-susceptibility classification.

## Solution

Using Table 2–6, we obtain the following results:

- Sample 1 to be designated as F4
- Sample 2 to be designated as F3

- Sample 3 to be designated as F3
- Sample 4 to be designated as F2

## PROBLEMS

**2–1.** Given the grain size distribution, the Atterberg limits, and the visual description of the three soils recovered from different sites, classify the following three frozen soils.

| | Percent Passing Sieve: | | | | | | |
|---|---|---|---|---|---|---|---|
| Soil | No. 4 | No. 10 | No. 40 | No. 200 | LL | PL | |
| 1 | 95 | 88 | 38 | 5 | — | — | NP[a] |
| 2 | 100 | 95 | 80 | 55 | 20 | 15 | |
| 3 | 40 | 30 | 15 | 4 | — | — | NP[a] |

[a]Nonplastic.

| Soil | Visual Description | Field Data |
|---|---|---|
| 1 | Visible ice coating on particles | Nonwoody tall grass (2–5 ft) and sloping ground |
| 2 | Visible stratified ice | Nonwoody, thick continuous mat and hummocky ground |
| 3 | Nonvisible poorly bonded ice | Woody birch trees (>5 ft tall) and plain area |

**2–2.** Classify the three soils in Problem 2–1 according to the frost-susceptibility classification given in Table 2–6.

**2–3.** Grain size distributions and plasticity data are given in Fig. 2–1 for four soil samples recovered from a borehole. The visual descriptions of frozen-soil samples and field data are given below. Classify the soil samples and comment on their suitability as "fill" material.

| Sample | Description |
|---|---|
| A | Ice nonvisible, poorly bonded; birch trees of height greater than 5 ft on a ridge |
| B | Ice nonvisible, no excess ice, birch trees of height less than 5 ft on a sloping area |
| C | Ice coatings on particles, nonwooded area, sledges up to 2 ft high on a plain area |
| D | Stratified ice on a continuous vegetative mat up to 4 in. high close to a pond within 50 ft |

**2–4.** Which of the soils in Problem 2–1 would you expect to be frost susceptible? Is this confirmed by Table 2–7?

**2–5.** The following data were obtained by mechanical analysis and plasticity tests of unfrozen-soil samples.

| Sample | Percent Passing Sieve: | | | | | | | LL | PI |
|---|---|---|---|---|---|---|---|---|---|
| | No. 4 | No. 10 | No. 40 | No. 200 | 0.05 mm | 0.02 mm | 0.002 mm | | |
| 1 | 100 | 90 | 80 | 7 | 3 | 0 | — | — | NP |
| 2 | 90 | 85 | 80 | 30 | 25 | 20 | 10 | 20 | 10 |
| 3 | 98 | 95 | 92 | 80 | 75 | 50 | 30 | 60 | 25 |
| 4 | 100 | 97 | 90 | 40 | 38 | 25 | 10 | 40 | 15 |

**(a)** Plot the grain size curve.

**(b)** Classify the soil. Assume no visible ice, a low shrub area, and sloping ground.

**(c)** Determine the frost susceptibility of the samples.

CHAPTER 3

# Physical, Mechanical, and Thermal Properties of Frozen Soils

## 3.1 INTRODUCTION

The effective use of frozen ground as an engineering foundation material will not be possible without a more precise understanding of the physical, mechanical, and thermal properties of frozen soils. In this chapter, these properties are examined and properties related to strength and deformation are analyzed. Due to the complex nature of frozen-soil composition, various factors, such as soil type, ice content, strain rate, and temperature, must be analyzed to gain a working understanding of the physical and mechanical properties of frozen soils. The relative influence of several important physical parameters on the thermal properties is also reviewed. The notation used in this chapter is given in Table 3.1.

## 3.2 PHYSICAL PROPERTIES OF FROZEN SOILS

To understand the basic physical properties of frozen soils, the interrelationships between the various soil components, such as soil particles, unfrozen water, ice, and air of frozen soils, must be identified. Anderson and Morgenstern (1973), Tsytovich (1975), Beskow (1935), Scott (1969), and others have recognized the interactions of these components and their distribution under different below-freezing temperatures. Stresses originating from several causes are vital to the properties and mechanics of frozen ground. The following physical properties of

**TABLE 3–1 Notation**

| Symbol | SI Unit | Definition |
|---|---|---|
| $a$ | mm$^2$/s | Thermal diffusivity (Eq. 3–60) |
| $A(t)$ | kN/m$^2$ | Modulus of nonlinear deformation (Eq. 3–22) |
| $C$ | kJ/m$^3$ K | Heat capacity (Eq. 3–56) |
| $C_t$ | kN/m$^2$ | Cohesion (Eq. 3–27) |
| $e$ | — | Void ratio (Eq. 3–13) |
| $E$ | kN/m$^2$ | Elastic modulus (Eq. 3–31) or dynamic elastic modulus (Eq. 3–47) |
| $G$ | kN/m$^2$ | Shear modulus (Eq. 3–50) |
| $G_s$ | — | Specific gravity of soil solid particles |
| $K$ | W/m K | Thermal conductivity (Eq. 3–52) |
| $L$ | J/m$^3$ | Latent heat of fusion (Eq. 3–61) |
| $S$ | % | Degree of saturation |
| $S_i$ | % | Degree of ice saturation (Eq. 3–12) |
| $\Delta u$ | kN/m$^2$ | Change in pore pressure |
| $\omega$ | % | Water content (Eq. 3–2) |
| $\omega_t$ | % | Total water content (Eq. 3–1) |
| $\omega_i$ | % | Ice water content |
| $\omega_{uw}$ | % | Unfrozen water content |
| $\rho$ | kg/m$^3$ | Unit weight or bulk density |
| $\rho_d$ | kg/m$^3$ | Unfrozen dry density |
| $\rho_f$ | kg/m$^3$ | Frozen density (Eq. 3–6) |
| $\rho_{df}$ | kg/m$^3$ | Frozen dry density (Eq. 3–7) |
| $\rho_w$ | kg/m$^3$ | Density of water |
| $\rho_i$ | kg/m$^3$ | Density of ice |
| $\mu$ | — | Poisson's ratio (Eq. 3–51) |
| $V$ | m$^3$ | Total volume |
| $V_A$ | m$^3$ | Volume of air |
| $V_{uw}$ | m$^3$ | Volume of unfrozen water |
| $V_i$ | m$^3$ | Volume of ice |
| $V_s$ | m$^3$ | Volume of solids (Eq. 3–13) or shear wave velocity (Eq. 3–48) |
| $V_v$ | m$^3$ | Volume of voids |
| $V_p$ | m/s | Compressional wave velocity (Eq. 3–47) |
| $\varepsilon$ | — | Total strain |
| $\varepsilon_1$ | — | Strain rate 1 hr after stress application |
| $\varepsilon_0$ | — | Elastic strain |
| $\varepsilon(t)$ | — | Nonrecoverable or plastic strain |
| $\dot{\varepsilon}$ | — | Axial strain rate |
| $\dot{\varepsilon}_e$ | — | Effective shear strain (Eq. 3–22b) |
| $\sigma$ | kN/m$^2$ | Axial stress |
| $\sigma_e$ | kN/m$^2$ | Effective shear stress |
| $\sigma_c$ | kN/m$^2$ | Compressive strength |
| $\sigma_n$ | kN/m$^2$ | Normal stress (Eq. 3–27) |
| $\sigma_{oct}$ | kN/m$^2$ | Octahedral normal stress |
| $\sigma_{ult}$ | kN/m$^2$ | Long-term strength (Eq. 3–43) |
| $\tau$ | kN/m$^2$ | Shear strength (Eq. 3–27) |

frozen soils are of primary interest and have particular engineering design relevance:

1. Total water content $\omega_t$ (%)
2. Ice water content $\omega_i$ (%)
3. Bulk density $\rho$ (kg/cm$^3$ or lb/ft$^3$)
4. Dry density $\rho_d$ (kg/cm$^3$ or lb/ft$^3$)
5. Ice saturation $S_i$ (%)

### 3.2.1 Frozen-Soil Constituents

Similar to three-phase unfrozen soil, frozen soil is also a multiphase, complex material. As shown in Fig. 3–1, frozen soil consists of four phases for a given volume. These phases are:

1. *Solid phase or soil particles:* These are mineral particles (silicates), organic, or both.
2. *Plastic-viscous phase or ice:* The viscosity of frozen soil bonded or cemented by the ice as well as the viscosity of ice is a highly plastic material. Their properties depend strongly on the internal below-freezing temperatures and the magnitude and duration of load applied.
3. *Liquid phase or unfrozen water:* Water molecules fill part or all of the voids or pores between the soil particles.
4. *Gaseous phase or air vapor:* Air vapor fills voids that are not occupied by liquid.

Also present are several interphases, the most important of these being ice-water, ice-silicates, water-silicate, air-water, and air-ice. The interrelationships of

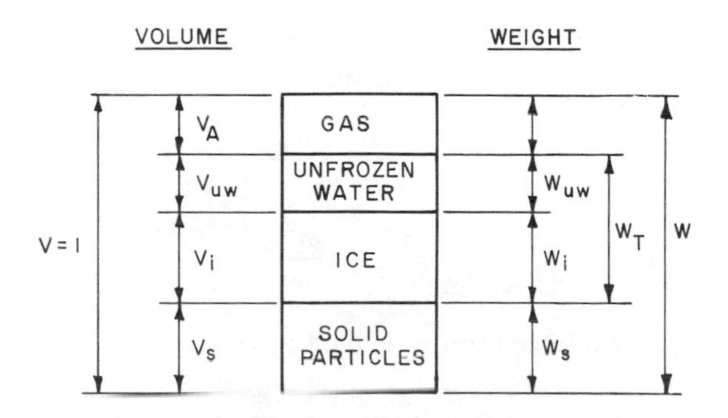


**Figure 3–1** Constitution of frozen soil.

these soil constituents depend on the properties of each phase as well as external influences such as temperature and stress. Physical relationships among the minerals, water, and air interphase are routinely described in most soil mechanics text (Lambe and Whiteman, 1969; Mitchel, 1976; Holtz and Kovacs, 1981). The size and composition of the mineral particles may have a considerable influence on the properties of frozen soil. Anderson and Morgenstern (1973) have reviewed the water–ice phase composition, interface characteristics, and applicable thermodynamic relationships. The phases may be determined in terms of the physical parameters defined below.

Neglecting the vapor phase, the total water content $\omega_t$ is given by

$$\omega_t = \omega_i + \omega_{uw} \tag{3–1}$$

where

$$\omega_i = \text{ice water content}$$
$$\omega_{uw} = \text{unfrozen water content}$$

All water content is defined as

$$\omega_t = \text{water content} = \frac{\text{weight of water}}{\text{weight of dry soil}} = \frac{W_w}{W_s} \tag{3–2}$$

From Fig. 3–1,

$$W_i = W_t - W_{uw} \tag{3–3}$$

The unfrozen water content $\omega_{uw}$ may be determined by various experimental methods (Anderson and Morgenstern, 1973).

Relative iciness $i_{\text{ice}}$ is defined as the ratio between the weight of ice and the weight of total water content in a given unit volume of frozen soil:

$$i_{\text{ice}} = \frac{W_i}{W_t} \tag{3–4}$$

or

$$\begin{aligned} i_{\text{ice}} &= \frac{\omega_i W_s}{\omega_t W_s} \\ &= \frac{\omega_t - \omega_{uw}}{\omega_t} \\ &= 1 - \frac{\omega_{uw}}{\omega_t} \end{aligned} \tag{3–5}$$

The bulk density $\rho_f$ of a frozen soil is defined as

$$\rho_f = \frac{\text{weight}}{\text{volume}} = \frac{W}{V} \tag{3–6}$$

The dry density $\rho_{df}$ of a frozen soil is the weight of dry soil divided by the total volume of frozen soil:

$$\rho_{df} = \frac{W_s}{V} \tag{3–7}$$

The relationship between $\rho_f$, $\rho_{df}$, and $\omega_t$ is given by

$$\begin{aligned}\rho_f &= \frac{W}{V} = \frac{W_s + W_{iuw}}{V} \qquad \text{(see Fig. 3–1)}\\ &= \frac{W_s}{V} + \frac{W_{iuw}}{V}\\ &= \rho_{df} + \omega_t \frac{W_s}{V} = \rho_{df}(1 + \omega_t)\end{aligned} \tag{3–8}$$

Assuming full saturation (no unfrozen water but with excess ice in soils), we have from Eq. (3–8);

$$\rho_f = \frac{G_s\rho_w(1 + \omega)}{1 + 1.09G_s\omega} \tag{3–9}$$

where $G_s$ is the specific gravity of soil solid particles and $\rho_w$ is the density of water = 62.4 lb/ft$^3$ or 1000 kg/m$^3$ or 1 g/cm$^3$.
The frozen dry unit weight $\rho_{df}$ is given by

$$\rho_{df} = \frac{\rho_f}{1 + \omega}$$

Using the relationship of Eq. (3–9), we obtain

$$\rho_{df} = \frac{G_s\rho_w}{1 + 1.09G_s\omega} \tag{3–10}$$

or

$$\rho_{df} = \frac{\rho_w}{(1/G_s) + 1.09\omega} \tag{3–11}$$

The degree of ice saturation $S_i$ is given by

$$\begin{aligned}S_i &= \frac{\text{volume of ice}}{\text{volume of pores in frozen soil}}\\ &= \frac{V_i}{V_v} = \frac{W_i}{\rho_i e V_s} = \frac{\omega G_s \rho_w}{\rho_i e} \qquad (\text{using } V_s = W_s/G_s\rho_w)\end{aligned} \tag{3–12}$$

Generally, the degree of saturation $S$ is defined as the ratio between the volume of water $V_w$ and the volume of voids $V_v$. If all voids are filled with liquid ($V_w = V_v$), the degree of saturation is 100%. The void ratio $e$ is given by

$$e = \frac{V_v}{V_s} \tag{3–13}$$

We have for the density of water,

$$\rho_w = \frac{\text{weight of water}}{\text{volume of water}} = \frac{W_w}{V_w}$$

or

$$W_w = \rho_w V_w = \omega W_s = \omega V_s \rho_s$$

But $S = V_w/V_v$ and $e = V_v/V_s$, so

$$\rho_w S V_v = \omega V_s \rho_S$$

or

$$\rho_w S e V_s = V_s \rho_s \tag{3–14a}$$

or

$$\rho_w S e = \omega \rho_s$$

*or*

$$Se = \omega G_s \tag{3–14b}$$

The relationship between void ratio $e$, degree of saturation $S$, water content $\omega$, and specific gravity of soil particles $G_s$ as given by Eq. (3–14b) is one of the most useful equations for phase problems.

**Example 3–1**

A silty frozen-soil sample has the following physical properties:

- Specific gravity $G_s = 2.67$
- Total water content $\omega_t = 30\%$
- Bulk density $\rho = 90$ lb/ft$^3$

Determine the void ratio and the degree of saturation.

### Solution

From Eq. (3–2) we have

$$\omega_t = \frac{W_t}{W_s} = 0.3 \tag{3–15a}$$

or

$$W_t = 0.3 W_s$$

Assume total volume $V = 1$. Given

$$\rho = \frac{W}{V} = \frac{W_t + W_s}{V}$$

$$= 90$$

or

$$W_t + W_s = 90 \tag{3–15b}$$

From Eqs. (3–15a) and (3–15b) we get

$$W_s = 69.23 \qquad \text{and} \qquad W_t = 20.77$$

We now have that the volume of solid

$$V_s = \frac{W_s}{G_s \rho_w}$$

$$= \frac{69.23}{(2.67)(62.4)} = 0.415$$

The void ratio

$$e = \frac{V_v}{V_s}$$

$$= \frac{1 - 0.415}{0.415}$$

$$= 1.409$$

The degree of saturation

$$S = \frac{V_w}{V_v} \times 100$$

$$= \frac{W_t/\rho_\omega}{V_v} \times 100$$

$$= \frac{20.77/62.4}{1 - 0.415} \times 100$$

$$= 56.9\%$$

**Example 3–2**

A frozen-soil sample weighs 1317.50 g in its natural state and 850 g after drying. The unfrozen water content of the sample is found to be 10% at the natural temperature of the sample. What are the void ratio, dry density, and ice saturation of the sample? Assume that $G_s = 2.66$ and that the volume of air is negligible.

### Solution

The weight of water = 1317.50 − 850 = 467.5 g. The total water content

$$\omega_t = \frac{467.5}{850} \times 100$$
$$= 55\%$$

The volume of solid

$$V_s = \frac{W_s}{G_s \rho_w}$$
$$= \frac{850}{(2.66)(1)} \qquad \text{(assuming that } G_s = 2.66\text{)}$$
$$= 319.55 \text{ cm}^3$$

The volume of water

$$V_w = \frac{467.5}{1} = 467.5 \text{ cm}^3$$

The volume of air is negligible ($V_w = V_v$), so

$$e = \frac{V_v}{V_s} = \frac{467.5}{319.55} = 1.463$$

The ice content

$$\omega_i = \omega_t - \omega_{uw}$$
$$= 55 - 10$$
$$= 45\%$$

The frozen dry density

$$\rho_{df} = \frac{\rho_w}{1/G_s + (1.09\omega_i + \omega_{uw})/S}$$
$$= \frac{1}{(1/266) + [1.09(0.45) + 0.10]/1}$$
$$= \frac{1}{0.376 + (0.49 + 0.1)}$$
$$= 1.035 \text{ g/cm}^3$$
$$= 1035 \text{ kg/m}^3$$

The ice saturation

$$S_i = \frac{\omega_i G_s \rho_w}{e \rho_i} \times 100$$

$$= \frac{(0.45)(2.66)(1)}{(1.463)(0.916)} \times 100$$

$$= 89\%$$

**Example 3–3**

Derive an expression of void ratio $e$ in terms of porosity $n$.

Solution

Porosity is defined as $n = V_v/V$ (see Fig. 3–1). Then the void ratio

$$e = \frac{V_v}{V_s}$$

$$= \frac{V_v}{V - V_v}$$

$$= \frac{V_v/V}{1 - V_v/V}$$

$$= \frac{n}{1 - n} \tag{3–16}$$

### 3.2.2 Unfrozen Water in Frozen Soils

As a result of various investigations, it is recognized that an unfrozen water phase that separates ice from the mineral and organic soil matrix exists even at $-10°C$. The amount of unfrozen water in fine-grained frozen soils at negative temperatures may be significant. Following are listed various factors that govern the presence of unfrozen water.

Primary factors:

1. Specific surface area of solid phase present in the frozen soil
2. Negative temperature of frozen soils
3. Confining pressure exerted or existing at soil–ice–water interfaces
4. Osmotic potential of the soil solution or composition of absorbed cations

Secondary factors:

1. Mineralogical composition of soil
2. Particle packing geometry
3. Surface charged density
4. Exchangeable absorbed ions

Various investigators (Anderson and Morgenstern, 1973; Andersland and Anderson, 1978; Williams, 1967; Johnston, 1981) have explained the presence of water–ice phase composition in frozen soils. It may be stated that this water–ice phase composition in frozen soils results from surface forces or intermolecular forces that limit the amount of unfrozen water surrounding the soil particles. The freezing-point depression or the temperature below 0°C (32°F) at which nucleation and ice crystalization begin will vary depending on the surface forces, which consist of capillary and absorbed forces. Unfrozen water can move within the water–ice interfacial films under electrical and osmotic as well as thermal gradients. Also, unfrozen water may be present and be the result of pressure melting of high-ice-content soils. The amount of water present at the water–ice interface and the variations associated with negative temperatures are of importance to the engineer for the design of foundations for structures in frozen ground.

The direct methods of dilatometry, adiabatic calorimetry, x-ray defraction, heat capacity, nuclear magnetic resonance, and differential thermal analysis may be used to identify and quantify the amount of unfrozen water present in a soil. Anderson and Morgenstern (1973) have discussed in detail several direct and indirect methods developed for the measurement of unfrozen water content in frozen soils.

Based on experimental results, several empirical formulas are also reported for the determination of unfrozen water content. Anderson and Tice (1973) reported the relationship for remolded frozen soils as

$$\omega_{uw} = \alpha\theta^{\beta} \tag{3–17}$$

where

$\alpha, \beta$ = characteristics of soil parameter (Table 3–2)
$\theta$ = temperature below freezing, °C

**TABLE 3–2 Soil Parameters $\alpha$ and $\beta$**

| Soil Type | $\alpha \times 10^{-2}$ | $\beta$ |
|---|---|---|
| Fairbanks silt | 4.8 | −0.33 |
| Limonite | 8.81 | −0.33 |
| Kaolinite | 23.80 | −0.36 |
| Wyoming bentonite | 55.99 | −0.29 |

*Source:* After Anderson and Tice (1972).

Dillon and Andersland (1966) and Anderson and Tice (1973) have published equations giving the unfrozen water content of frozen soil as a function of temperature. The equations of Dillon and Andersland incorporate values for the specific surface area as well as the Atterberg limits, the freezing-point depression of the pore water, the clay mineral type, and a defined activity ratio for the soil and other required values for surface area only.

From a regression analysis of phase composition data for representative soils having widely varying properties and characteristics, the following relationship was found:

$$\ln \omega_{uw} = a + b \ln S + CS^d \ln \theta \tag{3–18}$$

where $a, b, c, d$ = empirical constants

$S$ = specific surface area of the soil, $m^2/g$

$\theta$ = temperature below freezing, °C

Equation (3–17) given by Anderson and Tice (1973) is well suited for incorporation in mathematical models of the thermal behavior of frozen ground (Ho et al., 1969; Nakamo and Brown, 1972).

Figure 3–2 presents data for six representative soils. Tsytovich (1978) suggested the following equation for the determination of unfrozen water content:

$$\omega_{uw} = K_{PI}K_{H}$$

where $K_{PI}$ = plasticity index of soil

$K_H$ = soil constant which varies at different temperatures (see Table 3–3)

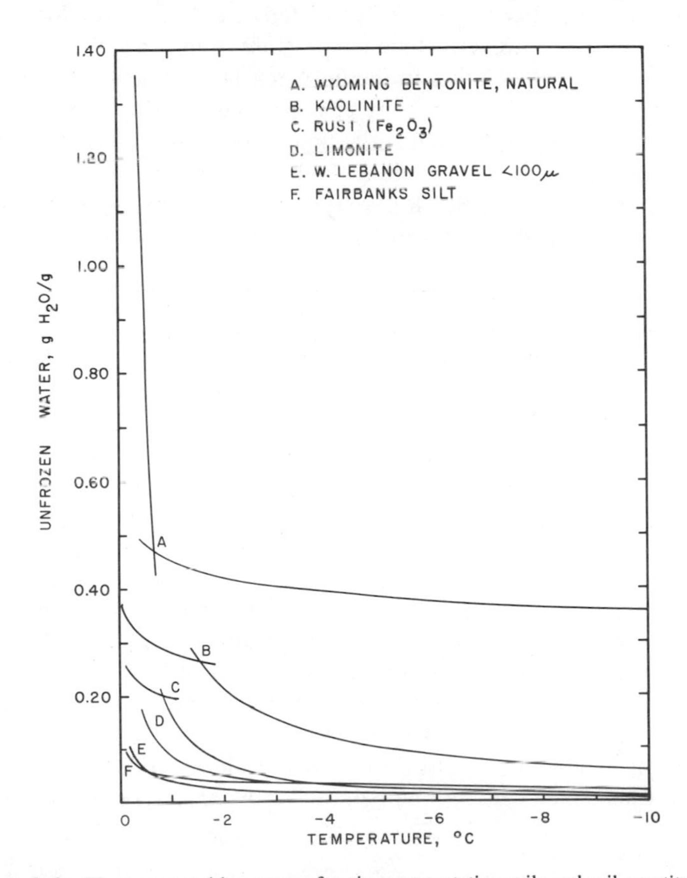


**Figure 3–2** Phase composition curves for six representative soils and soil constituents. (After Anderson and Morgenstern, 1973.)

**TABLE 3–3 Values of $K_H$**

| Soil | Plasticity Index, $W$ | $K_H$ at Soil Temperature (°C) of: | | | | | |
|---|---|---|---|---|---|---|---|
| | | −0.3 | −0.5 | −1 | −2 | −4 | −10 |
| Sand | $W < 1$ | 0 | 0 | 0 | 0 | 0 | 0 |
| Sandy loam | $1 \leq W \leq 2$ | 0 | 0 | 0 | 0 | 0 | 0 |
| Sandy loam | $2 < W \leq 7$ | 0.6 | 0.5 | 0.4 | 0.35 | 0.3 | 0.35 |
| Clay loam | $7 < W \leq 13$ | 0.7 | 0.65 | 0.6 | 0.5 | 0.45 | 0.4 |
| Clay loam | $13 < W \leq 17$ | [a] | 0.75 | 0.65 | 0.55 | 0.5 | 0.45 |
| Clay | $> 17$ | [a] | 0.95 | 0.9 | 0.65 | 0.6 | 0.55 |

[a]The pores contain unfrozen water only.

*Source:* After Tsytovich (1975).

The effect of solute in the water–ice composition must also be taken into account, and such efforts generally shift the unfrozen water content versus temperature curve relationship toward lower temperatures. This shift may be compared with the freezing-point depression of the soil solution corresponding to the osmotic potential. The freezing-point depression is generally low for frozen soils with low salt content. However, as the temperature of frozen soil is lowered, there is a concentration of salt solution in the unfrozen liquid fraction and the freezing point of the residual liquid film is further depressed. Tsytovich et al. (1978) reported that the chemistry of the interstitial fluid is very important, particularly with respect to frozen saline soils encountered both onshore and offshore.

Soviet investigators (U.S.S.R., 1973) consider a frozen soil to be salty if the salt content exceeds the following limits:

| Soil Type | Salt Content[a] (%) |
|---|---|
| Fine, medium, coarse gravelly sands | 0.10 |
| Silty sands | 0.05 |
| Sandy loam | 0.15 |
| Clay | 0.25 |

[a]Salt content is measured as the ratio between the weight of salts and the weight of dry soil.

A basic question may be asked. What is the significance of unfrozen water content in frozen soils? Results of many experimental studies have shown that the material properties (physical, mechanical, thermal) as well as the behavior of frozen soils is affected by the water–ice phase composition within the range of negative temperatures. The effect of overburden pressure of applied pressure, salt solution, and negative temperatures must be considered to determine the amount of unfrozen water. Such data will be helpful for the analysis of deformation characteristics as well as adfreeze strength of frozen soils. These aspects are discussed further in later sections of this chapter and in Chapter 6.

**Example 3–4**

Calculate the unfrozen water content of frozen fine-grained Fairbanks silt at a temperature of $-4°C$. The specific surface area of Fairbanks silt may be assumed as 40 $m^2/g$.

**Solution**

Using Eq. (3–18), we have

$$\ln \omega_{uw} = a + b \ln S + CS^d \ln \theta$$

The values of empirical constants $a$, $b$, $c$, and $d$ may be assumed to be (Anderson and Tice, 1973)

$$a = 0.2618 \qquad b = 0.5519 \qquad c = -1.449 \qquad d = -0.264$$

We obtain

$$\ln \omega_{uw} = 0.2618 + 0.5519 \ln (40) - 1.440(40)^{-0.264} \ln (4)$$

or

$$\omega_{uw} = 4.68$$

**Example 3-5**

Calculate the freezing-point depression for Fairbanks silt with an unfrozen water content of 2 g of $H_2O$ per gram of soil.

**Solution**

Rearranging Eq. (3–17), we obtain

$$\theta = \left(\frac{(\omega_{uw})^{1/\beta}}{\alpha}\right)$$

$$= \frac{2^{1/-0.33}}{0.048} = 1.236 \times 10^{-5}°C$$

($\alpha$ and $\beta$ are obtained from Table 3–1.) The freezing-point depression $\simeq$ 0°C.

Ice-phase properties are also temperature dependent. Deformation of ice and its interaction with the soil mineral skeleton must be understood for the determination of mechanical properties of frozen soils. The mechanical behavior of ice and frozen soils is discussed in the following sections.

## 3.3 MECHANICAL PROPERTIES OF FROZEN SOILS

Understanding the mechanical properties of frozen soil is of primary importance to frozen-ground engineering. The phenomena that control the mechanical behavior of frozen soils have been studied with both artificially frozen soil samples and naturally frozen soil samples for the last decade. It is to be recognized that the structure of frozen soil in sites differs significantly from that of artificially prepared material in the laboratory. The application of laboratory results to practical problems requires much caution and judgment, as there are a number of dominant variables influencing the way in which frozen soils behave with respect to their mechanical properties. Some factors, such as soil type, ice content, strain rate, and temperature, which control the mechanical properties of frozen soils must be fully understood before one can solve specific engineering problems.

The mechanical properties of frozen soils that are of importance to engineers in the design and construction of various structures in frozen ground are as follows:

1. Compression strength and tensile strength
2. Shear strength
3. Creep strength and long-term strength
4. Compressibility

The strength of frozen soils is highly temperature and time dependent. The strength of a soil is governed by cohesion and interparticle friction of its soil–ice matrix system. Generally, the strength of ice-poor frozen soil (ice content less than 30% by volume) is dominated by interparticle frictional forces. The strength of ice-rich frozen soils (ice content greater than 30% by volume) is given by cohesive forces between the soil–ice matrix system. Ice found in frozen ground will normally be hexagonal ice. Ice crystals within lenses are usually elongated in the direction that heat flowed during freezing, with the *c* axes being oriented randomly in a plane orthogonal to the direction of growth (Penner, 1960). The properties of ice are responsible for several aspects of unique behavior that characterize frozen soils. Their viscoplastic strength and deformation properties can be attributed largely to the presence of ice matrix and ice cementation bonds. Changes in temperature alter the phase composition of water in frozen soils, which in turn control the degree of ice cementation and total ice content.

### 3.3.1 Mechanical Behavior of Ice

The presence of ice as a matrix or internal bonding agent greatly affects the rheological (i.e., load–deformation) characteristics of frozen soils. Moreover, grain-to-grain contact has an influence on soil behavior, and in ice-rich materials, a significant portion of the particles are completely separated from each other by ice. With ice present as pore cement or discrete veins, ice-phase behavior will probably dominate any load–deformation relationships determined for frozen soil. Therefore, studies of mechanical behavior or creep in ice are a useful starting point in the development of appropriate stress–strain relationships for frozen soils.

Various investigators (Glen, 1952, 1955, 1974; Hult, 1966; Odquist, 1966; Hobbs, 1974) have studied various ice types under widely differing conditions. Early investigators assumed that ice would behave as a simple Newtonian viscous material, so that at any given temperature, strain rate would be linked to stress by a constant coefficient of viscosity. Attempts to determine the constant coefficient of viscosity led to results that varied by as much as six orders of magnitude. From other data, it was apparent that ice was a non-Newtonian material; therefore, an examination of deformation mechanisms was required.

Glen (1952, 1955, 1974) performed temperature-controlled laboratory creep tests by subjecting randomly oriented polycrystalline ice to uniaxial compression. He observed a rheological response, with the ice undergoing a small instantaneous deformation (elastic strain) at the outset and continuing to deform at a rate that gradually decreased with time (transient or primary creep). With a sufficient passage of time, a steady creep rate was usually obtained (constant or secondary creep). Under higher stresses, this eventually gave way to accelerated creep (tertiary) and ultimately to failure. Since ice is a viscoplastic crystalline material, its time-dependent plastic deformation can be described with three modes of creep: primary, secondary, and tertiary creep. Typical creep behavior of ice is presented in Fig. 3–3. The major factors that control the development of a particular creep mode are the magnitude and duration of the applied load and the ice temperature. The creep behavior of material is discussed further in Section 3.3.5. A flow law (Glen, 1952, 1955, and 1974) may be used for polycrystalline ice to describe its deformation characteristics under simple compression as

$$\dot{\varepsilon} = A\sigma^n \tag{3–19}$$

where $\dot{\varepsilon}$ = axial strain rate
$\sigma$ = axial stress
$A$ = constant for a given temperature and ice type
$n$ = exponent

Under low stresses, Eq. (3–19) is modified (Butkovich and Landaver, 1959) as

$$\dot{\varepsilon} = A\sigma + B\sigma^n \tag{3–20}$$

where $A$ and $B$ are constant for a given ice type and temperature and $n$ is a constant.

After a detailed review of the mechanics of ice creep (Barnes, 1966; Glen, 1974; Roggensack, 1977; Mellor and Cole, 1982), an Arrhenius type of equation could be used. This equation differs from Eq. (3–20) by including temperature effects and takes the form

$$\dot{\varepsilon} = A'\sigma^n \exp\left(\frac{-Q}{RT}\right) \tag{3–21}$$

where $Q$ = activation energy
$R$ = gas constant
$T$ = absolute temperature

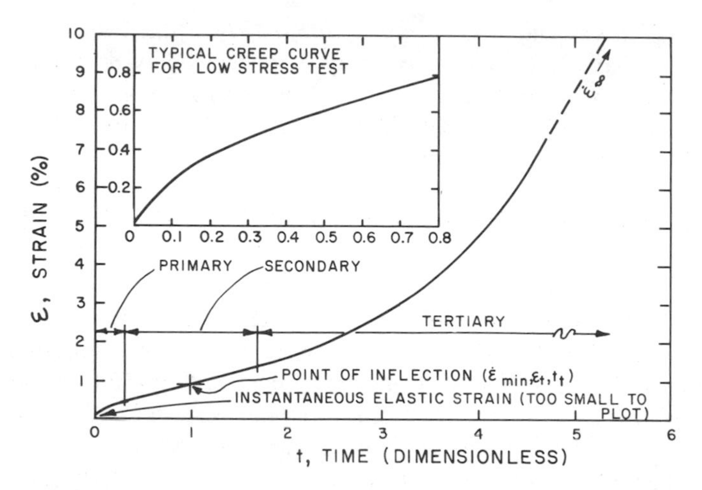


**Figure 3–3** Creep behavior of polycrystalline ice. (After Mellor and Cole, 1982.)

The constant $A$ in Eq. (3–19) has been replaced by $A'\exp(-Q/RT)$ in Eq. (3–21). From Eq. (3–21) it is clear that $\dot{\varepsilon}$ is proportional to $\sigma^n$, and a plot of log $\sigma$ against log $\dot{\varepsilon}$ would produce a line with a slope equal to $n$. Barnes et al. (1971) found this to be the case for stresses less than 1 $MN/m^2$ (150 psi) and obtained an exponent $n$, approximately equal to 3. At higher stresses, the exponent approached a value of 5.

At warmer temperatures, the creep process was found to accelerate as a result of pressure melting at intergranular contacts plus intergranular sliding. Glen (1974) has also shown that the creep rate of polycrystalline ice is strongly influenced by the average ice-crystal grain size.

The flow law for random polycrystalline ice under multiaxial conditions may be described in terms of the second invariants of the deviator stress and strain rate (Emery and Nguyen, 1974; Ladanyi, 1974) using the relationships

$$\sigma_e = \text{effective shear stress} = \sqrt{3}\sqrt{I_2} \tag{3–22a}$$

and

$$\dot{\varepsilon}_e = \text{effective shear strain rate} = \frac{2}{\sqrt{3}}\sqrt{J_2} \tag{3–22b}$$

where $I_2$ and $J_2$ denotes the second invariants of deviatoric stress and strain rate, respectively. The flow law for random polycrystalline ice becomes

$$\dot{\varepsilon}_e = B\sigma_e^n \tag{3–23}$$

where $B$ and $n$ are temperature-dependent constants.

Figure 3–4 summarizes the published data on ice creep for temperatures of approximately $-1$, $-2$, $-5$, and $-10$°C.

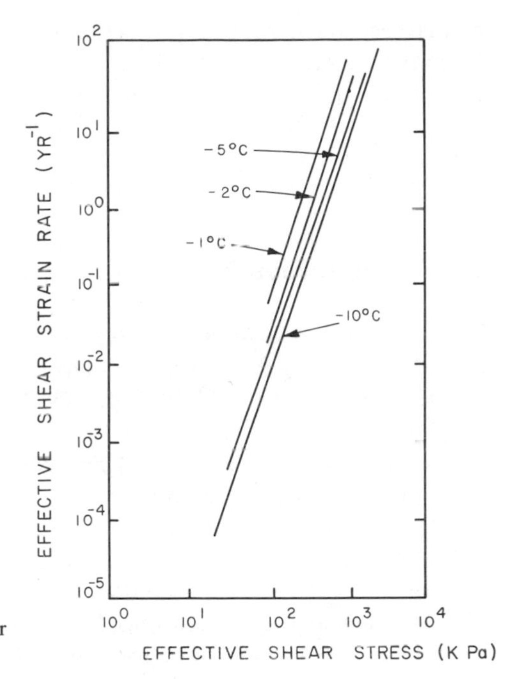


**Figure 3–4** Summary of creep data for ice. (After Morgenstern et al., 1980.)

Morgenstern et al. (1980) have reviewed the secondary creep rates in random polycrystalline ice and presented the following values for creep parameters $B$ and $n$:

| Temperature (°C) | $B$ (kPa$^{-n}$/yr) | $n$ |
|---|---|---|
| −1 | $4.5 \times 10^{-8}$ | 3.0 |
| −2 | $2.0 \times 10^{-8}$ | 3.0 |
| −5 | $1.0 \times 10^{-8}$ | 3.0 |
| −10 | $5.6 \times 10^{-9}$ | 3.0 |

## 3.3.2 The Deformation Process in Frozen Soils

As discussed in Section 3.2.1, the structure of frozen soil is complex. The presence of soil particles in a body of ice may modify or entirely change the indigenous stress–strain relationship. The degree of alteration is particularly sensitive to the bulk density of the frozen soil and the grain size of the soil particles.

In general, the presence of soil particles impedes the movement of dislocations within the ice crystals, thereby suppressing ice creep. However, the soil particle may also cause a reduction in the average grain size of the ice crystals, and at very low concentrations of soil (i.e., bulk densities less than 950 $kg/m^3$ or 60 $lb/ft^3$) the creep rate may be enhanced. At higher concentrations of solids, the suppressive mechanism dominates and the creep rate is reduced. The mechanical properties of frozen soils under uniaxial compressions, tension, and shear are discussed in the following sections. These sections are followed by a detailed discussion of the long-term creep characteristic of frozen soils.

### 3.3.3. Compressive Strength and Tensile Strength

The strength and deformation characteristics of frozen soils depend on several factors, of which temperature, ice content, and strain rate are the most dominant. Haynes (1978), Baker (1978), Kaplar (1971), Sayles and Epanchin (1966), Mellor and Smith (1967), Vyalov et al., (1966a), and Phukan (1980b) have reported on the dependence of the unconfined compressive strength on strain rate and temperature. In fine-grained frozen soils, the interstitial ice bonds the soil grains together, so that the strength of ice becomes an important factor. In addition, the unfrozen water surrounds the soil particles and this influences the strength and the type of failure to be anticipated. Sayles and Haynes (1974) have postulated that the intergranular friction can impede failure when the deformation permits grain-to-grain contact of the soil or ice. Generally, the ice matrix is strengthened with increasing strain rate and decreasing temperature. So the compressive strength of frozen soils increases at a higher strain rate and a decreasing temperature. The ice matrix accounts for 58% of this rate of increase (Haynes, 1978) and the strength increment is about 1.362 $MN/m^2$ per 0°C (32°F). At a high strain rate, 1.0 $S^{-1}$, and at low temperatures, $\theta < -40$°C, an elastic behavior is usually observed. Typical stress–strain behavior of frozen Fairbanks silt at various negative temperatures is presented in Fig. 3–5. Figure 3–6 illustrates the compressive strength $\sigma_c$ as a function of temperature. The following equations are derived from the relationships obtained between the compressive strength and the temperature of soil samples at a strain rate of 5% per minute (Phukan, 1980b) and the relationships may be used to determine the unconfined compressive strength of frozen silt.

$$\sigma_c = 0.7 - 0.45\theta \qquad \theta < -3.9°\text{C} \qquad (3\text{–}24)$$

$$= 0.55 - 0.4\theta + 0.1\theta^2 \qquad \theta > -3.9°\text{C} \qquad (3\text{–}25)$$

where $\sigma_c$ is in $MN/m^2$ and $\theta$ is in °C. At a given temperature, the compressive strength is related to the strain rate (Baker, 1978) by

$$\sigma_c = A\dot{\varepsilon}^{\,b} \qquad (3\text{–}26)$$

where $\sigma_c$ is in $MN/M^2$ and $\dot{\varepsilon}$ is in $s^{-1}$. $A$ and $b$ are constants dependent on soil type and temperature. They may be determined from the relationship between log $\sigma_c$ and log $\dot{\varepsilon}$.

Equation (3–26) may also be used in determining the tensile strength of frozen fine-grained soils. The tensile strength was found to be relatively insensi-

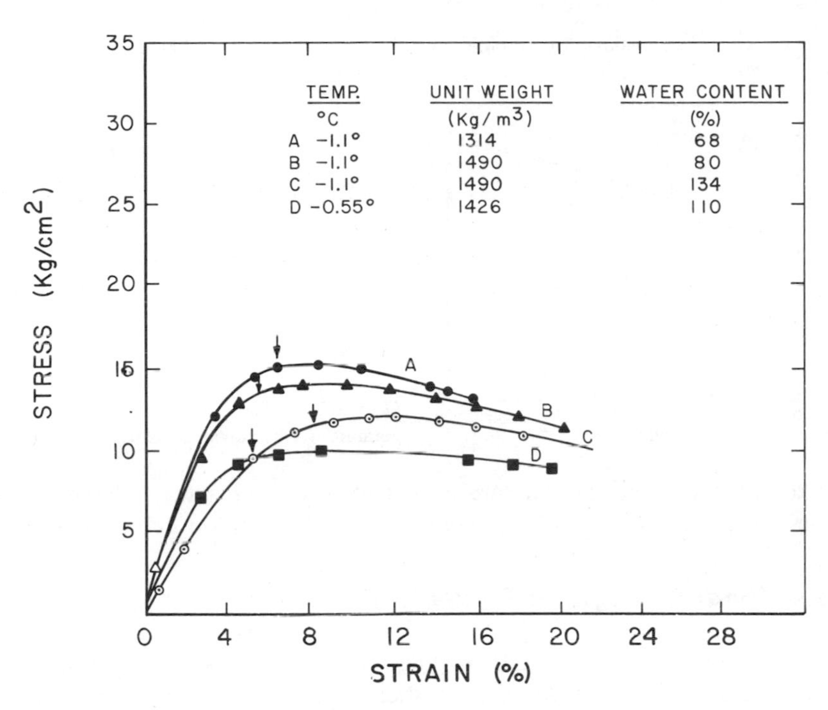


**Figure 3–5** Relationship between stress and strain at temperatures $-0.55°C$ and $-1.1°C$, Fairbanks silt.

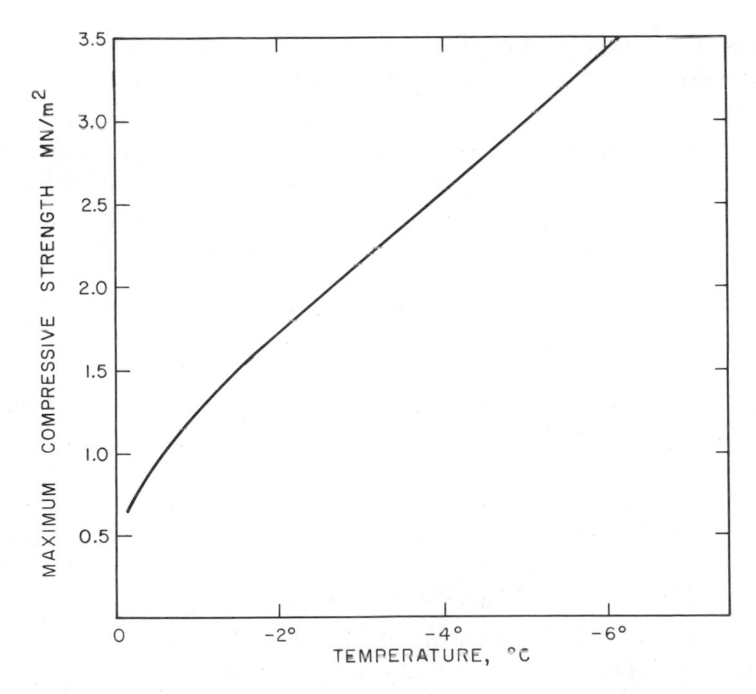


**Figure 3–6** Maximum compressive strength versus temperature, Fairbanks silt.

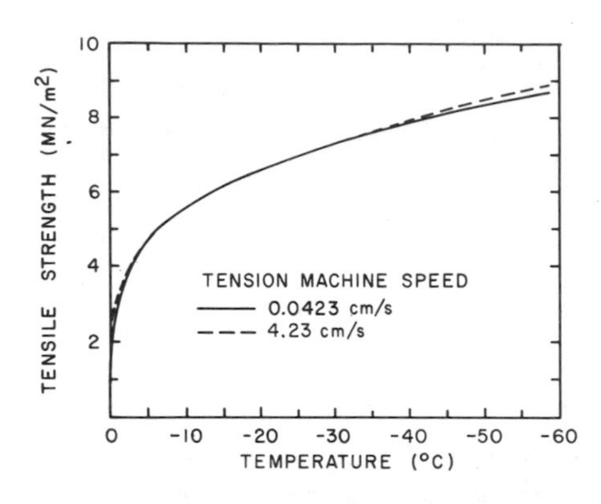


**Figure 3–7** Tensile strength versus temperature. (After Haynes, 1978.)

tive to strain rate and temperature except at low strain rates and temperatures above −6°C. A typical tensile strength relationship is shown in Fig. 3–7.

### 3.3.4 Shear Strength of Frozen Soils

The shear strength of frozen soils may be separated into two components (Ruedrich and Perkins, 1973; Vyalov and Shusherina, 1970): (1) the resistance of ice to deformation, and (2) the frictional response of soil grain contacts. These separate components can be well approximated by the following conditions:

1. The time-dependent strength of the ice matrix, taken as the unconfined compressive strength of the soil in question and obtained at an appropriate strain rate
2. The essentially time-independent strength associated with the mobilization of frictional resistance at particle contacts under the applied boundary stresses.

As such, the shear strength of frozen soils can be estimated by obtaining the uniaxial strength as a function of strain rate or time to failure, and adding to this the frictional response of the soil matrix for a particular confined pressure (Roggensack, 1977).

The following modified Mohr–Coulomb failure theory may be used to determine the shear strength of frozen soils (Vyalov, 1962):

$$\tau = C_t + \sigma_n \tan \phi_t \qquad (3\text{–}27)$$

where $\tau$ = shear strength
$\sigma_n$ = normal stress on shear plane
$C_t$, $\phi_t$ = functions of both temperature and time

Two components of shear strength are defined as follows: (1) resistance to "smooth shear" (i.e., cohesion) and (2) frictional resistance to shear as a function of normal stress.

The cohesive component $C_t$ can be attributed to molecular forces of attraction between particles, including physical or chemical cementing of particles together and cementing of particles by ice formation in the soil voids. Cohesion depends on the amount, strength, and area of ice in contact with the soil particles, each of which is temperature dependent. On the basis of experimental data, Vyalov and Shusherina (1970) concluded that the cohesion intercept $C_t$ would be given by

$$C_t = \frac{\beta}{\log (t/B)} \tag{3–28}$$

where $\beta$ and $B$ are constants obtained from a plot of $C_t$ against log $t$, $t$ being the time period. This equation has the same form as that proposed by Vyalov (1962) to estimate long-term uniaxial compressive strengths from the results of several short-term creep tests.

The frictional component represented by the angle of internal friction is dependent on the ice content and soil-grain arrangement, as well as size, distribution, shape, and the number of grain-to-grain contacts. Except for the ice content, each of these factors is independent of temperature.

Direct shear and triaxial tests may be performed to determine the shear strength of frozen soils. Schematic views of direct shear and triaxial test arrangements are shown in Figs. 10–4 and 10–5, respectively.

Results of typical triaxial tests at a constant strain rate and various confining pressures on frozen sand are presented in Fig. 3–8. Observations of test results suggest that the first peak is essentially the consequence of the strength of the ice matrix and the second peak is the development of friction between the grains of sand and/or ice crystals as the strain progresses (Sayles, 1973). The frictional resistance increases until the maximum dilatancy of the sand grains is reached and the second peak in the stress–strain curve is attained. Figure 3–8 illustrates Mohr's envelopes for the first peak resistance of frozen sand tested at different strain rates. Strength envelopes of frozen sand at different ice saturation are shown in Fig. 3–9. Figure 3–10 illustrates typical data from one of the direct shear tests on frozen clay.

From Eq. (3–27) and (3–28) it should be recognized that creep and shear strength are intimately related to the strain required to mobilize resistance, resistance being a direct function of the time required for the ice matrix to yield under an applied stress.

The shear strength of frozen soils is also a function of the confining pressure (Vyalov, 1962; Neuber and Wolters, 1970; Alkire and Andersland, 1973). Friction angles of frozen soils reported have been less than values for the same soils when unfrozen. Generally, considerable energy is required to overcome the strength of the ice matrix so that particle dilation can occur. The ice frost yields at a strain of about 1% and is followed by the gradual development of frictional strength. Strains of about 10% or more are required to mobilize full frictional shear strength under triaxial conditions. However, the full development of fric-

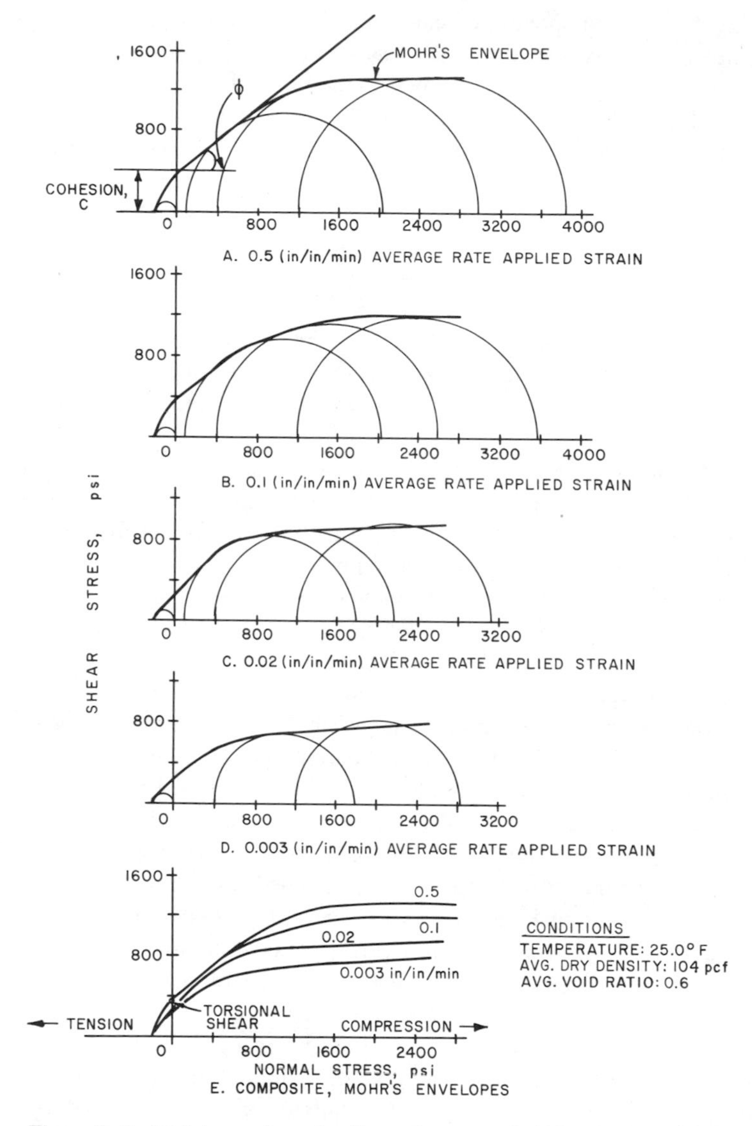


**Figure 3–8** Mohr's envelopes for frozen Ottawa sand. (After Sayles, 1973.)

tional strength may be inhibited as ice saturation is increased, as shown in Fig. 3–9. The modified Mohr–Coulomb equation (3–27) adequately describes the shear strength behavior of frozen soils. The observed friction values in Fig. 3–9 are 29.8° and 24.6°, which are less than the 37° value obtained for unfrozen sand, suggesting that ice does interfere with the development of full frictional strength.

Stress concentration produced by a small deviatoric stress may also cause melting of ice and the movement of water to regions of lower stress, where it

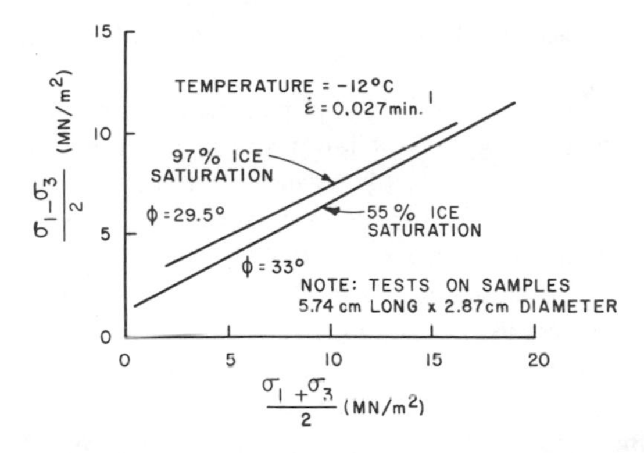


**Figure 3–9** Failure envelopes for frozen Ottawa sand. (After Alkire and Andersland, 1973, reproduced from the *Journal of Glaciology* by permission of the International Glaciological Society.)

refreezes. As such, frozen soils exhibit dilation and a frictional response when sheared with a confining pressure.

Direct shear and triaxial tests may be performed to determine the shear strength of frozen soils. A schematic view of a direct shear test arrangement is shown in Fig. 10–4. Based on various laboratory test results (Roggensack and Morgenstern, 1978; Sayles, 1973; Roggensack, 1977), it may be concluded that the shear strength of frozen fine-grained soils is purely cohesive. At slow strain rates, ice-poor frozen soils will exhibit frictional responses. The strength and deformation characteristics of ice-rich frozen soils are affected by the behavior of ice inclusion in soils.

### 3.3.5 Creep Behavior of Frozen Soils

The time-dependent deformation behavior of material under stress is called *creep*. Frozen soils exhibit creep behavior even at low stress levels at rates dependent on ice content, temperature, and loading rate. Vyalov (1966) explained the physical process of creep in frozen soils and they attribute the time-dependent deformation

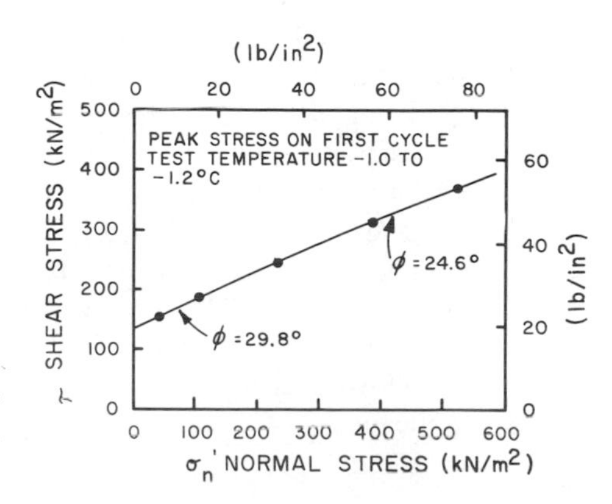


**Figure 3–10** Direct shear envelopes for frozen remoulded clay. (After Roggensack et al., 1978.)

to (1) pressure melting of ice in the soil at points of soil-grain contact, (2) migration of unfrozen water to regions of lower stress, (3) breakdown of the ice and structural bonds to the soil grain, (4) plastic deformation of pore ice, and (5) a readjustment in the particle adjustment. During the creep process, the structural deformation leads to a denser packing of the soil particles. This denser packing in turn causes a strengthening of the materials due to the increased number of firm contacts between soil grains, hence an increase in internal friction between grains. During the creep process there is also a working of the structural cohesion and possibly an increase in the amount of unfrozen water in the frozen soil (particularly in fine-grained soils). All of this action is time dependent. If the applied load does not exceed a critical stress called the *long-term strength* of the frozen soil, the weakening process is compensated by the strengthening process.

Deformation is damped; that is, the rate of deformation decreases with time, referred to variously as *primary creep, transient creep, decaying,* and *alternating creep* (Vyalov et al., 1966; Sayles and Haynes, 1974; Hult, 1966; Ladanyi, 1972) (Fig. 3–11a). However, if the applied load exceeds the critical stress of the frozen soils, the weakening process will exceed the strengthening process and the rate of deformation will increase with time, resulting in an undamped deformation stage which is called *secondary creep, steady-state,* or *nonalternating creep,* as shown in Fig. 3–11a. Steady-state creep will eventually develop into a plastic flow or accelerated creep rate which is called *tertiary creep* and normally leads to the failure of frozen soils. As shown in Fig. 3–11b, the primary creep stage is characterized by a continuous decreasing creep rate, the secondary creep stage is characterized by a constant creep rate, and the tertiary creep stage is characterized by an accelerated creep rate. Idealized time-dependent behaviors of frozen soils are illustrated in Fig. 3–12.

Further to the discussion in Section 3.3.1, the shape of the creep-deformation behavior curve of frozen soils is influenced by various factors, such as temperature, the magnitude and duration of applied stress, the soil type, and ice saturation. Generally, the creep behavior of ice-rich frozen soils is dominated by secondary creep, with a short time interval for the primary creep stage, whereas ice-poor frozen soils may exhibit the dominant stage of primary creep. At high stress levels, the soil may reach the tertiary stage without well-defined primary and secondary stages and fall under a short duration of time.

Engineering design requires that the loads transmitted to the supporting foundation materials should not exceed the strength and deformation or settlement of the foundation materials allowable within the design life of the structure. As the strength and deformation of frozen soils are time dependent, theories are required to predict the expected deformation during the service life of the structure. Vyalov (1959, 1962, 1973), Hult (1966), Sayles (1968, 1973), and Ladanyi (1972) have presented various theoretical concepts for the creep of frozen soils, and in the following sections these theories are discussed as they relate to engineering applications.

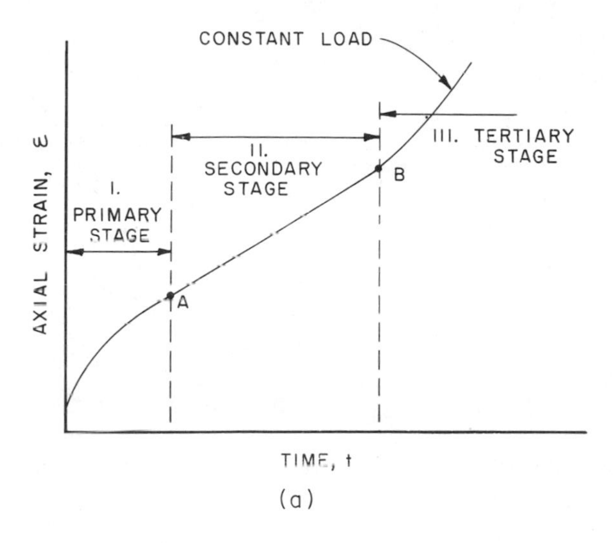


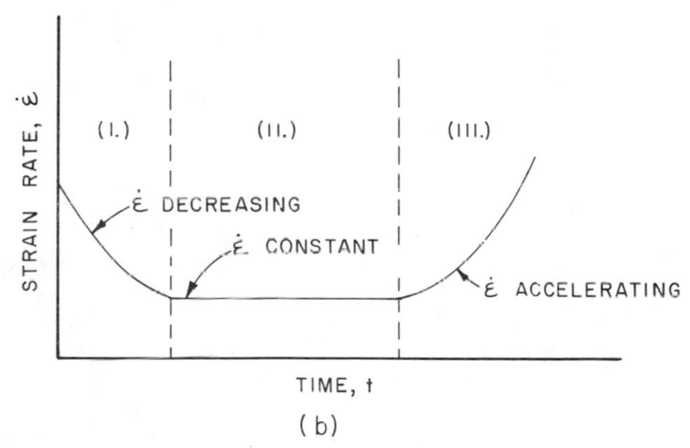


**Figure 3–11** Basic creep behavior.

### *3.3.5.1 Primary Creep Theory*

Vyalov (1962) reported the primary creep theory for frozen soils. The deformation $\varepsilon$ of frozen soil at any time is given by

$$\varepsilon = \varepsilon_0 + \varepsilon(t) \tag{3–29}$$

where $\varepsilon_0$ is elastic and

$$\text{recoverable strain} = \frac{\sigma}{E}$$

where $\sigma$ = applied constant stress in this case
$E$ = instantaneous elastic modules

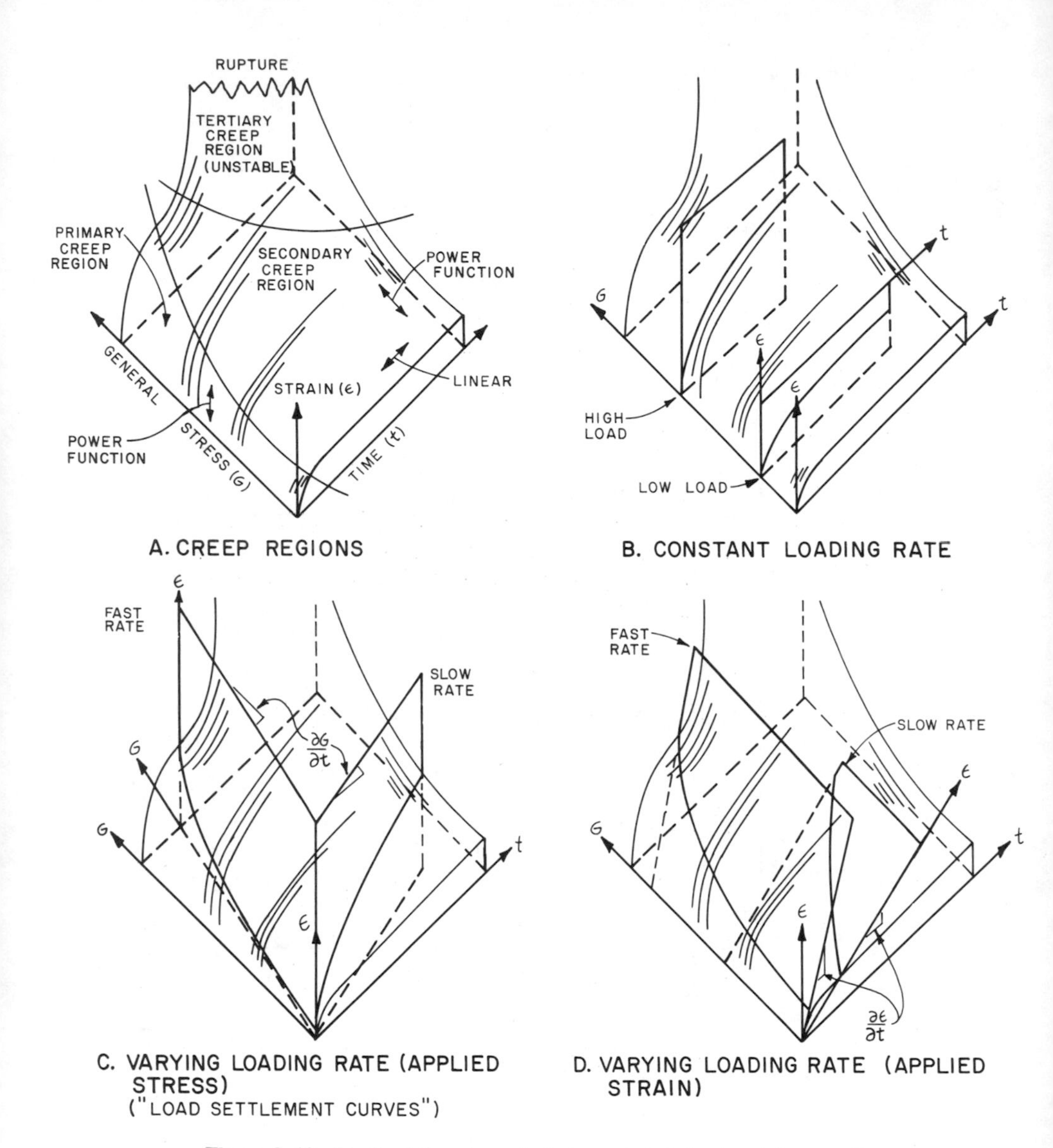


**Figure 3–12** Idealized time-dependent stress–strain behavior of frozen soils.

$\varepsilon(t)$ is elastic and nonrecoverable or plastic:

$$\varepsilon(t) = \int_0^t \sigma(t)\, K(t - t_n)\, dt_n \tag{3–30a}$$

where $t$ = time elapsed

$t_n$ = times at which stresses ($\sigma$) are applied

$K(t - t_n)$ = function describing the strain taking place after stress application at time $t_n$

For constant stress $\sigma$ at an initial time equal to zero ($t_n = 0$), Eq. (3–30a) becomes

$$\varepsilon(t) = \sigma\int_0^t K(t)\,dt \tag{3–30b}$$

So Eq. (3–29) becomes

$$\varepsilon = \frac{\sigma}{E} + \sigma\int_0^t K(t)\,dt \tag{3–31}$$

The first term on the right-hand side of Eq. (3–31) represents the instantaneous strain caused by the stress. The second term represents the increase in strain with time produced by stress.

Using the time-hardening relationship between stress, strain, and time, we have

$$f(\sigma, t, \varepsilon) = 0$$

and the power function for strain

$$\sigma = A(t)\varepsilon(t)^m$$

or

$$\varepsilon(t) = \left[\frac{\sigma}{A(t)}\right]^{1/m}$$

where $A(t)$ = modulus of nonlinear deformation
$m$ = constant

Substituting the $\varepsilon$ value above in Eq. (3–31), we get

$$\varepsilon = \frac{\sigma}{E} + \left[\frac{\sigma}{A(t)}\right]^{1/m} \tag{3–32}$$

These parameters vary from the largest, relatively instantaneous value, corresponding to the infinitely fast application of the load ($\Delta t \rightarrow 0$), to the terminal value, corresponding to the continued application of each stage of the load until deformation completely disappears. Disappearance of load deformation may take place under a critical stress level.

Values of $A$ and $m$ may be obtained from log plots of $A$ stress against strain for various time intervals after application of stress to the test specimens (Fig. 3–13). As shown in Fig. 3–13, slopes of lines are represented by $m$, which are the characteristics of the frozen soil tested. The value of $A$ (unit of stress) is given by the ordinate for log strain equal to unity.

Equation (3–32) may be replaced generally by a temperature- and time-dependent expression:

$$\sigma = f(t,\Theta)\varepsilon^m$$

or (3–33)

$$\varepsilon^m = \frac{\sigma t \Lambda}{A(t)}$$

The temperature effect on the frozen soil is accounted for by the equation

$$A(t) = \omega(\Theta + 1)^k$$

Considering stress–strain with time, Eq. (3–33) becomes

$$\varepsilon = \left[\frac{\sigma t^{\Lambda}}{\omega(\Theta + 1)^k}\right]^{1/m} \tag{3–34a}$$

where $\sigma$ = applied constant stress, kg/cm$^2$
$t$ = time elapse after application of load, hr
$\Theta$ = temperature below the freezing point of water, °C
$\omega, \Lambda, m, k$ = constants that are characteristics of the material (all the constants are dimensionless except that $\omega$ depends on units)

Equation (3–34a) conceals the fact that in $(\Theta + 1)$, the 1 has the temperature unit. Assur (1963) suggested that Eq. (3–34a) be written as

$$\varepsilon = \left[\frac{\sigma t^{\Lambda}}{\omega\,(\Theta + \Theta_0)^k}\right]^{1/m}$$

or

$$\varepsilon = \left[\frac{\sigma t^{\Lambda}}{\omega\Theta_o^k\,(\Theta/_{\Theta_o} + 1)^k}\right]^{1/m} \tag{3–34b}$$

Here the numerical values of $m$ and $\Lambda$ do not vary with the units employed, but $\omega$ and $\Theta_o^k$ do. Vyalov assumed that $\Theta_o = 1$°C, a constant reference temperature greater than zero, so that log $(1 + \Theta)$ has a meaning at $\vartheta = 0$. Typical values of these constants are presented in Table 3–4 for several soil types.

*Sayle's Method.* Vyalov's equation (3–25) demands precise measurements of the absolute values of strain (i.e., deformation) and also requires several tests for the evaluation of the formula constant. To overcome these difficulties, and to provide a more flexible approach, Sayles (1968) developed a strain equation using strain rates where only the difference in strain (i.e., deformation) needed to be measured. Typical creep tests (where the applied stress is less than the long-term strength, i.e., primary creep) on Ottawa sand were plotted as the reciprocal of the time log $t$ against the log strain rate ($\dot{\varepsilon}$) (Fig. 3–14a).

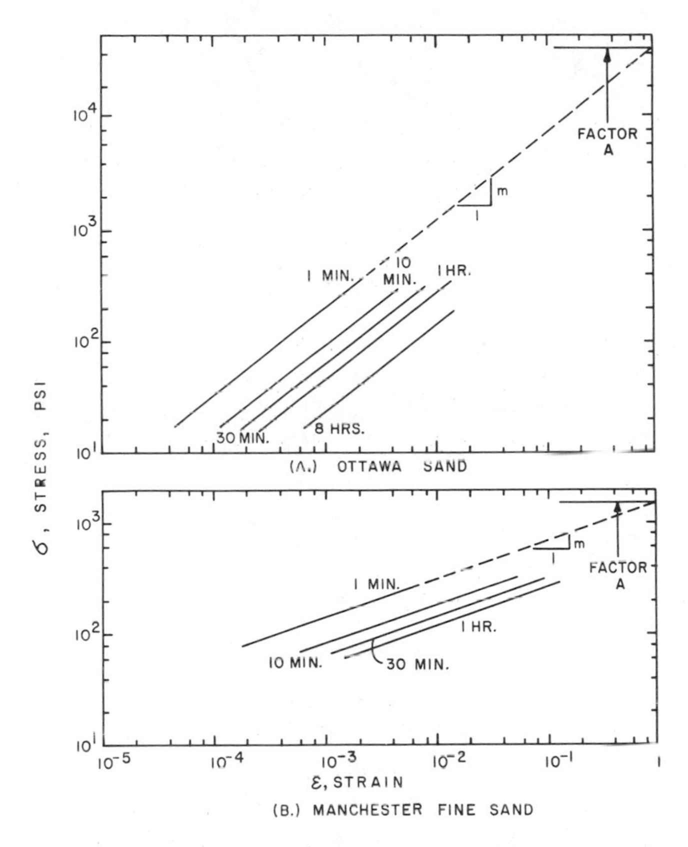


**Figure 3–13** Values of $A$ and $m$. (After Sayles, 1968.)

**TABLE 3–4 Constants for Eq. (3–42)**

| Frozen Soil | $m$ | $\lambda$ | $k$ | $\omega$ MPa(h$^\lambda$)/°C$^k$ | $\omega$ psi(hr)$^\lambda$/°F$^k$ |
|---|---|---|---|---|---|
| Saturated Ottawa sand[a] | 0.78 | 0.35 | 0.97 | 44.72 | 5500 |
| Manchester fine sand[a] | 0.38 | 0.24 | 0.97 | 2.20 | 285 |
| Suffield clay[a] (saturated clay,[a] CL, LL 35, PL 20, no visible ice lenses, $w < 35\%$) | 0.42 | 0.14 | 1.00 | 0.73 | 93 |
| Callovian sandy loam[b] (sandy silt, ML) | 0.27 | 0.10 | 0.89 | 0.88 | 90 |
| Bat-Baioss clay[b] | 0.40 | 0.18 | 0.97 | 1.25 | 130 |

[a]Data from Sayles (1968).
[b]Data from Vyalov (1962).

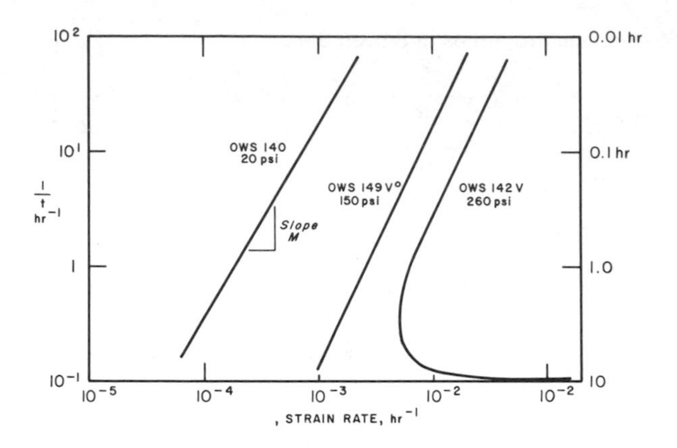


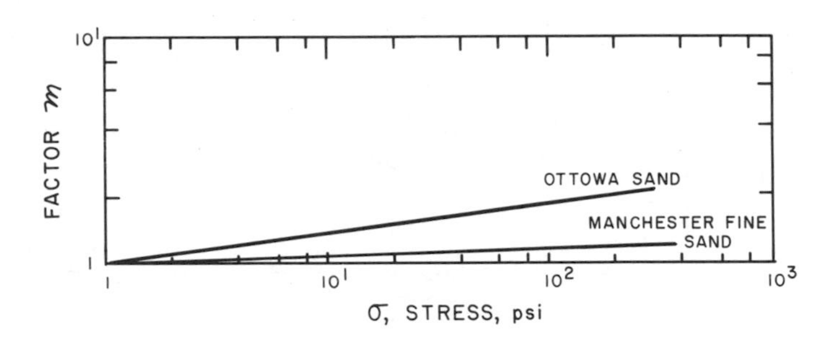


**Figure 3–14** (a) Creep rate and reciprocal of time, Ottawa sand; (b) factor *m* and stress. (After Sayles, 1968.)

The straight-line portion of the log (1/*t*) versus log strain-rate curves for damped creep tests on Fig. 3–14a can be represented by

$$\frac{1}{t} = \left[\frac{\dot{\varepsilon}}{\dot{\varepsilon}_1}\right]^m$$

or

$$\varepsilon = \dot{\varepsilon}_1 \, t^{-1/m} \tag{3–35}$$

By integration,

$$\varepsilon = \dot{\varepsilon}_1 \left(\frac{m}{m - 1}\right) t^{(m - 1)/m} + \varepsilon_0 \tag{3–36}$$

where $\dot{\varepsilon}_1$ = strain rate 1 hr after the stress is applied
$\varepsilon_0$ = instantaneous strain as stress applied (generally neglected)
$t$ = time after stress is applied
$m$ = slope of log $(1/t)$ versus log $\dot{\varepsilon}$

$m$ is found to be more dependent on stress. The values of $m$ as a function of stress are presented in Fig. 3–14b. Assuming that $m$ is independent of temperature and is a straight-line function of stress on logarithmic coordinates (Fig. 3–14), the equation of $m$ becomes

$$m = C\sigma^{1/W} \tag{3–37}$$

where $W$ is the slope of the log $\sigma$ versus log $m$ curve treated as a constant for each material. Sayles (1968) determined the values of $W$ for Ottawa sand and Manchester fine sand and they are 9 and 35, respectively. $C$ is almost equal to unity for each sand.

#### *3.3.5.2 Secondary Creep Theory*

Hult (1966) proposed methods for processing test data which deal with secondary creep phenomena. This section presents Hult's methods for secondary creep in frozen soils.

*Hult's Method.* In this method, the creep curves are approximated by straight lines having intercepts $\varepsilon^{(i)}$ and constant slope $\dot{\varepsilon}^{(c)}$ as shown in Fig. 3–15.

Using the straight-line relationships, the creep strain $\varepsilon$ at a particular stress and temperature is given by

$$\varepsilon = \varepsilon^{(i)} + \dot{\varepsilon}^{(c)} t \tag{3–38}$$

where $\varepsilon^{(i)}$ = instantaneous strain
$\dot{\varepsilon}^{(c)}$ = constant strain rate

These parameters are a function of stress and temperature; in mathematical form Eq. (3–38) can be written as

$$\begin{aligned} \varepsilon &= X(\sigma, T) + \int_0^t Y(\sigma, T)dt \\ &= X(\sigma, T) + tY(\sigma, T) \end{aligned} \tag{3–39}$$

But the pseudo-instantaneous strain $\varepsilon^{(i)}$ consists of both elastic and plastic strains; therefore, we have

$$\begin{aligned} \varepsilon^{(i)} &= \varepsilon^{(ie)} + \varepsilon^{(ip)} = X(\sigma, T) \\ &= \frac{\sigma}{E} + \varepsilon_k\left(\frac{\sigma}{\sigma_k}\right)^k \end{aligned} \tag{3–40}$$

where $E$ = time-dependent Young's modulus
$\varepsilon_k$ = arbitrary small strain for normalization
$\sigma_k$ = temperature-dependent deformation modulus

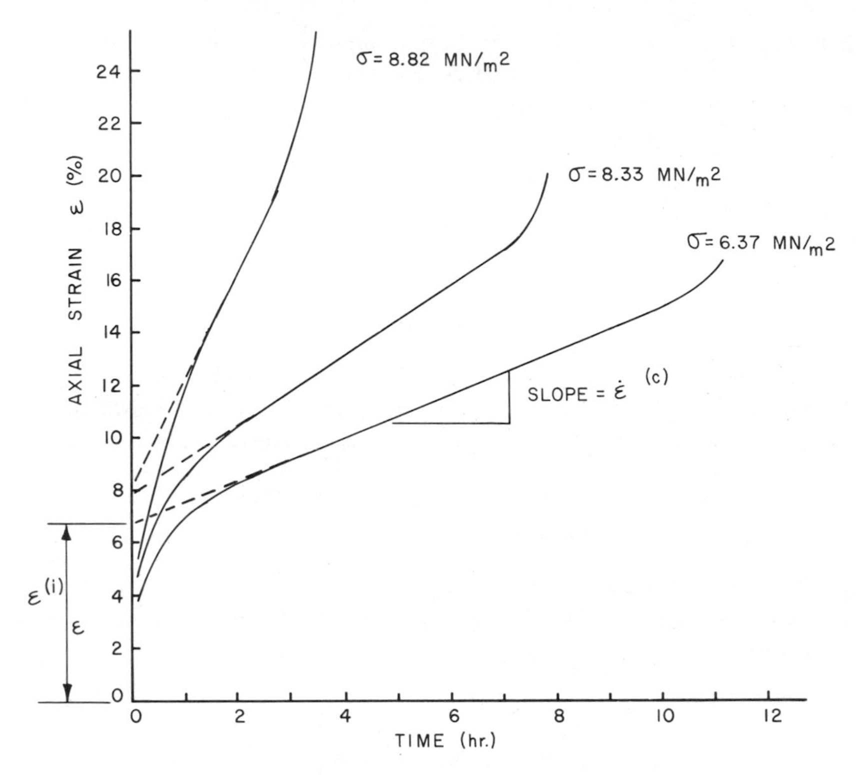


**Figure 3–15** Creep curves for frozen Callovian sandy loam at $-20°C$ in uniaxial compression. (Data from Vyalov, 1962.)

The creep function $Y(\sigma, T)$ may be written as

$$(\sigma, T) = \dot{\varepsilon}_c\left(\frac{\sigma}{\sigma_c}\right)^n \tag{3–41}$$

where $\sigma_c$, $n$ = temperature-dependent creep parameters

$\dot{\varepsilon}_c$ = small strain rate for normalization

from eqs. (3–39), (3–40), and (3–41), we get

$$\varepsilon = \frac{\sigma}{E} + \varepsilon_k\left(\frac{\sigma}{\sigma_k}\right)^k + t\,\dot{\varepsilon}_c\left(\frac{\sigma}{\sigma_c}\right)^n \tag{3–42}$$

The first two terms in Eq. (3–42) contribute less than 10% of the total creep strain if the time interval is greater than 1 day (Vyalov, 1962). So for engineering purposes, the third term only in Eq. (3–42) may give good approximations of the total strain.

### 3.3.6 Long-Term Strength of Frozen Soils

The long-term strength of frozen soils is defined as the stress level, after a finite time, at which accelerated strain or deformation leading to rupture or extremely large deformation without rupture occur. In Fig. 3–11, the long-term strength may be identified with the beginning of tertiary creep, point B. Frozen fine-grained soils may undergo large deformations without rupture near 0°C, especially when they are subjected to triaxial stresses. The strength of those soils must be considered in terms of the degree of strain or deformation that a given structure can tolerate. In laboratory testing of soil specimens, a specimen is generally considered to have failed when strain reaches 20%.

Vyalov (1959) proposed that the strength of frozen soils is an exponential function in the form

$$\sigma_{ult} = \frac{\beta}{\log_e(t/B)} \tag{3–43}$$

where $\beta$ and $B$ are parameters dependent on soil type, its properties, and temperatures. Time to failure and failure stress are represented by $t$ and $\sigma_{ult}$, respectively. The method suggested by Vyalov for determining these parameters from a plot of $1/\sigma_{ult}$ versus log time is shown in Fig. 3–16. The disadvantages of this method are the extensive test program involving several tests, and allowing sufficient time

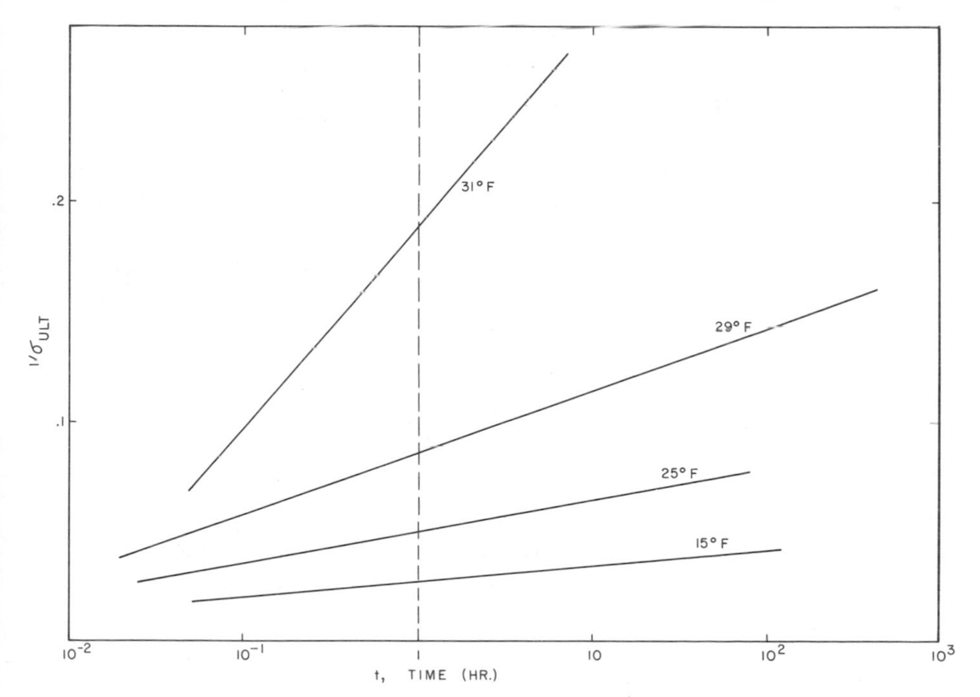


**Figure 3–16** Time and reciprocal of ultimate stress, Suffield clay. (After Sayles and Haynes, 1974.)

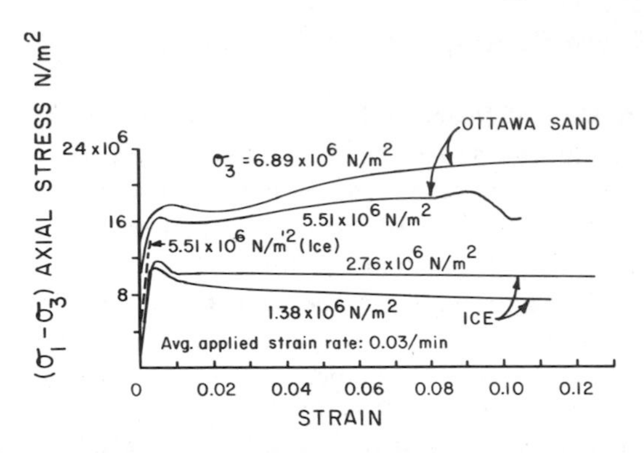


**Figure 3–17** Axial stress–strain curves for Ottawa sand and ice. (After Sayles, 1973.)

of several days for the tests at the lower stress. Using the results of short-term creep tests for predicting long-term strength has the advantage of requiring only two simple creep tests lasting less than 8 hr to be performed at two different stress levels. Values of $\beta$ and $B$ can be determined by solving Eq. (3–43) using the results of two short-term tests. Vyalov (1962) reported that the long-term strength of frozen soil ranged from 18 to 37% of the instantaneous cohesion for a frozen sandy silt. Sayles and Haynes (1974) found that the long-term strength of frozen silt and clay is less than 45% of the unconfined compression instantaneous strength.

Sayles (1973) modified Eq. (3–43) for triaxial condition as

$$\sigma_{\text{ult}} = \sigma_1 - \sigma_3 \quad \text{or} \quad \sigma_{\text{ult}} = \frac{\beta}{\log_e (t/B)} \tag{3–44}$$

He also studied the effects of confining pressure and strain rate on the long-term strength of Ottawa sand. Typical test results are shown in Fig. 3–17. The initial nonlinear relationships found on the test results indicate that the ice behavior dominates the strength before the angle of internal friction between grain particles develops. Those results are consistent with the explanation given regarding the shear strength of frozen soils in Section 3.3.4.

**Example 3–6**

Creep data for frozen Svea clay at $-5$°C in uniaxial compression are given in Fig. 3–18. Determine the creep parameters $K$, $\sigma_k$, $n$, and $\sigma_c$ and compute the secondary creep strain at an axial load of 100 kN/m² after 100 hr.

### Solution

1. Find the intercept $\varepsilon^{(i)}$ for the straight-line extensions shown in Fig. 3–18.
2. Estimate the creep rates $\dot{\varepsilon}^c$ from the curves at different loadings; the values are:

| $\sigma$ (kN/m²) | 200 | 300 | 395 |
|---|---|---|---|
| $\dot{\varepsilon}^c$ (min⁻¹) | $1.67 \times 10^{-3}$ | $6.67 \times 10^{-3}$ | $1.12 \times 10^{-2}$ |

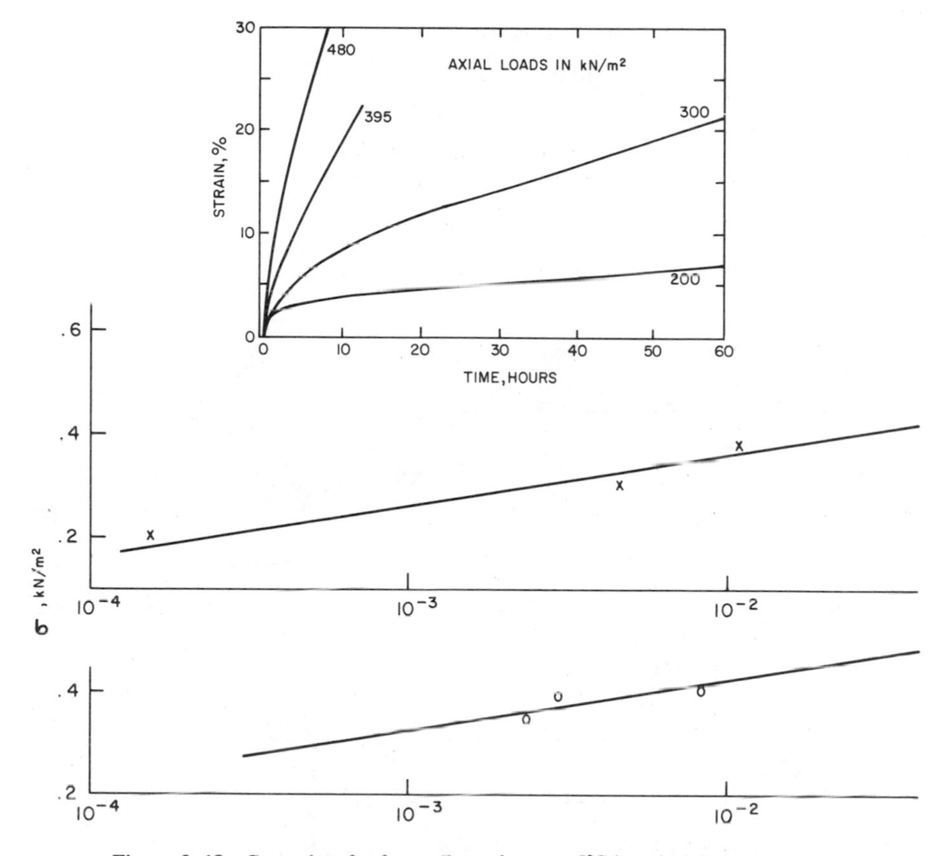


**Figure 3–18** Creep data for frozen Svea clay at $-5°C$ in uniaxial compression.

3. Plot the values of $\varepsilon^{(i)}$ and $\dot{\varepsilon}^c$ against the applied stress as shown in Fig. 3–18.
4. Draw the best-fit straight lines through the plotted points and determine the slopes.

   The slope obtained from the relationship between lσg and log $\varepsilon^{(i)}$ will give $K$, which is in this case, $K = \cot 79.7° = 0.18$. The slope $n$ is determined from the relationship between $\sigma$ and log $\dot{\varepsilon}^c$ and is found to be cot 20.5 = 2.67.
5. Selecting $\varepsilon^{(i)} = 10^{-2}$ and $\dot{\varepsilon}^c = 10^{-5}$ min$^{-1}$, determine $\sigma\zeta o_k$ and $\sigma_c$, respectively, from Fig. 3–18. In this case,

$$\sigma_k = 50 \text{ kN/m}^2$$
$$\sigma_c = 30 \text{ kN/m}^2$$

The secondary strain

$$\varepsilon = \frac{\sigma}{E} + \varepsilon_k \left(\frac{\sigma}{\sigma_k}\right)^k + t\dot{\varepsilon}^c \left(\frac{\sigma}{\sigma_c}\right)^n$$

Assuming that $\sigma/E$ is negligible, we get

$$\varepsilon = 10^{-2}\left(\frac{100}{50}\right)^{0.18} + (100 \times 60)(10^{-5})\left(\frac{100}{30}\right)^{2.67}$$

$$= 1.54$$

**Example 3–7**

Determine the long-term strength for Suffield clay at 25°F after 20 years. Use the data given in Fig. 3–16.

## Solution

1. From Eq. (3–43), rearranging, we obtain

$$\frac{1}{\sigma_{ult}} = \frac{1}{\beta}\ln\left(\frac{t}{B}\right)$$

$$= \frac{1}{\beta}(2.303)\log\frac{t}{B}$$

The slope of the relationship between $1/\sigma_{ult}$ and time $t$ in Fig. 3–16 equals $2.303(1/\beta)$.

2. From Fig. 3–16 at 25°F, we obtain

$$\text{slope} = 2.303\left(\frac{1}{\beta}\right) = \frac{6.49 \times 10^{-2}}{\log 10} - \frac{5.08 \times 10^{-2}}{\log 1}$$

Solving for $\beta$, we get $\beta = 1.63$ psi.

3. Take any time period, say 1 hr, and read the intercept $1/\sigma_{ult}$ from Fig. 3–16. In this case,

$$\frac{1}{\sigma_{ult}} = 5.08 \times 10^{-2}\ (\text{psi})^{-1} \text{ at } 25°\text{F}$$

4. Solve for $B$ using

$$2.303\log\frac{1}{B} = \frac{\beta}{\sigma_{ult}} = (1.63)(5.08 \times 10^{-2})$$

or

$$B = 0.92$$

5. Determine the long-term strength at 20 years.

$$t = 20 \times 365 \times 24 = 1.75 \times 10^5 \text{ hr}$$

$$\sigma_{ult} = \frac{1.63}{\log\dfrac{1.75 \times 10^5}{0.92}} = 0.31 \text{ psi}$$

### 3.3.7 Compressibility of Frozen Soils

Tsytovich (1975) measured the coefficient of volume compressibility ($m_v$) of several frozen soils and the values are presented in Table 3–5. Mathematically, the coefficient of volume compressibility is given by

$$m_v = \frac{\Delta v / v}{\Delta \sigma}$$

where $\Delta v$ = change in volume
$v$ = original volume
$\Delta \sigma$ = change in stress

Frozen soils are usually considered incompressible, and therefore deformations resulting from volume change can be neglected in comparison with creep deformation, as discussed in the preceding section. However, some frozen soils at different freezing temperature may show significant volume compressibility (Tsytovich, 1975; Brodskaia, 1962). The engineering application of volume compressibility is discussed further in Chapter 5.

### 3.3.8 Interaction between Strength and Deformation

When a soil, either frozen or unfrozen, is subjected to an increasing load, it responds by deformation and eventually will fail depending on the strength of the soil. Considering a soil element located in the centerline of a vertically loaded footing, the soil element will undergo an increase in both vertical and horizontal principal stresses. The former, however, will be much larger than the latter. Since such a stress increment implies an increase in both the spherical and deviatoric components of the stress tensor, the resulting deformation will be composed of compression and distortion, the latter usually associated with the dilatancy.

In a frozen soil, the distortional creep is considered to be the main source of the delayed response to a stress increase, while the consolidation is usually thought to be negligible. Some investigators (Vyalov, 1959; Brodskaia, 1962) assumed that the consolidation can be neglected only for relatively cold frozen soils which contain practically no unfrozen water, whereas for relatively warm frozen soils containing large amounts of unfrozen water, consolidation may be quite substantial. Ladanyi (1974, 1975) presented a theoretical treatment that takes into account the processes of creep, consolidation, and shear in frozen soils. As discussed in Section 2.5, an arbitrary boundary has been defined between hard-frozen and plastic-frozen soils on the basis of texture and temperature. For the analysis of strength and deformation in frozen soils, one can consider the frozen soil as a quasi-single-phase medium with mathematically well-defined creep properties, neglecting the fact that one portion of the observed creep is actually due to volumetric strain. Another approach will be to consider the consolidation

**TABLE 3–5 Coefficient of Volume Compressibility ($m_v$) of Frozen Soils**

| Material | $\omega$ (%) | $\omega_{uw}$ (%) | Temp. (°C) | $m_v$ (cm²/kg × $10^{-4}$) with a Load (kg/cm²) of: 0–1 | 1–2 | 2–4 | 4–6 | 6–8 |
|---|---|---|---|---|---|---|---|---|
| Medium sand | 21 | 0.2 | −0.6 | 12 | 9 | 6 | 4 | 3 |
| | 27 | 0.2 | −4.2 | 17 | 13 | 10 | 7 | 5 |
| | 27 | 0.2 | −0.4 | 32 | 26 | 14 | 8 | 5 |
| Silty sand, massive structure | 25 | 5.2 | −3.5 | 6 | 14 | 18 | 22 | 23 |
| | 27 | 8.0 | −0.4 | 24 | 29 | 26 | 18 | 14 |
| Medium silty clay, massive structure | 35 | 12.3 | −4.0 | 8 | 15 | 26 | 28 | 24 |
| | 32 | 17.7 | −0.4 | 36 | 42 | 37 | 21 | 14 |
| Medium silty clay, reticulate structure | 42 | 11.6 | −3.8 | 5 | 10 | 18 | 42 | 32 |
| | 38 | 16.1 | −0.4 | 56 | 59 | 39 | 24 | 16 |
| Medium silty clay, layered structure | 104 | 11.6 | −3.6 | 54 | 54 | 59 | 44 | 34 |
| | 92 | 16.1 | −0.4 | 191 | 137 | 74 | 36 | 18 |
| Varved clay | 36 | 12.9 | −3.6 | 15 | 22 | 26 | 23 | 19 |
| | 34 | 27.0 | −0.4 | 32 | 30 | 25 | 20 | 16 |

*Source:* After Tsytovich (1975).

and the deviatoric creep as two simultaneous but separate phenomena whose relative amounts depend on the applied stress and the elapsed time. Of the two lines of approach, the latter is clearly more advanced and appropriate, but few attempts have been made to investigate separately these two phenomena in frozen soils, as basic data are still missing. Using an approach based on the unique relationship between water content contour and effective stress paths, Ladanyi (1974) examined stress changes in frozen soils with the aid of a Rendulic (1937) plot. This format permits the convenient separation of any stress into its normal and deviatoric components.

#### *3.3.8.1 Frozen Soil Response to a Stress Increment*

Following Ladanyi (1974), Fig. 3–19 illustrates schematically how the separation of applied stresses into effective stresses and pore pressure can be performed in case of a normally consolidated clay under triaxial compression test conditions with $\sigma_1 > \sigma_2 = \sigma_3$, where $\sigma_1$ denotes the principal axial stress and $\sigma_2$ and $\sigma_3$ denote the confining pressures.

A "normally consolidated" triaxial specimen (in equilibrium at point O′) is subjected to a stress increment that will be carried by the mineral particles and the

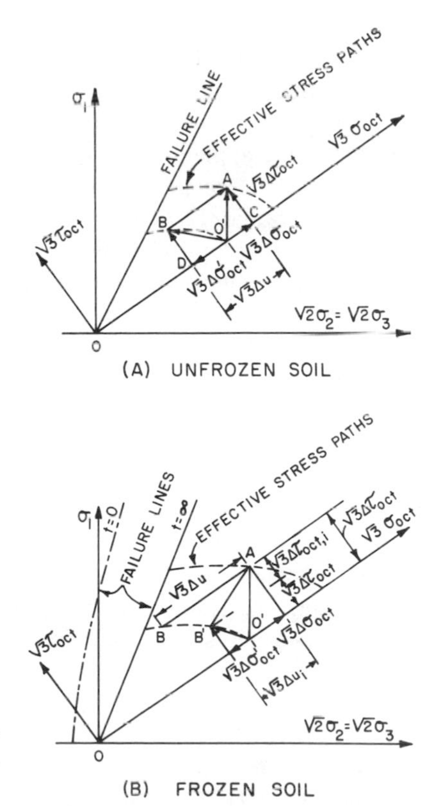


**Figure 3–19** Redulic plot for normally consolidated soil in compression. (After Ladanyi, 1974.)

pore-filling matrix. Some of the applied stress is immediately assumed by the soil skeleton and can be likened to an effective stress in unfrozen soils. In the case where the applied stress increment $\Delta\sigma_1$ (O′A) does not exceed the long-term strength, it can be separated into two components, $\sqrt{3}\Delta\sigma_{oct}$ and $\sqrt{3}\Delta\tau_{oct}$, which are the octahedral normal and octahedral shear stress changes, respectively. As the soil deforms without drainage, unfrozen water in the soil will carry only the hydrostatic stress. Shear stresses can be temporarily supported by the ice matrix, and it will assume those not immediately taken up by the soil skeleton. A portion of the shear stress is initially supported by the ice, with the result that the soil skeleton will be less strained than it would be under the same stress in the unfrozen state. Effective shear strength is then mobilized initially to B′ instead of B and the hydrostatic pore pressure generated by this straining (assumed to be the same in the ice and unfrozen water) is given by

$$\Delta u = \Delta\sigma_{oct} - \Delta\sigma^{1}{}_{oct} \tag{3–45}$$

This will be smaller than the pore pressure change that would be generated in the same soil if it were unfrozen. Shear stresses are shared by the ice matrix and soil skeleton; similarly, the following can be written:

$$\Delta\tau_{oct,i} = \Delta\tau_{oct} - \Delta\tau'_{oct} \tag{3–46}$$

where $\Delta\tau_{oct,i}$ and $\Delta\tau_{oct}$ are the shear stresses assumed by the ice and soil, respectively. Ladanyi suggests that the creep relaxation of the ice can be likened to the dissipation of the excess pore pressures during consolidation since it results in a gradual transfer of the applied stress to the soil skeleton. Eventually, under conditions of closed-system creep, $\Delta u_i \rightarrow \Delta u$ and $\Delta\tau_{oct,i} \rightarrow 0$ (the long-term strength of ice being zero), so that the point B′ will move toward B.

By opening the system when the effective stresses are at point B′, consolidation will occur simultaneously with creep of the ice matrix and B′ will move toward A. In this case, the long-term conditions will have $\Delta u_i \rightarrow 0$, $\Delta u \rightarrow 0$, and $\Delta\tau_{oct,i} \rightarrow 0$. With point A below the long-term failure line, both creep and consolidation processes assume an attenuating character. Therefore, deformation rates will decrease continuously with time.

Two features distinguish frozen-soil behavior from that commonly associated with normally consolidated soils:

1. Although the soil appears to be very stiff at the outset and creep and consolidation attenuate with time, ultimate deformations can still be quite large.
2. By mobilizing the short-term strength of the ice matrix, frozen soils can temporarily sustain stress increments which exceed the soil's long-term strength.

This second characteristic is depicted in Fig. 3–19. As before, the effective strength of the soil skeleton is initially mobilized to B′ under closed-system conditions. The ice matrix then creeps to mobilize full soil strength as B′ moves toward B. Simultaneously, progressive straining brings ice closer to failure. Loss of strength with time can be likened to a gradual shrinking of the delayed failure surface so that with point A lying outside the long-term strength envelope, failure will occur at some time $t'_A$. Permitting consolidation by opening the system at B′, the stress path will move toward A. The relative rates of consolidation will determine whether A′ is reached before failure occurs. In this case, straining consists of simultaneous consolidation and steady-state creep, and would be characterized by parabolic time–deformation curves.

An approach similar to that shown in Figs. 3–19 and 3–20 can be used for hard-frozen soils, with the difference being that these will behave as closed systems under all circumstances since the possibility of consolidation is eliminated by the virtual absence of mobile unfrozen water.

Figure 3–20 illustrates a similar consideration of the probable behavior of an "overconsolidated" frozen clay or dense granular soil assuming a differently shaped effective stress path. Point B represents the peak undrained strength and may lie slightly beyond the long-term strength envelope. A steady-state creep response will be initiated, with failure at some finite time being the inevitable result.

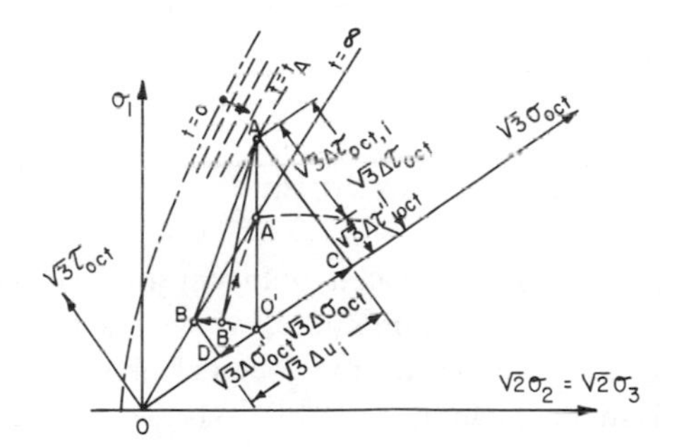


(a) NORMALLY CONSOLIDATED PLASTIC FROZEN SOIL

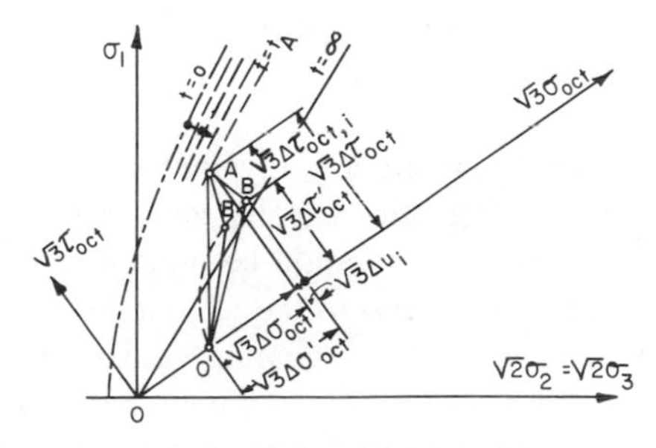


(b) OVER-CONSOLIDATED FROZEN SOIL

**Figure 3–20** Frozen soil stressed beyond long-term strength. (After Ladanyi, 1974.)

A précis of Ladanyi's (1974) excellent conceptual description of the response of frozen soil to a stress increment has been included here to clarify soil behavior and demonstrate the interaction between strength and deformation properties. It is apparent that an estimate of the position of the applied stress with respect to the long-term strength is an essential first step in deciding the design approach to be adopted in either limit equilibrium or deformation analysis. Ladanyi's approach suggests that the long-term strength for ice-rich materials will be zero, whereas for massive soils, this strength can be approximated by the drained strength of the unfrozen soil. Between these extremes and the short-term strength lie a number of delayed strength envelopes, each having a shape that may be different from the straight-line Mohr–Coulomb failure envelope. These envelopes must be defined in terms of total stresses, so it seems entirely possible that the apparent friction angles obtained could be different from the drained friction angle found for the same soil in an unfrozen state. The extreme sensitivity of strength to strain rate or time to failure can be directly attributed to the extremely nonlinear behavior of the ice matrix that is described by a flow law such as Eq. (3–19).

Experimental determination of the various strength and deformation components is an enormous task, but Ladanyi's (1974) model is certainly a useful framework within which a better understanding of observed behavior can be obtained.

#### 3.3.8.2 Consolidation of Frozen Soils

It is essential that deformations associated with consolidation have shown that frozen soils, hitherto considered incompressible, were capable of sustaining appreciable volume changes, with the controlling factors being soil texture, temperature, and ice content. Low permeabilities and the viscous nature of ice prolonged deformations so that even though the consolidation process did attenuate, equilibrium conditions may not be reached in any laboratory tests. Based on this observation, there is good reason to suspect that the compressibility coefficients for frozen ground reported by Brodskaia (1962) and given elsewhere in the Soviet literature may be nonconservative. Some experimental studies have reported finite soil permeabilities and confirmation of water flow in frozen soil (Burt and Williams, 1976).

Kent et al. (1975) have described a study that set out to verify the state variables required to described deformation mechanisms in frozen soils. They proposed that two volume-change processes would be produced by the application of an isotropic stress: first, steady-state creep in the ice due to intrinsic shear stresses imposed at the microscale; and second, attenuating consolidation. $(\sigma - u_w)$ was identified as a state variable and by plotting volume change against log time, a curve resembling the conventional consolidation relationship was obtained. This finding suggests that the rate $\sigma_i$ volume change in frozen soil is governed by the dissipation of excess pore water pressures. Similar results were obtained for volume changes produced by an incremental change in temperature. Volumetric strains resulting from a temperature increase exceeded the amount that could be explained by phase change alone. The authors attribute these volume changes to additional creep in the ice matrix as it yielded to conform to its new equilibrium ice porosity. Behavior observed in this study is in general agreement with the conceptual model for volume change in frozen soils proposed by Ladanyi (1974).

### 3.3.9 Dynamic Properties of Frozen Soils

The study of the dynamic characteristics of frozen soils is very important to cope with a number of very specific problems, chiefly, the response of frozen soil to vibrating machinery, the "blastibility" of frozen ground for excavation, and the identification of the presence of permafrost from geophysical surveys such as seismic reflection and seismic refraction. The response to vibratory loads and the identification of permafrost from seismic survey records may be evaluated from the dynamic moduli or compressional and shear wave velocities of frozen soils at low strain. Generally, elastic solutions are used to solve dynamic problems. The analysis and understanding of the response of frozen ground to earthquake excitation is at an elementary stage. The dynamic response of frozen ground has been discussed by several authors (Finn and Young, 1978; Vinson, 1978). In this section, parameters influencing the dynamic properties of frozen soils are discussed. Also, some of the basic dynamic properties of frozen soils, such as Young's modulus and the shear modulus, are presented.

#### *3.3.9.1 Parameters*

Parameters influencing dynamic properties of frozen soils may be divided into two groups (Vinson, 1977):

1. Field and/or test condition parameters
2. Material parameters

Group 1 includes temperature, strain or stress amplitude of loading, frequency of loading confining pressure, and duration of loading. Group 2 consists of material type and composition, material density or void ratio, ice content or degree of ice saturation, unfrozen water content, and anisotrophy.

Based on field and laboratory investigations, it may be concluded that:

1. Coarse-grained soils generally have higher compressional wave velocities than those of fine-grained soils.
2. The compression wave velocities of materials in the frozen state are much greater than those in the unfrozen state.
3. The dynamic stress–strain properties of frozen soils are directly proportional to the degree of ice saturation.

4. Damping properties increase with increasing temperature and increasing axial strain amplitude.
5. The dynamic stress–strain properties are not affected by the frequency of loading.
6. The dynamic properties of coarse-grained soils are affected only by the confining pressure.

### *3.3.9.2 Dynamic Properties*

Various investigators (Kaplar, 1969; Stevens, 1975; Vinson, 1978) have reported on the dynamic properties of frozen soils. The following relationships are commonly used to calculate the elastic constants:

$$\text{dynamic Young's modulus } E = \rho V^2_s \frac{3V^2_p - 4V^2_s}{V_p^2 - V^2_s} \quad (3\text{–}47)$$

$$\text{Poisson's ratio } \nu = \frac{[1/2\,(V_p/V_s)^2] - 1}{(V_p - V_s)^2 - 1} \quad (3\text{–}48)$$

$$\text{bulk modulus } K = \rho(V_p^2 - 4/3\, V_s^2) \quad (3\text{–}49)$$

$$\text{shear modulus } G = \rho V_s^2 \quad (3\text{–}50)$$

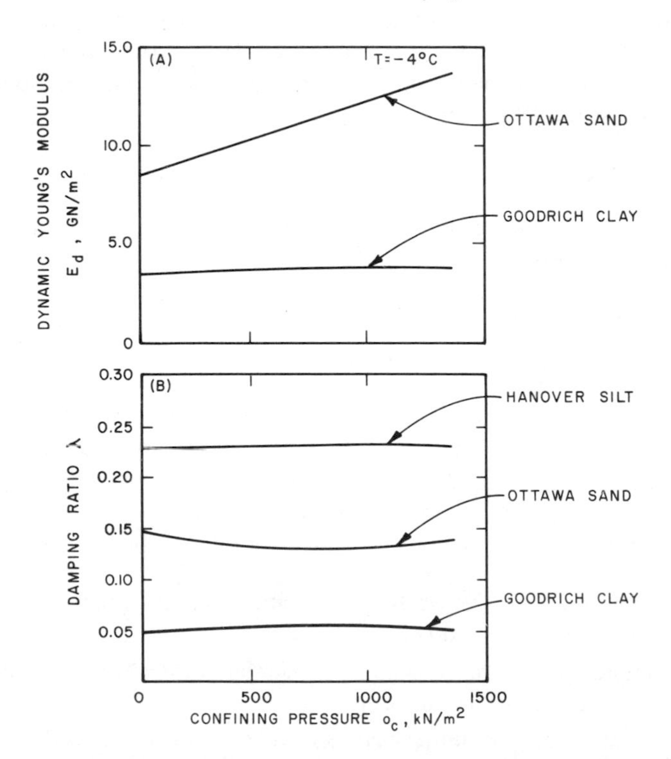


**Figure 3–21** Dynamic Young's modulus of frozen soils. (After Vinson, 1977.)

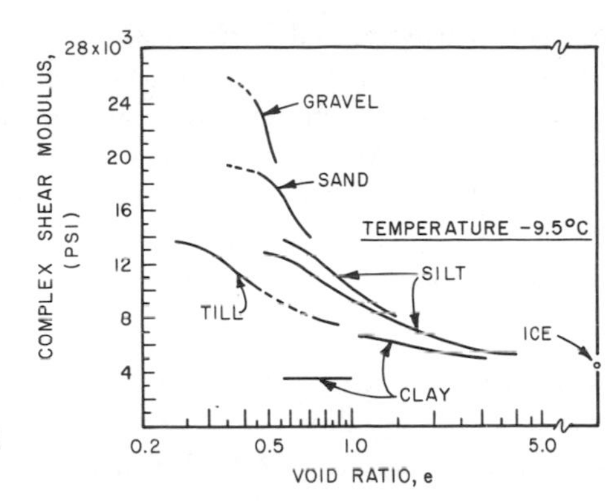


**Figure 3–22** Shear modulus versus void ratio relationship. (After Stevens, 1975.)

where $V_p$ = compressional wave velocity
$V_s$ = shear wave velocity
$\rho$ = bulk density of the material

The formulas above are strictly applicable to isotropic elastic material complying with Hooke's law. The dynamic Young's modulus $E$ can also be found knowing $V_p$, $\nu$, and $\rho$ by the following relationship:

$$E = V_p^2 \rho \frac{(1 + \nu)(1 - 2\nu)}{1 - \nu} \tag{3–51}$$

where $\nu$ is Poisson's ratio.

Kaplar (1969) found that the dynamic moduli ($E$ and $G$) increase significantly between 0 and −6.7°C (32 and 20°F). Some typical results for $E$ are presented in Fig. 3–21. Stevens (1975) studied the effect of void ratio on the shear modulus and the results indicate that the shear modulus decreases as the void ratio increases (Fig. 3–22). The use of dynamic properties of frozen soils for the detection of permafrost is discussed in Chapter 10.

## 3.4 THERMAL PROPERTIES OF SOILS

The thermal properties of soil are of great importance in many engineering projects, such as buildings, roads, airports, pipelines in frozen ground as well as underground cables, hot water pipes, and cold gas pipelines in unfrozen ground. It is also important to determine the position of the interface boundary between the thawed and frozen soil with respect to the ground surface for a given surface temperature. The position of thaw or frost penetration is dependent on the thermal properties of soils. This section deals with the thermal properties of soils and the factors influencing them. Also, thermal properties of different soils are included for use in different analytical methods for the solution of ground heat-transfer problems. The basic thermal properties of soils are thermal conductivity, heat capacity, diffusivity, and latent heat. Kersten (1949), Johansen (1975), and Farouki (1981) have published information on various aspects of thermal properties of soils. The thermal properties of soils vary with phase composition as discussed in Section 3.2.1. As such, temperature, soil type, water content, degree of saturation, dry density, and organic contents are some of the most predominant factors that affect a soil's thermal properties. Application methods related to ground thermal analyses are presented in Chapter 4.

### 3.4.1 Thermal Conductivity

The thermal conductivity of a soil is defined as the amount of heat passing in unit time through a unit cross-sectional area of the soil under a unit temperature gradient applied in the direction of the heat flow. Considering a prismatic element of soil having a cross-sectional area $A$ at right angles to the heat flow $q$, the thermal conductivity $K$ is given by

$$K = \frac{q}{A\,(T_2 - T_1)/L} \tag{3–52}$$

where the temperature drops from $T_2$ to $T_1$ over a length $L$ of the element. Its units are

$$\frac{\text{Btu}}{\text{hr ft}^2\ ^\circ\text{F/ft}} \quad \text{or} \quad \frac{\text{cal}}{\text{s cm }^\circ\text{C}} \quad \text{or} \quad \frac{\text{W}}{\text{m K}}$$

The relationship between thermal conductivity and amount of water in soil was explored by Kersten (1949), and Figs. 3–23 to 3–25 present his experimental data on the thermal conductivity of both unfrozen and frozen soils at different dry densities and water contents. His empirical equations showed that the thermal conductivity is linearly related to the logarithm of the moisture content at a constant dry density. For unfrozen silt and clay soils containing 50% or more of silt and clay, the equation for the thermal conductivity $K$ is

$$K = (0.9 \log \omega - 0.2)10^{0.01\rho_d} \qquad \text{for } w \geq 7\% \tag{3–53a}$$

and for unfrozen sandy soils (clean sand), it is

$$K = (0.7 \log \omega + 0.4)10^{0.01\rho_d} \qquad \text{for } w \geq 1\% \tag{3–53b}$$

where $K$ is in Btu in./ft$^2$ hr °F, the moisture content $w$ is in percent, and the dry density $\rho_d$ is in lb/ft$^3$. Equations (3–53a) and (3–53b) can be rewritten as

$$K = 0.00144 \times 10^{1.373\rho_d} + 0.01226 \times 10^{0.499\rho_d}\omega \qquad \text{for silt and clay} \tag{3–54a}$$

and

$$K = 0.01096 \times 10^{0.8116\rho_d} + 0.00461 \times 10^{0.9115\rho_d}\omega \qquad \text{for sandy soils} \tag{3–54b}$$

where $K$ is in W/m K, $\rho_d$ in g/cm$^3$, and $\omega$ in percent.

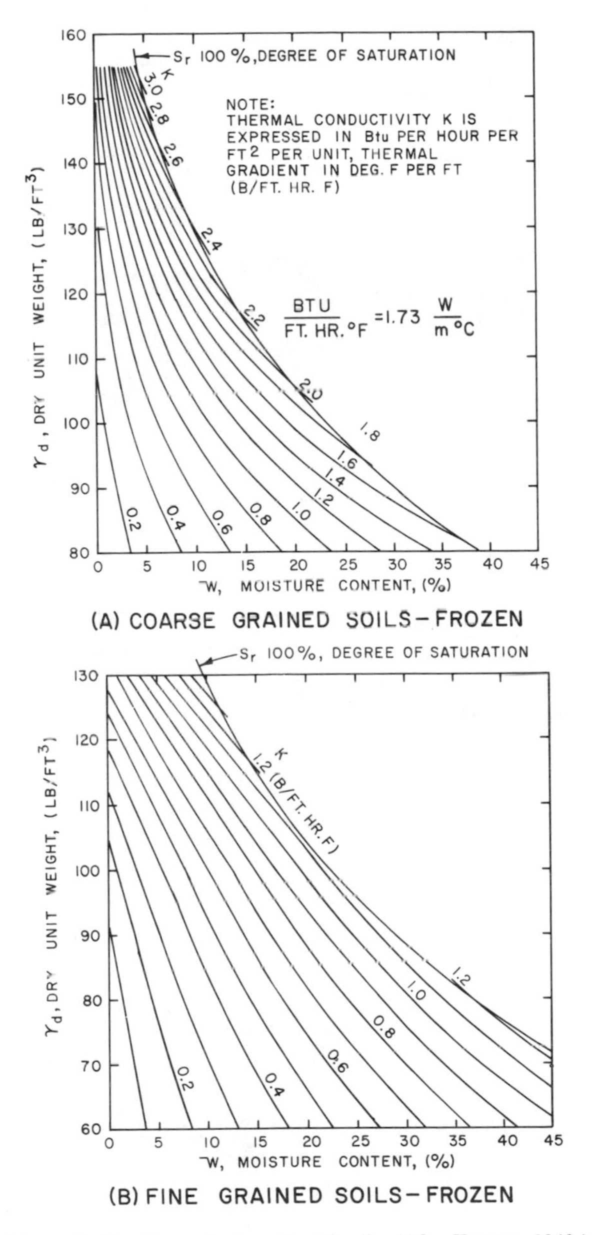


**Figure 3–23** Thermal properties of soils. (After Kersten, 1949.)

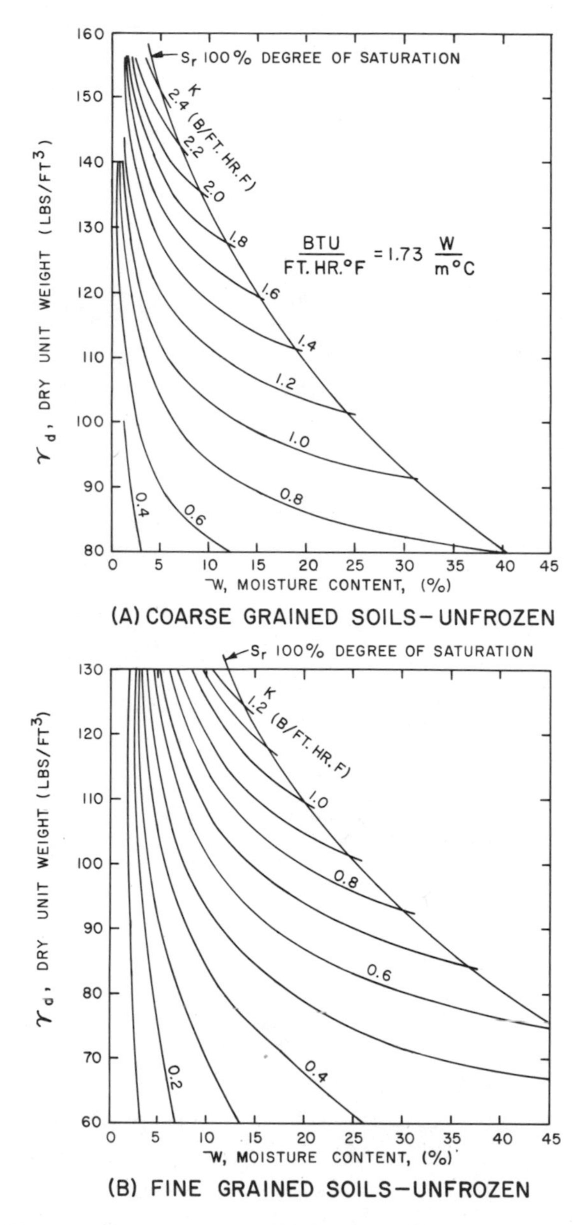


**Figure 3–24** Thermal properties of soils. (After Kersten, 1949.)

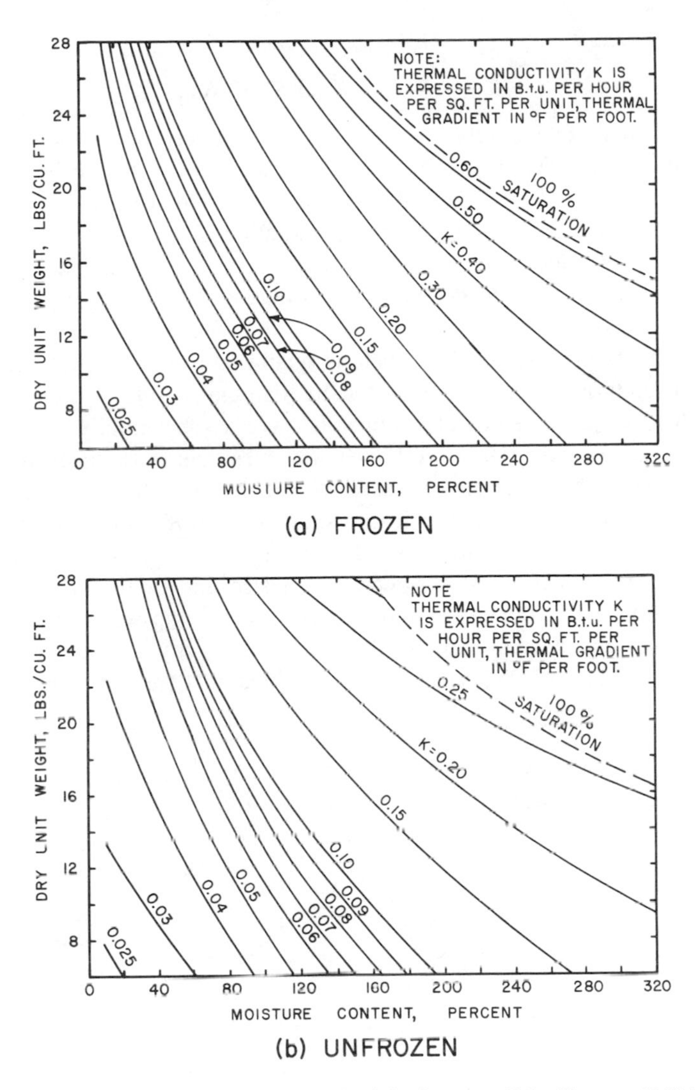


**Figure 3–25** Average thermal conductivity for peat. (After Kersten, 1949.)

Johansen (1975) introduced the concept of the Kersten number $K_e$ to calculate the thermal conductivity $K$ of a soil at partial saturation from known values of the conductivity in the saturated state ($K_{sat}$) and that in the dry state ($K_{dry}$) by the following relationship:

$$K_e = \frac{K - K_{dry}}{K_{sat} - K_{dry}}$$

or

$$K = (K_{sat} - K_{dry}) K_e + K_{dry} \tag{3–55}$$

The thermal conductivity of ice is about four times that of water. Therefore, the thermal conductivity of frozen soils ($K_f$) may be expected to be greater than that of unfrozen soil ($K_u$). This is the case for saturated soils and soils with high saturation. For soils with a low degree of saturation, $K_f$ may be less than $K_u$.

Figure 3–26 illustrates the relationships between $\alpha$, $\mu$ and $\lambda$, and these relationships are useful in the determination of thaw and frost penetration in soil discussed in Chapter 4.

The effects of other factors, such as quartz and micaeous minerals, on the thermal conductivity have been reported by Farouki (1981). Quartz has a relatively high $K$ value, whereas clay minerals have a substantially lower $K$ value. This is the reason sandy soils have higher $K$ values than those of clayey soils.

Equation (3–55) may also be used to calculate the thermal conductivity of frozen unsaturated soil in terms of frozen saturated soil $K_{sat}$ and that of frozen dry soil $K_{dry}$ using the Kersten number $K_e$. The occurrence of ice in soils can be quite complex, as ice crystals vary considerably in size and orientation. Ice banding may decrease particle contacts and air pores may develop within the ice. Figure 3–27 illustrates the decrease in the thermal conductivity as the ice content increases. Slusarchuk and Watson (1975) attributed this trend to the presence of

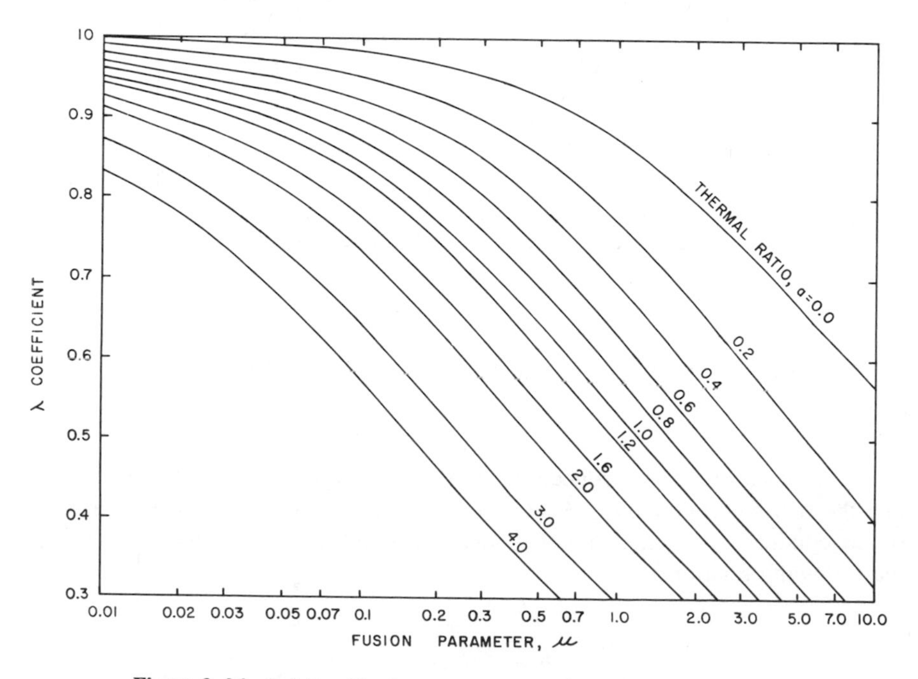


**Figure 3–26** Relationships between $\alpha$, $\mu$, and $\Lambda$. (After Kersten, 1949.)

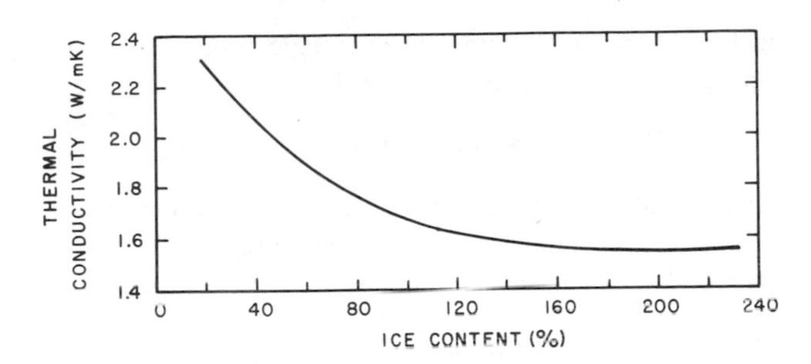


**Figure 3–27** Thermal conductivity as function of ice content of undisturbed permafrost samples. (After Farouki, 1981.)

many small air bubbles and discontinuities in the naturally occurring ice-rich permafrost. Typical values of thermal conductivity for a variety of materials are presented in Appendix C.

### 3.4.2 Heat Capacity

Heat capacity is defined as the quantity of heat necessary to raise the temperature of a given material by 1°C. When expressed on a unit volume basis, the quantity is known as the *volumetric heat capacity,* and when expressed on a unit weight basis, the quantity of heat is referred to as the *specific heat capacity*. The units are

$$\frac{\text{Btu}}{\text{ft}^3\ {}^\circ\text{F}} \quad \text{or} \quad \frac{\text{cal}}{\text{cm}^3\ {}^\circ\text{C}} \quad \text{or} \quad \frac{\text{kJ}}{\text{m}^3\ \text{K}}$$

The heat capacity of a frozen soil can be determined by adding the heat capacities of its constituents (e.g., solid, ice, unfrozen water, and air components) present in a unit of mass of frozen soil. Therefore,

$$CW = C_s W_s + C_i W_i + C_{uw} W_{uw} + C_a W_a + \frac{1}{\Delta\Theta} \int_{\Theta}^{\Theta + \Delta\Theta} L \frac{\partial \omega_{uw}}{\partial \Theta} d\Theta \tag{3–56}$$

where $C_s$, $C_i$, $C_{uw}$, $C_a$ = heat capacity of solid particle, ice, unfrozen water, and air phases, respectively

$L$ = latent heat of fusion of ice

$W_s$, $W_i$, $W_{uw}$, $W_a$ = corresponding weight factors

$\Theta$ = temperature

$\omega_{uw}$ = unfrozen water content

The last term in Eq. (3–56) is incorporated due to latent heat involved in the ice–water phase transformation (Hoekstra, 1969). But the term can be neglected for most engineering applications, as the dependence of latent heat on the unfrozen water content is significant only for temperatures below about −20° C (Anderson and Morgenstern, 1973).

Dividing Eq. (3–56) by volume $V$ and neglecting small term $C_a$, the volumetric heat capacity of a frozen soil is given by

$$C_f = C\rho_f = \rho_{df}(C_s + C_i\omega_i + C_{uw}\omega_{uw}) \tag{3–57}$$

Typical values of the specific heat capacity for mineral soil constituents, organic material, and ice near freezing temperature are 0.17, 0.40, and 0.50, cal/gm °C respectively. Equation (3–57) can be rewritten for frozen soils as

$$C_f = \rho_{df}[0.17 + 0.5(\omega - \omega_{uw}) + \omega_{uw}]$$

or

$$C_f = \rho_{df}(0.17 + 0.5\omega) \qquad \text{(assuming that } \omega_{uw} = 0) \tag{3–58}$$

For unfrozen soils ($\omega_i = 0$, $\omega = \omega_{uw}$) the volumetric heat capacity is given by [from Eq. (3–57)]

$$C_u = \rho_d\,(0.17 + 1.0\omega) \tag{3–59}$$

### 3.4.3 Thermal Diffusivity

Where an unsteady state exists, the thermal behavior of a soil is governed not only by its thermal conductivity $K$ but also by its heat capacity $C$. The ratio of these two properties is termed the thermal diffusivity $a$, which becomes the governing parameter in such a transient state and is given by

$$a = \frac{K}{C} \tag{3–60}$$

Its units are $ft^2/hr$ or $mm^2/s$.

A high value for the thermal diffusivity implies a capacity for rapid and considerable changes in temperature. The diffusivity of ice is about eight times that of liquid water, and consequently the diffusivity of frozen soil is much higher than that of the same soil in a thawed condition. In frozen soils, temperature can therefore change much more rapidly and to a greater extent than in unfrozen soils.

### 3.4.4 Latent Heat

The volumetric latent heat of fusion $L$ is the amount of heat required to melt the ice or freeze the water in a unit volume of soil without a change in temperature. Its units are $Btu/ft^3$ or $J/m^3$. The latent heat of the soil is dominated by the water content. We have

$$L = \rho_d\omega(1 - \omega_{uw})L' \tag{3–61}$$

or

$$L = \rho_d \omega L' \qquad \text{for } \omega_{uw} = 0 \tag{3–62}$$

where $L' = 79.6$ cal/g or 143.4 Btu/lb.

Theoretically at 0° C at atmospheric pressure, with no temperature change during crystallization, the amount of heat energy removed is equal to 79 cal/g or 143.4 Btu/lb. When ice melts, it absorbs heat from the air with which it is in contact and the same amount of heat must be added to ice to transform it to water without a change in temperature.

## PROBLEMS

**3–1.** The dry density of a frozen soil is 1.60 $Mg/m^3$ and the solids have a density of 2.65 $Mg/m^3$. Find the **(a)** total water content, **(b)** void ratio, and **(c)** porosity.

**3–2.** A natural deposit of frozen fine-grained soil was found to have a total water content of 40% and to be 100% saturated. What are the porosity and void ratios of this soil? Assume that $G_s = 2.77$.

**3–3.** In Problem 3–2, if the unfrozen water content was found to be 10%, what is the frozen dry density of the solids? What is the ice saturation?

**3-4.** The total volume of a thawed soil specimen is 80 $cm^3$ and it weighs 140 g. The dry weight is 120 g. Find the **(a)** water content, **(b)** void ratio, **(c)** degree of saturation, and **(d)** wet density. Assume that $G_s = 2.70$.

**3–5.** The total water content of a frozen clayey sample taken from a subsoil investigation was found to be 55%. Assuming that the ice lenses contribute one-third of the total water content, determine the frozen dry density of the sample **(a)** if the sample is fully saturated and **(b)** if the sample is 90% saturated.

**3–6.** If the specific surface area of the sample in Problem 3–5 was found to be 60 $m^2/g$, determine the unfrozen water content at $-2$, $-3.5$, and $-5$ °C.

**3–7.** Calculate the unfrozen water content of a frozen sandy soil at a temperature of $-1.5$° C. Assume the necessary data.

**3–8.** The following information is given for Fairbanks silt:

- specific surface area = 40 $m^2/g$
- unfrozen water content = 5%
- dry density = 1440 $kg/m^3$

What will be the freezing point of depression?

**3–9.** Determine the range of unfrozen water content of a clayey soil at different below freezing temperatures. Assume that the specific surface area is 900 $m^2/g$. If the predominant clay minerals (i.e., kalonite, illite, and montmorillonite) vary in the soil samples, compute the variation of unfrozen water content of soil samples according to the predominant clay minerals.

**3–10.** What is the freezing-point depression of soil? Discuss its relevance with regard to the data presented in Fig. 3–2.

**3–11.** **(a)** From Fig. 3–15, determine the creep parameters $K$, $\sigma_k$, $n$, and $\sigma_c$ for frozen Callovian sandy loam. **(b)** Determine the creep strain after 30 years for a stress of 100 kN/m$^2$.

**3–12.** Using the creep parameters, find the primary creep strain after 20 years to a stress of 2 tons/ft$^2$ applied on Ottawa sand and Manchester fine sand at 31° F.

**3–13.** Explain the difference between primary and secondary creep. Discuss the creep behavior of ice-poor and ice-rich frozen soils.

**3–14.** Using the data from Example 3–6, estimate the seondary creep strain for frozen Svea clay after 10 years with an axial load of 50 kN/m$^2$.

**3–15.** Using the creep parameters and observing that the failure strain was close to 10% for the frozen Callovian sandy loam (Fig. 3–15), estimate the 30-year strength at −20° C using the relationship

$$\sigma_f = \sigma_c \left( \frac{\varepsilon_f}{\dot{\varepsilon}_c t_f} \right)^{1/n}$$

**3–16.** A silty soil with a dry density $\rho_d$ of 1440 kg/m$^3$ and a water content of 25% changes temperature from −7°C to 5°C. Discuss quantitatively the changes in thermal conductivity, volumetric heat capacity, and latent heat of the soil.

**3–17.** Work Problem 3–16 in U.S. customary units given that $\rho_d$ = 90 lb/ft$^3$. The temperature changes from 20°F to 41°F.

CHAPTER 4

# Foundation Design Philosophy and Considerations

## 4.1 INTRODUCTION

Basic foundation design concepts (i.e., allowable soil pressure limiting short- and long-term settlement) that are applied to the stability of structures on unfrozen soils are also applicable to frozen ground. However, the design approach to foundations in frozen ground is more complex. Factors such as type and nature of structure, heated or unheated, ground thermal regime, specific site conditions, load–deformation characteristics of foundation soils, construction schedule, and environmental constraint will govern the design and construction of foundations in frozen ground. As the foundation is an integral part of the superstructure and transfers all loads acting on the superstructure to the supporting materials, the interactions between the structure–foundation and the foundation materials must be understood. Generally, site conditions dictate the type of suitable foundations that can be constructed to support the superstructure and its loads. Either shallow or deep foundations may be designed and they are discussed in Chapters 5 and 6, respectively. In this chapter the basic foundation design philosophy and design considerations that are usually used in frozen ground are presented.

As discussed in Chapter 1, frozen ground may be perennial or seasonal. The upper layer, called the *active layer* of frozen ground, undergoes a seasonal thaw–frost cycle. This cycle causes various phenomena to occur, such as thaw settlement, down-drag force, frost-heave force, and differential settlement. To ensure proper foundation design, the temperature changes, uplift forces, deformation, and reduction of long-term strength of soils due to phase changes must be considered. Foundations should be designed in a manner that provides an adequate factor of safety against the stability of the structure and establishes limits of settlement within an acceptable range during the design life of the structure.

As frozen soils are particularly sensitive to temperature change and their strength–deformation characteristics are governed mainly by below-freezing or negative temperatures, the thermal regime at the site must be assessed to evaluate the interactions between the foundation and the supporting frozen ground. Design parameters such as dead and live loads of structure, structural rigidity or flexibility, dynamic loads (due to vibrations or earthquakes), structural loads on the floor, and temperatures to be maintained must be considered in the selection of particular foundation types.

The notation introduced in this chapter is given in Table 4–1.

**TABLE 4–1 Notation**

| Symbol | SI Unit | Definition |
|---|---|---|
| $a$ | mm$^2$/s | Thermal diffusivity |
| $A$ | — | Thaw strain (Eq. 4–26) |
| $A'$ | — | Cross-sectional area |
| $A_0$ | °C | Temperature amplitude (Eq. 4–7) |
| $a_f$ | mm$^2$/s | Diffusivity of frozen soil |
| $a_u$ | mm$^2$/s | Diffusivity of unfrozen soil |
| $C_{av}$ | kJ/m$^3$ K | Average heat capacity |
| $C_f$ | kJ/m$^3$ K | Volumetric heat capacity of frozen soil |
| $C_u$ | kJ/m$^3$ K | Volumetric heat capacity of unfrozen soil |
| $C_v$ | m$^2$/yr | Coefficient of consolidation (Eq. 4–31) |
| $d$ | m | Layer thickness |
| $e_f$ | — | Initial void ratio of frozen sample (Eq. 4–27) |
| $e_t$ | — | Initial void ratio of thawed sample (Eq. 4–27) |
| $erf$ | — | Error function |
| $F_+$ | J/m$^2$ | Heat fluxes in the thawed zone (Eq. 4–9) |
| $F_-$ | J/m$^2$ | Heat fluxes in the frozen zone (Eq. 4–9) |
| $H$ | m | Total thaw settlement (Eq. 4–28) |
| $I_f$ | degree-days | Freezing index (Eq. 4–20) |
| $I_t$ | degree-days | Thawing index (Eq. 4–21) |
| $K_a$ | W/m K | Average thermal conductivity |
| $K_f$ | W/m K | Thermal conductivity of frozen soil |
| $K_u$ | W/m K | Thermal conductivity of unfrozen soil |
| $k$ | cm/s | Coefficient of permeability |
| $L$ | J/m$^3$ | Volumetric latent heat of soil |
| $m_v$ | — | Coefficient of volume compressibility (Eq. 4–34) |
| $P$ | kN/m$^2$ | Frost-heaving pressure (Eq. 4–23) |
| $P_o$ | kN/m$^2$ | Overburden pressure |
| $P_w$ | kN/m$^2$ | Pore water pressure (Eq. 4–37) |
| $R$ | m$^2$ K/W | Thermal resistance of layer |
| $R_n$ | m$^2$ K/W | Thermal resistance of $n$ layers |
| $r$ | m | Radius of sphere (Eq. 4–23) |
| $r_i$ | m | Radius of curvature of ice–water interface (Eq. 4–23) |
| $S$ | m | Settlement |
| $S_{max}$ | m | Total consolidation settlement (Eq. 4–45) |
| $S_t$ | m | Consolidation at any time $t$ |
| $T$ | °C | Temperature |

TABLE 4–1 *(Cont.)*

| Symbol | SI Unit | Definition |
|---|---|---|
| $T_a$ | °C | Mean air temperature |
| $T_g$ | °C | Mean ground temperature |
| $T_s$ | °C | Surface temperature |
| $t$ | — | Time |
| $U$ | — | Pore water pressure |
| $V$ | — | Total volume |
| $\Delta V$ | — | Volume of pore water (Eq. 4–32) |
| $V_0$ | °C | Mean annual site temperature |
| $X_f$ | m | Frost depth |
| $X_t$ | m | Depth of thaw |
| $X_{(t)}$ | m | Thaw penetration after time $t$ |
| $Z$ | — | Depth |
| $\Delta\sigma$ | — | Change in effective stress |
| $\sigma_{is}$ | — | Ice–solid interfacial energy (Eq. 4–23) |
| $\sigma_{iw}$ | — | Ice–water interfacial energy (Eq. 4–23) |
| $\sigma_{ws}$ | — | Water–solid interfacial energy (Eq. 4–23) |
| $\alpha$ | — | Thermal ratio |
| $\mu$ | — | Fusion parameter |
| $\Lambda$ | — | Constant |
| $\rho$ | kg/m$^3$ | Bulk density |
| $\rho_d$ | kg/m$^3$ | Dry density |
| $\rho_b$ | kg/m$^3$ | Buoyant weight |
| $\rho_w$ | kg/m$^3$ | Density of water |

## 4.2 FOUNDATION DESIGN APPROACH IN FROZEN GROUND

Frozen ground potentially provides an excellent foundation support in areas where thermal degradation cannot occur. In extreme cases where the frozen ground consists of ice-rich soils at warm temperature, it may undergo deformation even at a much lower design load. So, depending on the nature and composition of frozen ground as well as the type of structure, the following design approaches are generally used in frozen ground:

METHOD A. Maintain existing thermal regime or keep in frozen state.

METHOD B. Accept thermal regime changes caused by construction and operation or allow thermal degradation of existing ground support materials.

METHOD C. Improve or modify foundation support material before construction.

METHOD D. Use a conventional foundation design approach for unfrozen soils.

Method A is most commonly used in perennially frozen or continuous-permafrost regions as well as in discontinuous-permafrost regions when the frozen ground consists of thaw-unstable materials. This passive method is specifically suitable where ice-rich frozen soils are encountered at shallow depths. Method B can also be used in continuous or discontinuous permafrost regions where *suitable* foundation support materials are encountered at shallow depths. The anticipated total settlement or differential settlement due to changes in thermal regime should not exceed the acceptable limit of the structure. Method C is applied primarily in discontinuous-permafrost regions where the thickness of the frozen layer may be shallow enough to use soil improvement techniques. Method D may be used only in subsoil conditions where thaw-stable dense sands and gravels or ice-free bedrock are encountered. Each method is discussed in more detail in the following sections.

### 4.2.1 Method A: Maintaining Frozen Ground in Its Frozen State

The foundation design approach of maintaining frozen ground in its frozen state is a passive method. This passive method should be used at site conditions where the thermal degradation of existing foundation materials is not acceptable. Various subgrade cooling systems (see Chapter 5) and thermal pile support systems (see Chapter 6) may be used to retain the existing frozen conditions. Mechanical refrigeration systems consisting of a grid of cooling pipes through which the refrigerated fluid is circulated may also be used to maintain the frozen state of the ground. Generally, the implementation of a mechanical system is expensive, involving high construction costs as well as long-term expenses in maintaining equipment, refrigeration loads, and a steady fuel supply.

Heat tubes or self-refrigerated systems are becoming more popular devices to keep the ground frozen. A typical two-phase heat-tube system is shown in Fig. 4–1. This device works only in the winter, when air temperatures are colder than ground temperatures. The device pumps heat from the ground by a two-phase convective process that works as follows. When the ground temperature is higher than the air temperature, film liquid evaporation occurs along the evaporating section embedded in the ground. The resulting vapor rises to the upper portions, where it releases heat to the air through a radiator section. The vaporized gas condenses as it comes into contact with the colder air temperatures and falls back, causing a cycle as shown in Fig. 4–1. Many authors have described this heat-tube device (Heuer, 1979; Water, 1973). Similar systems, such as the long pile (Long, 1966, 1978) and the air convection system (Reed, 1966), may also be used to maintain the existing thermal regime.

The application of this passive method to a heated structure is more complex than its use with an unheated structure. The thermal interactions between the heated structure and the frozen ground must be analyzed in detail, considering the

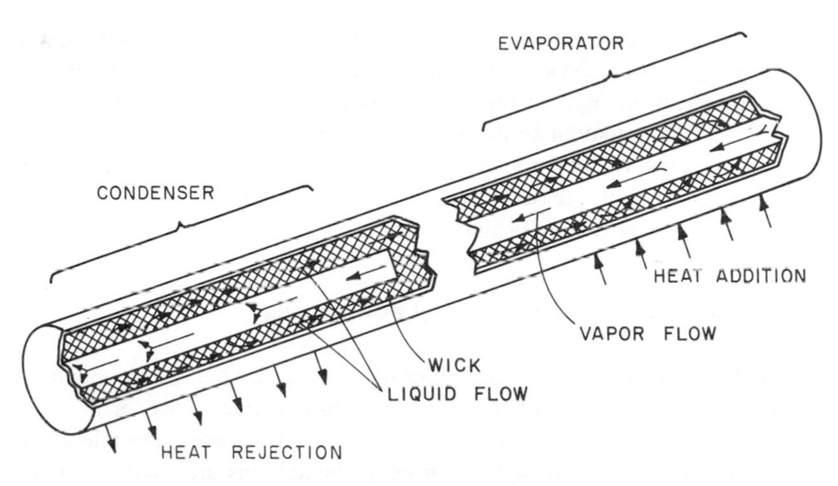


**Figure 4–1** Basic heat pipe. (After Waters, 1973.)

most suitable device to maintain underlying frozen ground. A combination of gravel pad, insulation material, and thermal heat-removal device may be considered under specific conditions for the design of a foundation for heated structures, as discussed in Chapter 5.

### 4.2.2 Method B: Allowing Frozen Ground to Undergo Thermal Changes

This method of foundation design allows the frozen soil to thaw under the thermal-regime changes caused by the structure or other environmental conditions. This method may be applied if such thermal-regime changes maintain adequate bearing capacity of foundation supporting soils and if the total settlement or differential settlement is within acceptable limits. This active method is used only if the foundations can be placed on a dense, thaw-stable soil stratum or competent bedrock at shallow depth. Extensive site investigation, rate of thaw, and thaw-settlement predictions (see Section 4.5) are required for the application of this method.

### 4.2.3 Method C: Improvement of Foundation Support Conditions

By this method of foundation design, undesirable foundation materials may be excavated and backfilled with non-frost-susceptible (NFS) material. In other cases, prethawing existing frozen soils using steam or cold water and then compacting the resulting altered soils may be used to improve the site conditions. Sanger (1969) quoted a Soviet criterion in which the prethawed method was used when the thickness of the frozen layer was less than 60% of the computed thaw penetration in 10 years. Prethawing precludes the anticipated thaw settlements and improves the shear strength of the soils by compaction near the structure. However, some problems may arise during the prethawing period, including metastability of soils, unworkable ground surface, poor drainage, poor stability under vibration, excessive pore pressure generation, and loss of shear strength, and must be considered when analyzing design data.

### 4.2.4 Method D: Conventional Techniques

Conventional techniques for foundation design are used when the foundation supporting materials consist of either competent bedrock free of ice or dense thaw-stable sands and gravels. Based on the bearing capacity theory, the allowable soil pressures are determined and shallow or deep foundations are used depending on the magnitude of the load to be supported on the foundations.

## 4.3 FOUNDATION DESIGN CONSIDERATIONS

The two basic aspects of foundation design in frozen ground are rheological and thermal. The former defines the effects of stress on the supporting material, consisting of deformation and flow of materials, and the latter considers the thermal aspects, consisting of heat flow and thermal analyses between the structure and the foundation material. Chapter 3 discusses the rheological aspects and the thermal aspects. The thermal aspects are explained in the following section. Rheological and thermal aspects must be combined with the engineering judgments for the most appropriate design of foundations.

Some of the major factors related to the design of shallow and deep foundations are discussed in Chapters 5 and 6, respectively. The upper layer of frozen ground may undergo a freeze–thaw cycle that reduces soil strength. The phase changes may cause frost heave and thaw weakening, both of which must be considered in the design of foundations.

### 4.3.1 Thermal Analysis

Thermal analysis between the structure and ground can be made by considering boundary conditions and thermal properties of the ground. The basic heat transfer of energy due to conduction, convection, and radiation must be understood in doing heat-flow calculations. Detailed thermal analysis is beyond the scope of this book and basic references on heat transfer by Carslaw and Jaeger (1959), Ingersoll et al. (1954), and Kreith (1973) may be used for rigorous solutions of two- and three-dimensional problems.

In frozen ground, the temperature tends to vary seasonally with the below-ground surface to a depth of about 10 to 30 m (Fig. 1–7). Below this depth, the temperature increases gradually under the geothermal gradient.

As discussed in Chapter 1, the presence or absence of frozen ground in a given region depends primarily on the persistence of ground temperatures below 0°C for 2 or more years. There is a difference between the average annual air and ground-surface temperatures. This difference is due to the effects of such near-surface factors as snow cover, vegetation, time-varying ground thermal properties, relief, slope and orientation of the surface, and surface and subsurface drainage. Each temperature change at the ground surface imposes a response that moves downward to a depth below the ground surface. This movement is dependent on the thermal properties of the ground. Lunardini (1981) discussed the heat-transfer problems in cold regions.

From an engineering point of view, the depth of thaw and frost penetration due to heated or unheated structures on the ground is of prime importance.

*Steady-State Conditions.* Conduction is the dominant ground heat-transfer process, and the governing equation of heat transfer assuming no phase change is

$$\frac{\partial T}{\partial t} = a \frac{\partial^2 T}{\partial z^2} \tag{4–1}$$

where $T$ = temperature
$t$ = time
$z$ = depth
$a$ = thermal diffusivity = $K/C$ ($K$ and $C$ are the thermal conductivity and volumetric heat capacity, respectively)

For steady-state conditions, that is, conditions in which the temperature distribution within a body is tending toward no change and the boundary conditions remain unchanged, one-dimensional steady-state equation (4–1) becomes (setting $\partial T/\partial t = 0$)

$$\frac{\partial^2 T}{\partial z^2} = 0 \tag{4–2}$$

Equation (4–2) can be integrated directly to give the temperature profile:

$$T = Mz + N \tag{4–3}$$

where $M$ and $N$ are constants obtained from the upper and lower boundary conditions.

In two dimensions, Eq. (4–1) may be written as

$$\frac{\partial T}{\partial t} = a\left(\frac{\partial^2 T}{\partial z^2} + \frac{\partial^2 T}{\partial x^2}\right) = 0 \tag{4–4}$$

Equation (4–4) is the well-known Laplace equation. A variety of mathematical and graphical methods have been developed to solve the Laplace equation depending on the boundary conditions.

The steady temperature beneath heated or cooled areas on the ground surface may be written as

$$T - T_g = \frac{T_s - T_g}{\pi} \tan^{-1} \frac{z}{x} \tag{4–5}$$

where $T$ = desired temperature
$T_g$ = mean ground surface temperature outside heated or cooled area
$T_s$ = temperature of heated or cooled area
$z$ = depth
$x$ = horizontal direction

For a long strip (such as roads) on the ground surface of width $2a$, the temperature field can be written as

$$T - T_g = \frac{T_s - T_g}{\pi} \tan^{-1} \frac{2az}{x^2 + z^2 - a^2} \tag{4–6}$$

Equation (4–6) is useful in estimating the temperature regime beneath roads, rivers, and heated or cooled structures on frozen or unfrozen ground.

*Time-Dependent Surface Temperature with No Phase Change.* If thermal properties are constant and there is no freezing or thawing, the ground temperature can be described approximately by the periodic steady-state solution for one-dimensional heat flow in a semi-infinite homogeneous medium as

$$T - T_g = A_0 \exp\left(-x \frac{\sqrt{\pi}}{365a}\right) \sin\left(\frac{2\pi t}{365} - x \frac{\sqrt{\pi}}{365a}\right) \tag{4–7}$$

where $A_0$ denotes the temperature amplitude.

When buildings, roads, and airstrips are constructed, the mean annual temperature in the ground beneath the structures will adjust to new equilibrium values and the thermal design must evaluate whether such changes will lead to thermal degradation of frozen ground or unacceptable foundation conditions.

Equation (4–7) may be used to calculate the full thickness required for permafrost protection by setting $T = 0$ at some depth $x = X$ and the sine function equal to 1 (maximum value).

In Eq. (4–4) we have

$$X = -\left(\frac{365a}{\pi}\right)^{1/2} \ln \frac{-T_g}{A_0} \tag{4–8}$$

Harlan and Nixon (1978) noted that Eq. (4–8) predicts large fill thickness in warmer permafrost areas where the ratio $T_g/A_0$ is small.

Thawing and freezing of soil significantly change the ground thermal regime in permafrost areas. The conductive heat flow, Eq. (4–1), must be accompanied by an equation to account for the liberation or absorption of latent heat ($L$). The approximate boundary condition at the moving interface separating the frozen and thawed phase is

$$L \frac{dX}{dt} = F_- - F_+ \tag{4–9}$$

where $X$ = position of the moving interface
$F_-$ = heat fluxes in the frozen phase
$F_+$ = heat fluxes in the thawed phase

A good approximation to one-dimensional freezing is provided by the Stefan formula, which assumes a material initially at the isothermal condition at the freezing point $T_f$, whose surface temperature is suddenly lowered to a freezing value $T_s < T_f$.

Neglecting the heat flow from the underlying thawed material, the boundary condition at the freezing point, Eq. (4–9), becomes

$$L \frac{dX_f}{dt} = F_- = -K_f \frac{\partial T}{\partial z} = X_f \tag{4–10}$$

If the temperature profile in the frozen zone is assumed to be linear, Eq. (4–10) becomes

$$L \frac{dX_f}{dt} = -K_f \frac{T_s - T_f}{X_f}$$

or

$$X_f = \sqrt{\frac{2K_f(T_f - T_s)}{L}} \tag{4–11}$$

where $X_f$ denotes the frost depth at time $t$, $K_f$ the thermal conductivity of the frozen material, and $L$ the volumetric latent heat.

In the derivation of the Stefan equation (4–11), both the heat capacity of the frozen layer and the heat flow from the underlying thawed zone are neglected. Since both quantities are responsible for the progress of frost or thawed penetration, the Stefan equation will overestimate frost or thaw penetration depths.

#### *4.3.1.1 Depth of Thaw and Frost Penetration in Soils*

Estimation of depth of thaw or frost penetration in soils or under engineering structures is one of the most important thermal calculations in frozen ground engineering. Such solutions are needed to estimate the rate and depth of the active layer in permafrost areas, rates of thermal degradation or deepening of the active layer following surface disturbances, depth of frost or thaw under highways and airstrips, rates of frost or thaw advance beneath chilled or warm buried pipelines, and many other engineering areas.

Neumann (in about 1860) derived the solution for determining the depth of thaw or frost, and Carslaw and Jaeger (1959) have presented Neumann's solution. As illustrated in Fig. 4–2, a uniform homogeneous frozen soil is subjected to a step increase in temperature from $T_g$ in the ground to $T_s$ at the surface. Assuming that properties of the frozen and thawed zones are independent of temperature and are homogeneous, the Neumann formulation leads to

$$X_t = \alpha \sqrt{t} \tag{4–12}$$

where $X_t$ denotes the depth of thaw and $\alpha$ is a constant that is determined from the equation

$$L\pi^{1/2} = \frac{2(a_u)^{1/2} C_u T_s e - \alpha^2/4a_u}{\text{erf}\,[\alpha/2(a_u)^{1/2}]} - \frac{e - \alpha^2/4a_f}{1 - \text{erf}\,[\alpha/2(a_f)^{1/2}]} \frac{T_g K_f}{T_s K_U} \left(\frac{a_u}{a_f}\right)^{1/2} (2a_u^{1/2}\, C_u T_s) \tag{4–13}$$

where erf = error function
$K_u$, $K_f$ = thermal conductivities of unfrozen and frozen soil
$C_u$, $C_f$ = volumetric heat capacities of unfrozen and frozen soil
$a_u$, $a_f$ = diffusivities of unfrozen and frozen soil
$L$ = volumetric latent heat of soil
$T_g$ = initial ground temperature below freezing
$T_s$ = applied constant surface temperature $> T_g$

So $\alpha$ is a function of seven variables:

$$\alpha = f(K_u, K_f, C_u, C_f, T_g, T_s, L)$$

To apply these solutions to practical field cases where the surface temperature varies, the average surface temperature for the thawing period is $T_s$ and $T_a$ is taken as the mean air or ground temperature $T_g$. With these assumptions, Eq. (4–12) is referred to as the *Berggren equation*. The average surface temperature may be estimated from the air-temperature thawing index using an *n*-factor correlation, as discussed in Section 1.4.

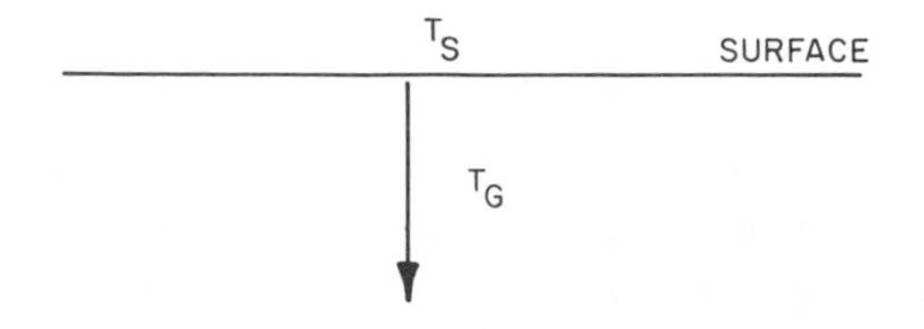


**Figure 4–2** Uniform homogeneous frozen soil subjected to step increase in temperature.

Using the average thawed and frozen values for the thermal properties and three dimensionless parameters as follows:

$$\alpha = \text{thermal ratio} = \frac{T_a - T_f}{T_s - T_f} = \frac{(T_a - T_f)t}{I_t} = \frac{(T_a - T_f)}{V_s} \tag{4–14}$$

where $T_f$ = 0°C or 32°F

$V_s$ = surface thawing index divided by the length of the thawing season

$$\mu = \frac{C}{L}V_s \tag{4–15}$$

$$\Lambda = \alpha \sqrt{\frac{L}{K_a(T_s - T_f)}} \tag{4–16}$$

where $K_a$ is the average thermal conductivity of soil, Eq. (4–12) becomes

$$X_t = \Lambda \sqrt{\frac{2K}{L} I_t} \tag{4–17}$$

which is the equation recommended by Aldrich and Paynter (1953) and is known as the *modified Berggren equation*.

For freezing problems, Eqs. (4–14) and (4–15) must be redefined in terms of the freezing index $I_f$ as

$$\alpha = \frac{T_a - T_f}{T_f - T_s} = \frac{(T_a - T_f)t}{I_f} \tag{4–18}$$

$$\mu = \frac{C(T_f - T_s)}{L} = \frac{C\,I_f}{L\,t} = \frac{C}{L}V_s \tag{4–19}$$

$V_s$ = freezing index divided by the length of freezing season

Equation (4–17) then becomes

$$X_f = \Lambda \sqrt{\frac{2KI_f}{L}} \tag{4–20}$$

where $X_f$ denotes the depth of frost penetration. Equations (4–17) and (4–20) may be rewritten as

$$X = \Lambda \sqrt{\frac{48K_a n I_t(I_f)}{L}} \quad \text{in U.S. customary units} \tag{4–21}$$

or

$$X = \Lambda \sqrt{\frac{7200K_nI_t(I_f)}{L}} \quad \text{in SI units} \tag{4–22}$$

(Aldrich and Paynter, 1953)

where $X$ = thaw or (frost) penetration
$K_a$ = average thermal conductivity of soils, Btu/ft hr °F or J/sm °C
$L$ = latent heat of fusion of soil, Btu/ft$^3$ or J/m$^3$
$n$ = factor
$I_t$ = thawing index, degree-days
$I_f$ = freezing index, degree-days
$\Lambda$ = thermal constant which is a function of $\alpha$ and $\mu$
$\alpha$ = thermal ratio = $V_0/V_s$
$\mu$ = fusion parameter = $(C/L)V_s$
$V_0$ = mean annual site temperature minus 0°C or 32°F
$V_s$ = surface thawing (or freezing) index divided by the length of thawing (or freezing) season
$C$ = volumetric heat capacity of the soil, Btu/ft$^3$ °F or J/m$^3$ °C

The thawing index $I_t$ is defined as the number of degrees above freezing integrated over the time during which the surface is above freezing. Similarly, the freezing index $I_f$ is defined as the number of degrees below freezing integrated over the time during which the surface is below freezing. For most engineering applications, the thawing or freezing indexes can be calculated by determining the area between the average monthly temperature curve and the 0°C line and time taken over the appropriate season. The U.S.Army/Air Force (1966) recommended

**TABLE 4–2 Some Examples of *n* Factors**

| Surface | Thawing Factor, $n_t$ | Freezing Factor, $n_f$ |
|---|---|---|
| Snow | — | 1.0 |
| Pavement free of snow and ice | — | 0.9 |
| Sand and gravel | 1–2 | 0.9 |
| Turf | 1.0 | 0.5 |
| Spruce | 0.35–0.53 | 0.55–0.9 |
| Spruce trees, brush | 0.37–0.41 | 0.28 |
| Willows | 0.82 | — |
| Weeds | 0.86 | — |
| Gravel fill slope | 1.38 | 0.7 |
| Gravel road | 1.99 | — |
| Concrete road | 2.03 | 0.8 |
| Asphalt road | 1.96–1.25 | 0.8 |
| White paint surface | 0.98–1.25 | — |

*Source:* After Lunardini (1978).

that for design purposes, the design freezing index or thawing index is taken as the average air-freezing or thawing index for the three coldest winters or warmest summers in the last 30 years of record.

As discussed in Section 1.4, the ratio between the surface index and air index, called the $n$ factor, is influenced by many factors, such as latitude, cloud cover, time of year or day, wind speed, surface characteristics, and subsurface thermal properties. Some typical values of $n$ factors are presented in Table 4–2.

**Example 4–1**

Determine the depth of thaw penetration into a homogeneous frozen silt for the following conditions:

- Mean annual air temperature (M.A.T.) = 15°F
- Surface-thawing index $nI$ = 1200 degree-days
- Length of thaw season = 110 days
- Soil properties: dry unit weight $\rho_d$ = 100 lb/ft$^3$; water content $\omega$ = 15%

## Solution

Volumetric latent heat of fusion:

$$L = 144\,(100)(0.15)$$
$$= 2160 \text{ Btu/ft}^3$$

Average volumetric heat capacity:

$$C_{av} = 100(0.17 + 0.75 \times 0.15)$$
$$= 28.2 \text{ Btu/ft}^3\ ^\circ\text{F}$$

Average thermal conductivity:

$$\text{Unfrozen: } K_u = 0.72 \text{ Btu/ft hr }^\circ\text{F} \qquad \text{(see Fig. 3–23)}$$

$$\text{Frozen: } K_f = 0.80 \text{ Btu/ft hr }^\circ\text{F} \qquad \text{(see Fig. 3–22)}$$

$$K_a = 1/2(K_u + K_f) = 0.76 \text{ Btu/ft hr }^\circ\text{F}$$

Average surface-temperature differential:

$$V_s = \frac{nI}{t} = \frac{1200}{110} = 10.91^\circ\text{F}$$

Initial temperature differential:

$$V_0 = \text{M.A.T.} - 31 = 15 - 32 = 17^\circ\text{F (below } 32^\circ\text{F)}$$

Thermal ratio:

$$\alpha = \frac{V_0}{V_s} = \frac{17}{10.91} = 1.56$$

Fusion parameter:

$$u = V_s\, C_{av}/L$$

$$= 10.91 \left(\frac{28.2}{2160}\right)$$

$$= 0.14$$

Lambda coefficient:

$$\Lambda = 0.79 \qquad \text{(see Fig. 3–25)}$$

Estimated depth of thaw penetration:

$$X = \Lambda \sqrt{\frac{48\, K_a\, nI_t}{L}} \qquad \text{(from Eq. 4–21)}$$

$$= 0.79 \sqrt{\frac{48(0.76)(1200)}{2160}}$$

$$= 3.6 \text{ ft}$$

**Example 4–2**

Determine the depth of frost penetration into a homogeneous deposit for the following conditions:

- Mean annual air temperature (M.A.T.) = 2.5°C
- Surface covered with snow
- Air freezing index = 1350 degree-days
- Length of freezing season = 150 days
- Soil conditions: silt deposit; unit weight of silt = 1600 kg/m$^3$; saturated water content = 30%

## Solution

The frost penetration depth is given by the modified Berggren equation (Aldrich and Paynter, 1953) (4–22) as

$$X = \Lambda \left(\frac{7200 K_a I_f}{L}\right)^{1/2}$$

where $X$ = frost penetration, m
$K_a$ = average thermal conductivity, J/s m °C
$L$ = latent heat, J/m$^3$
$t$ = time, hr
$I_f$ = freezing index, degree-days

$$\rho_d = \frac{1600}{1 + 0.3} = 1230.77 \text{ kg/m}^3$$

$$L = (334)\left(\frac{1230.77}{1000}\right)(0.3) = 123.32 \text{ MJ/m}^3$$

$$V_0 = 2.5 - 0 = 2.5°\text{C}$$

$$V_s = \frac{1350}{150} = 9°\text{C}$$

$$\alpha = \frac{V_0}{V_s} = \frac{2.5}{9} = 0.28$$

$$C_{av} = \frac{1230.77}{1000}[0.17 + (0.75)(0.30)] = 0.486 \text{ MJ/m}^3\ °\text{C}$$

$$\mu = V_s\frac{C_{av}}{L} = \frac{9(0.486)}{123.32} = 0.04$$

$$\Lambda = 0.94 \qquad \text{(see Fig. 3–25)}$$

Estimated depth of frost penetration:

$$X = 0.94\sqrt{\frac{7200(24)(1.21)(1)(1350)}{123.32 \times 10^6}}$$

$$= 1.42 \text{ m}$$

**Example 4–3**

Determine the depth of thaw penetration beneath an asphalt concrete pavement for the following conditions:

- Mean annual air temperature (M.A.T.) = 15°F
- Surface thawing index $(nI)$ = 1560 degree-days
- Length of thaw season = 105 days

Soil boring log:

| Layer | Depth (ft) | Material | Dry Unit Weight (lb/ft$^3$) | Water Content (%) |
|---|---|---|---|---|
| 1 | 0.0–0.4 | Asphalt concrete | 138 | — |
| 2 | 0.4–2.0 | GW-GP | 130 | 2.0 |
| 3 | 2.0–5.0 | GW-GP | 125 | 3.0 |
| 4 | 5.0–6.0 | SM | 110 | 7.0 |
| 5 | 6.0–8.0 | SM-SC | 105 | 5.0 |
| 6 | 8.0–9.0 | SM | 105 | 8.0 |

## Solution

1. The $V_s$, $V_0$, and $\alpha$ values are determined in the same manner as that for the Example 4–1.

$$V_s = \frac{1560}{105} = 14.8°\text{F}$$

$$V_0 = 15 - 32 = 17°\text{F below freezing}$$

$$\alpha = \frac{17}{14.8} = 1.15$$

2. The thermal properties $K$, $C$, and $L$ of the respective layers are obtained using the procedure described in Example 4–1.
3. The tabular arrangement shown in Table 4–3 facilitates a solution of the multilayer problem, and in the following discussion, layer 3 is used to illustrate quantitative values. Columns 9, 10, 12, and 13 are self-explanatory. Column 11, $L$, represents the average value of $L$ for a layer and equal to $\Sigma\, Ld/\Sigma d = 443.8$. Column 14, $C_{av}$, represents the average value of $C$ and is obtained from $\Sigma\, Cd/\Sigma d = 24.1$. $L$ and $C$ represent the weighted values of a depth of thaw penetration given by $\Sigma d$, which is the sum of all layer thicknesses to that depth.

   The fusion parameter $\mu$ for each layer is determined from

$$V_s\frac{C}{L} = 14.8 \times \frac{24.1}{443.8} = 0.80$$

   The $\Lambda$ coefficient is equal to 0.55 (from Fig. 3–25).
4. Column 18, $Rn$, is the ratio $d/K$ and for layer 3 equals 3.0/0.95 or 3.16.

   Column 19, $\Sigma\, R$, represents the sum of the $R_n$ values above the layer under consideration.
5. Column 20, $\Sigma\, R + (R_n/2)$, equals the sum of the $R_n$ value of the layer being considered. For layer 3 this is

$$2.14 + \frac{3.16}{2} = 3.72$$

6. Column 21, $nI$, represents the number of degree-days required to thaw the layer considered and is determined from

$$nI = \frac{Ld}{24\Lambda^2}\left(\Sigma R + \frac{R_n}{2}\right)$$

   For layer 3,

$$nI_3 = \frac{540(3)}{24(0.3)}(373) = 837$$

**TABLE 4-3 Multilayer Solution**

| (1) Layer | (2) Dry Density, $d$ | (3) Water Content, $\omega$ | (4) Depth, $d$ | (5) $\Sigma d$ | (6) Volumetric Heat Capacity, $C$ | (7) Thermal Conductivity, $K$ | (8) Latent Heat, $L$ | (9) $Ld$ | (10) $\Sigma Ld$ | (11) $AV.$ $L$ |
|---|---|---|---|---|---|---|---|---|---|---|
| 1 | 138 | — | 0.4 | 0.4 | 23.5 | 0.86 | 0 | 0 | 0 | — |
| 2 | 130 | 2 | 1.6 | 2.0 | 24.1 | 0.95 | 374.4 | 599.0 | 599.0 | 299.5 |
| 3 | 125 | 3 | 3.0 | 5.0 | 24.1 | 0.95 | 540.0 | 1620.0 | 2219.0 | 443.8 |
| 4 | 110 | 7 | 1.0 | 6.0 | 24.5 | 1.02 | 1108.8 | 1108.8 | 3327.8 | 554.6 |
| 4a | 110 | 7 | 0.5 | 5.5 | 24.5 | 1.02 | 1108.8 | 554.4 | 2773.4 | 504.3 |
| 4b | 110 | 7 | 0.3 | 5.8 | 24.5 | 1.02 | 1108.8 | 332.6 | 3106.0 | 535.5 |

| (12) $C_d$ | (13) $\Sigma C_d$ | (14) $C_{av}$ | (15) $\mu$ | (16) $\Lambda$ | (17) $\lambda^2$ | (18) $R_n$ | (19) $\Sigma R$ | (20) $\Sigma R + \frac{R_n}{2}$ | (21) $nI$ | (22) $\Sigma nI$ |
|---|---|---|---|---|---|---|---|---|---|---|
| 9.4 | 9.4 | — | — | — | — | 0.46 | 0 | 0.23 | — | — |
| 38.6 | 48.0 | 24.0 | 1.2 | 0.49 | 0.24 | 1.68 | 0.46 | 1.30 | 55 | 135 |
| 72.3 | 120.3 | 24.1 | 0.80 | 0.55 | 0.30 | 3.16 | 2.14 | 3.72 | 837 | 972 |
| 24.5 | 144.8 | 24.1 | 0.64 | 0.59 | 0.35 | 0.98 | 5.30 | 5.79 | 764 | 1736 |
| 12.25 | 132.55 | 24.1 | 0.71 | 0.58 | 0.34 | 0.49 | 5.30 | 5.55 | 377 | 1349 |
| 7.35 | 139.90 | 24.1 | 0.67 | 0.59 | 0.35 | 0.29 | 5.30 | 5.45 | 216 | 1565 |

7. The summation of the number of degree-days required to thaw layers 1 through 4 is 1736, leaving (1736 − 1560) = 176 degree-days to thaw layer 4. A trial-and-error method is used to determine the thickness of layer 4 thawed. First, it is assumed that 0.5 ft of layer 4 is thawed (designated as layer 4a). Calculations indicate that 377 degree-days are needed to thaw 0.5 ft of layer 4a or 1560 − 1349 = 211 degree-days less than that available. A new layer, 4b, is then selected as follows:

$$\left(\frac{211}{377}\right)(0.5) = 0.28 \text{ ft} \qquad \text{say, } 0.3 \text{ ft}$$

This new thickness results in 216 degree-days required to thaw layer 4b or (1565 − 1560), 5 degree-days more than available. Further trial and error is unwarranted and the estimated thaw penetration is 5.8 ft.

(Note: A similar technique is used to estimate penetration in a multilayered soil profile.)

## 4.4 FROST HEAVE

Vertical displacement of foundations is commonly seen in cold regions where the soils are frost susceptible (Fig. 4–3). Uplift resulting from the freezing process in frost-susceptible soils is a serious stability problem for various structures, such as unheated building foundations, transmission towers, utility poles for telephone and power lines, and transformer and distribution stations. Uplift forces caused by frost heaving in the active layer of permafrost areas have been shown to be substantial and various investigators (Penner, 1972, 1974; Hoekstra et al., 1965; Crory and Reed, 1965; Kinoshita and Ono, 1963; Vyalov, 1959; Tsytovich, 1975) have studied the uplift or heaving forces that result from adfreezing on different foundations in frost-susceptible soils.

An understanding of the basic frost-heave phenomena is essential to determine the magnitude of forces that are expected from frost heaving. The mechanism of ice-lens formation is explained in the following paragraphs. Because of the complexities of boundary conditions of heat and mass transfer associated with various parameters of soils, a precise engineering model to predict frost heave and associated frost-heave forces is still lacking.

When a soil mass is frozen, it increases in volume, often by an amount considerably greater than that which could be accounted for by the normal increase in the volume of water as it changes from the liquid to the solid state. When water crystallizes, it expands by approximately 9% of its original volume. When one takes into account the normal void ratios and the commonly encountered soil-water content, an overall expansion of about 2 to 3% of the original soil mass would seem to be the most reasonable expectation. This expansion would correspond to a vertical expansion on the order of 1 or 2 in. in typical temperate climates. In observations dating back more than 100 years in many soil types, the vertical expansions are termed *frost heaving*.

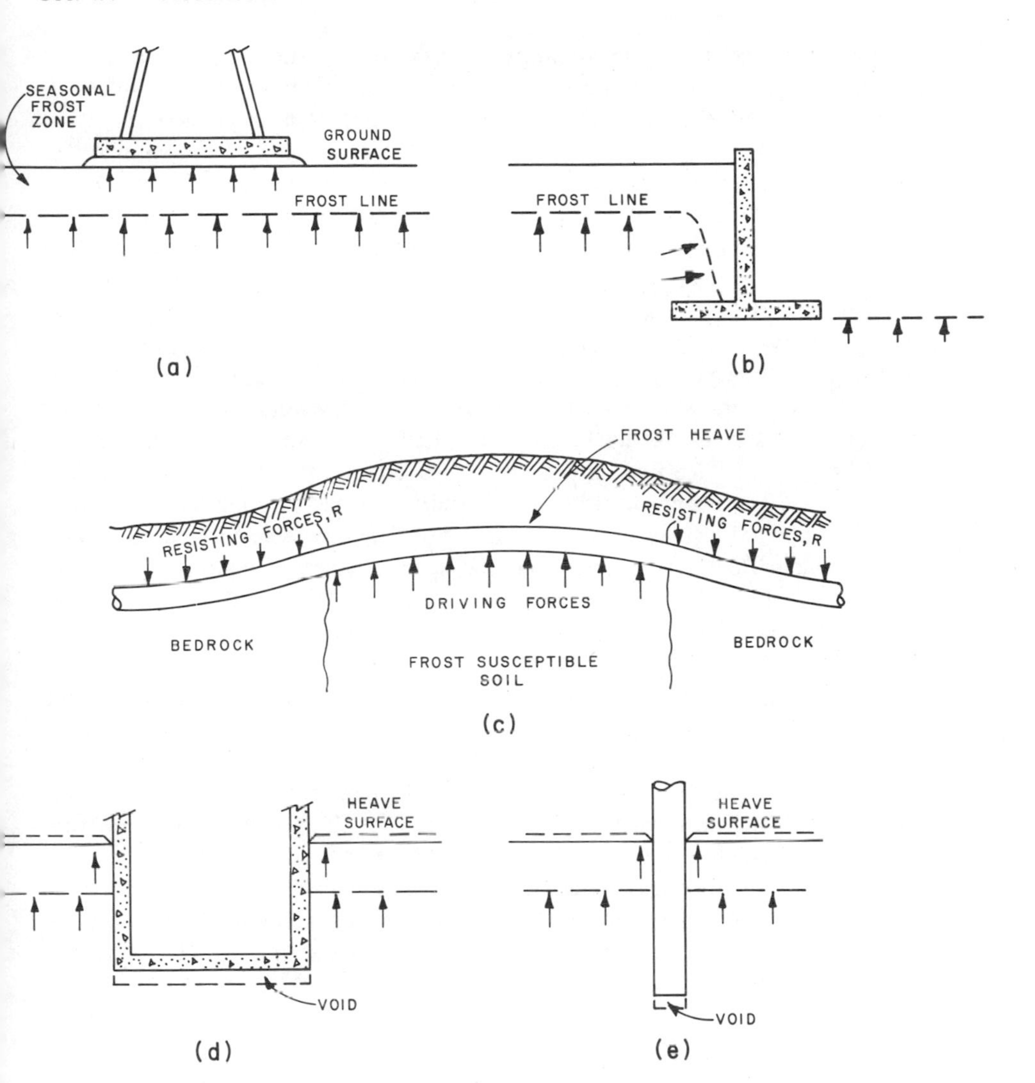


**Figure 4–3** Frost-heave effects.

A close examination of the frozen soil reveals that the pore water is frozen but that ice also appears in discontinuous layers. These are often referred to as *ice lenses*. The ice lenses appear to be more or less segregated from the soil grains. The total volume of the ice is often considerably greater than the volume of the original pore water, and the magnitude of vertical displacement has been observed to be approximately equal to the volume of segregated ice.

Frost heaving was first subjected to detailed study by Taber (1916, 1929, 1930). He assembled experimental apparatus and conducted freezing experiments under carefully controlled conditions. He showed that when soil is frozen from the top down under conditions permitting the upward migration of pore water from an artificial water table, continued exposure to freezing temperatures resulted in continuous vertical displacement with the appearance of a succession of segregated ice lenses as the freezing isotherm penetrated deeper and deeper into the soil material. Taber could observe no marked change in the rate of vertical displacement as one ice lens ceased to enlarge and a new one appeared below it.

The conditions necessary for frost heaving are represented in Fig. 4–4 by a schematic diagram of the process. Freezing air temperatures set up a thermal gradient that induces heat flow upward. As heat is extracted, freezing is initiated and growing ice crystals coalesce into planar ice lenses. Ice-lens enlargement is made possible by the transport of soil water from below. The enlargement in growth of the ice lenses occurs wherever the temperature is below freezing and the rate of evolution and dissipation of the latent heat of freezing does not exceed the upward flow of soil water. Ice-lens growth produces an upward displacement of the ground surface and eventually leads to the buildup of heaving pressures. Ice-lens growth and upward displacement continues until a favorable balance exists among the four principal governing factors: (1) the nature of the soil matrix, (2) the rate of heat removal, (3) the upward movement of soil water, and (4) the confining pressure. When all four occur, a new ice lens is initiated at a more favorable site just behind the descending 0°C isotherm.

The ground surface depicted in Fig. 4–4 shows the following important points: (1) flow lines for heat conduction exit perpendicular to the soil surface; (2) flow lines for soil-water movement in general are parallel to the flow lines for heat conduction; and (3) planar ice lenses tend to be perpendicular to the soil water and thermal-flux flow lines. Thus it is also possible for the process of frost heaving to produce horizontal displacements. The general rule, therefore, is: Displacements due to frost heaving are usually perpendicular to the ground surface. In general, the nearer the water table is to the surface, the more readily is water transported through the unsaturated capillary zone to enlarging ice lenses that are located just beyond the 0°C isotherm.

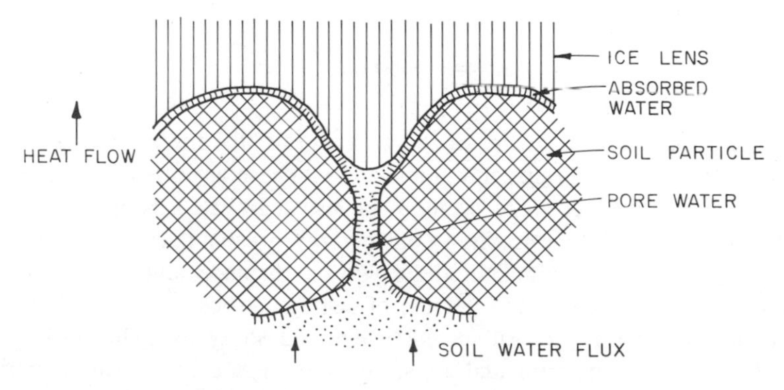


**Figure 4–4** Schematic diagram of ice lens.

In addition to deducing these basic processes and relationships, Taber recognized that the ice lenses must be separated by some sort of unfrozen boundary zone from the solid particles making up the soil fabric if the clear, segregated ice lenses he observed were to thicken and enlarge. He reasoned that as heat was conducted upward, a portion of this boundary layer supporting the ice lens would solidify, thickening the lens and reducing the boundary layer thickness. The boundary layer would tend to regain its former thickness by imbibing additional water from the moist, unsaturated soil below. Continued extraction of moisture from this region would, in turn, set up a hydraulic gradient that would induce the continual flow of water upward necessary to perpetuate the process. Shortly afterward, Beskow (1935) enlarged on this idea and proposed that the "capillary" behavior of soils, determined primarily by grain size distribution, determines the extent to which soils may heave when subjected to freezing conditions. These early observations and insights were subsequently confirmed, expanded, and elaborated upon by others. For example, frost susceptibility was extensively investigated by Kaplar (1968); the role of surface tensions, particularly the surface tension of the ice–water interface, was examined by Gold (1957), Everett (1961), Miller (1966, 1972, 1978), and Penner (1977), among others.

In recent years several international symposia were held dealing with ice segregation and frost heave in soils. These include:

- The International Symposium on Frost Action in Soils, University of Lulea, Sweden, 1977
- The First, Second, and Third International Symposia on Ground Freezing, held at the University of Bochum, Federal Republic of Germany, 1978; Norwegian Institute of Technology, Trondheim, June 1980; and Cold Regions Research and Engineering Laboratory, Hanover, June 1982
- The International Permafrost Conferences

The proceedings of these symposia present good reviews of recent developments on the subject.

In the preceding discussion we have explained how the theory behind the phenomena of frost heaving has been developed from either capillary theory or thermodynamic theory. Actually, the two theories are not truly independent: capillary theory can be derived from thermodynamical considerations. It is, therefore, not surprising that they should yield comparable predictions. As illustrated in Fig. 4–4, the dynamic aspects of frost heaving depend on both the rate of heat extraction and the rate of water movement to the freezing zone. Thermodynamic relationships and predictions are valid for static equilibrium conditions; applications to transient states are less firmly based. Thus it appears that thermodynamic theory is not fully adequate to describe the occurrence and processes associated with frost heaving.

To be complete, the theory of ice segregation and frost heaving should make possible accurate and reliable predictions of rates of heave, cumulative heave, cumulative ice–water contents, the rise and fall of heaving pressures, and ultimate (maximum) heaving pressures. The theory would be constructed from a set of equations that contain expressions for the heat and soil-water fluxes induced by prevailing thermal and water content gradients and one or more parameters to characterize the soil matrix, such as particle size, void ratio, interfacial curvature and dimensions, and so on. Such a theory is not yet in hand (Anderson et al., 1982).

Recently, attention had shifted to numerical modeling techniques. The basic aim of this approach is to devise algorithms that encompass the frost-heaving processes and phenomenology and, by empirical approaches utilizing numerical techniques, ultimately constructing a comprehensive simulation model. Once a suitable model or set of models have been constructed and shown to describe adequately frost heaving under fully controlled circumstances, they would be employed to predict ice segregation and frost heaving. This would be accomplished by extrapolation from fully verified circumstances in which the simulation has been shown to be accurate to other circumstances. Both finite-element and finite-difference techniques have been investigated and employed for this purpose (Guymon and Luthin, 1974; Berg et al., 1977; Guymon et al., 1980, 1981; Hopke, 1980; Miller and Koslow, 1980; O'Neill and Miller, 1980; Outcalt, 1980).

In practical terms, the most troublesome aspects of frost heaving are associated with the very large increases in total soil-water content, which leads to weakening of the soil on thawing; the displacements accompanying the growth of ice lenses within the soil accompanying the growth of ice lenses within the soil mass; and the buildup of pressure whenever the normal displacements accompanying ice-lens formation and growth are restrained. Some soils are more liable than others to exhibit extreme behavior in this regard. When seasonal freeze–thaw cycles are involved, it has been found that the most frost-susceptible soils are those comprised of particles of silt size. Gravels are not susceptible to any of the undesirable characteristics of frost heaving. This can be explained on the basis of the processes illustrated in Fig. 4–4. Ice segregation and frost heaving are dependent on a favorable balance between the removal of heat and the influx of soil water, usually by capillary flow. In addition, the soil particle sizes must be small enough to be rejected from growing ice crystals in much the same manner that dissolved salts and other solutes are excluded and rejected as liquids solidify. Crystalline solids form as individual atoms or molecules of geometry of the crystal lattice characteristic of that material. Ice molecules attach one by one during the freezing process, forming a nearly perfect crystal and in the process displace and thus reject all foreign substances, including colloidal materials such as clays and even silts and fine sands.

Gravels are not frost susceptible because of the relatively large sizes of the pebbles, cobbles, and rocks that make up gravel. Silts, on the other hand, have particles small enough to be easily rejected by slowly growing ice crystals. At the same time the hydraulic permeability of silt is sufficiently high to maintain a steady supply of water if it is available from a nearby source. Clay soils, at the other extreme, are not highly frost susceptible even though the clay particles are, if anything, more easily rejected by the enlarging ice lenses than are silt-size particles (Jackson et al., 1966; Hoekstra, 1969; Romkens and Miller, 1973). The governing factor in this case is the very low hydraulic permeability. If the water flow is restricted, the balance between heat removal and the influx of water may be shifted drastically. Under there conditions the freezing front can rapidly invade the freezing clay. The ice lenses formed usually are very thin and the displacements due to frost heaving tend to be very low. Thus, in naturally occurring seasonal freeze–thaw, experience has shown that gravels are the least troublesome and offer the greatest stability; silts are the most likely to exhibit large displacements due to freeze–thaw phenomena and are therefore classified as the most frost-susceptible soil materials. Clays, on the other hand, although they usually exhibit some vertical displacement and give rise to significant heaving pressures when confined, nevertheless are regarded as less frost susceptible than silts (Kaplar, 1968; Penner and Ueda, 1979; Jones, 1980).

Frost heaving occurs inevitably whenever the following three conditions exist: (1) the presence of frost-susceptible soil materials, (2) ground temperatures falling below 0°C (32°F), and (3) a source of capillary or groundwater sufficient to form and supply accumulating ice lenses in the frozen soil. Ice lenses form and enlarge steadily when the rate of heat removal due to freezing air temperatures above the ground surface creates a thermal gradient which, in turn, draws capillary water to the freezing front. Upon freezing, the water releases its characteristic latent heat. This heat is a relatively large quantity compared to the heat capacities of the soil constituents and is therefore the major factor in determining the balance between the two interconnected fluxes: (1) the thermal flux upward to the freezing air temperatures above, and (2) the capillary water flow from warm, nearly saturated soil below to the freezing front above. The freezing front tends to be stationary, and the ice lenses formed there tend to enlarge as long as this balance is favorable. During winter the freezing front tends to descend in incremental stepwise fashion as air temperatures steepen the thermal gradient and depletion of soil water below leads to a series of loci where ice-lens formation is initiated. Ice-lens growth proceeds for a time, and then is terminated as the freezing front descends to a new, more favorable location, where the process is repeated. Various authors (Konrad and Morgenstern, 1980; Penner, 1982) have carried out both theoretical and laboratory studies to predict the location of ice lenses in fine-grained soils.

A knowledge of the deformation behavior of frozen soil around a foundation and of the nature of the bond between the soil and the foundation structure is required when calculating the forces that may be imposed.

### 4.4.1 Model Studies and Theory

Various investigators (Everett, 1961; Everett and Haynes, 1965; Penner, 1966, 1967; Williams, 1977; Sutherland and Gaskin, 1973; Edlefson and Anderson, 1943) have used thermodynamic considerations in studying the freezing of water in unconsolidated porous materials. From these considerations, relationships have been developed to determine heaving pressure for a close-packed array of uniform-size spherical glass beads in the presence of an imposed thermal gradient. The heaving pressure $P$ is given by

$$P = \frac{2\sigma_{iw}}{r_i} + \frac{2}{r}(\sigma_{is} - \sigma_{ws}) \tag{4–23}$$

where $r_i$ = radius of curvature of ice–water interface
$\sigma_{iw}$ = ice–water interfacial energy
$r$ = radius of sphere
$\sigma_{is}$ = ice–solid interfacial energy
$\sigma_{ws}$ = water–solid interfacial energy

The first term on the right-hand side of the equation gives the pressure drop across the ice–water interface in the pores; the second term is referred to as the *flotation effect* of the ice on the sphere (Everett and Haynes, 1965). Equation (4–23) gives the maximum heaving pressure when the ice lens is of finite thickness, and the rate at which the molecules of water leave the ice is exactly balanced by the rate of molecule attachment. Experimental results supporting the Everett–Haynes theory are shown in Fig. 4–5.

By introducing the Young–Dupre equation, Eq. (4–23) is rewritten as

$$P = \frac{2\sigma_{iw}\cos\Theta\,(1 + B')}{r} \tag{4–24}$$

where

$$\sigma_{is} = \sigma_{ws} - \sigma_{iw}\cos\Theta$$

$$B' = \frac{r}{r_i}\frac{1}{\cos\Theta}$$

Penner (1967) discussed the use of Eq. (4–24) for the determination of maximum pressure produced by a growing ice lens and mentioned that the use of this equation is limited in that it applies to a close array of spherical particles of uniform size. The second component of pressure on Eq. (4–23) is small compared with the first. As no precise solution has yet been found for a complex internal geometry, Penner (1967) used the following equation:

$$\Delta P = \frac{2\sigma_{iw}}{r_i} \tag{4–25}$$

Sutherland and Gaskin (1973) reported the experimental results of heaving pressures developed in partially frozen soils and concluded that the measured heaving pressure for soils consisting of spherical particles [$r/r_i$ = 6.46 and $r$ = $0.5(D_{50})$] agreed closely with that predicted from Eq. (4–23) (Fig. 4–5).

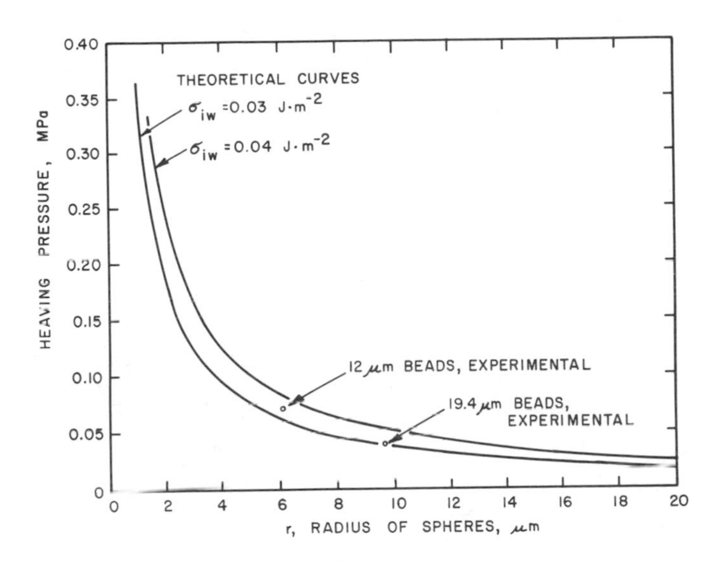


**Figure 4–5** Everett–Haynes equation. (After Everett and Haynes, 1965.)

## 4.4.2 Field Studies

Penner (1974) investigated the uplift forces of columns made of various types of materials (steel, concrete, and wood) and sizes (3, 6, and 12 in. in diameter) and a block concrete wall. The adfreeze strength was highest for the steel column, followed by that for columns made of concrete and wood. These results were attributed mostly to the influence of temperature on adfreeze strength. The maximum uplift forces were also higher on the steel columns. This was partially accounted for by the higher adfreeze strengths and the longer column length over which the adfreeze force was acting. The suppression of ground heaving by the fixed concrete block foundation wall and the resultant deformation pattern was used to predict the maximum total uplift force. Penner reported that the peak values of adfreeze strength occurred during the early freezing period when heaving rates were high, but maximum uplift forces often occurred near the time of maximum frost penetration late in the winter season.

Crory and Reed (1965) have reported on the frost-heaving forces on piles. The heave forces recorded during tests conducted in 1962–1963 are shown in Figs. 4–6 and 4–7. The heave force generally lagged behind ground freezing, with the heave force starting to increase after the 32°F isotherm had penetrated a few inches into the ground. The heave force increased to a maximum value when the 32°F isotherm had penetrated 80 to 100% of the seasonally thawed layer. The maximum unit adfreeze bond test (1962–1963) was about 41 psi for the steel pipe pile and 12 psi for the creosoted timber pile.

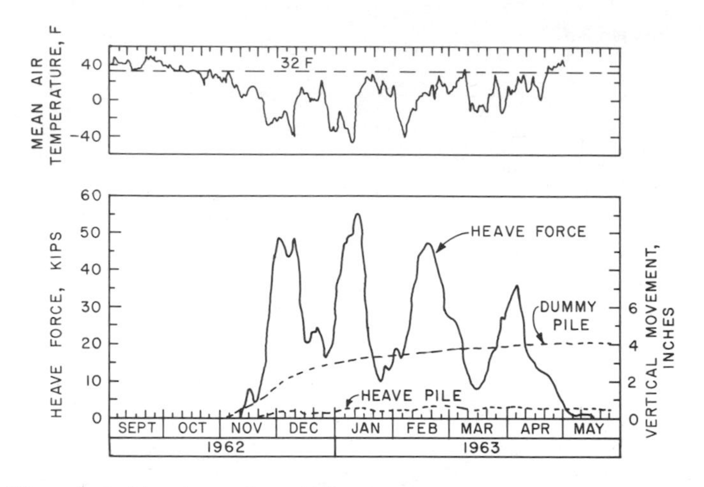


**Figure 4–6** Test observations, 1962–1963. (After Crory and Reed, 1965.)

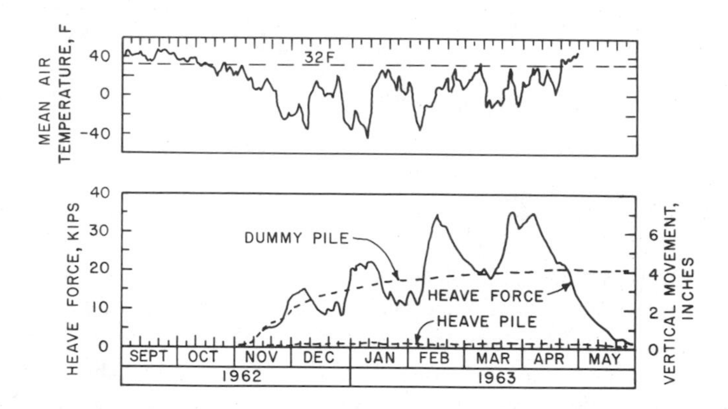


**Figure 4–7** Test observations, 1962–1963, creosoted timber pile, average diameter test pile 14 in., dummy pile 12 in. (After Crory and Reed, 1965.)

### 4.4.3 Frost-Heaving Forces on Foundations

From the discussion in Section 4.4.1, frost-heaving factors may be categorized into the following main divisions:

| Intrinsic Factors | Extrinsic Factors |
|---|---|
| Moisture conditions of soils | Climate |
| Soil structure and density | Groundwater level |
| Physical and chemical composition of soils | Surface condition of the soil |
| Permeability of soils | |
| Capillarity of soils | |

Intrinsic factors are those which belong to or are properties of the soil, whereas extrinsic factors are those that are outside the soil but condition it. The intrinsic and extrinsic factors interact to establish thermal and hydraulic conditions in the soils that are either favorable or unfavorable to frost heave and associated forces.

The frost-penetration depth is rarely the same around a foundation, as the depth of penetration is highly influenced by variations in building heat or insulation, shading, snow cover, and soil conditions, particularly soil moisture. Fine-grained soils with higher moisture contents, and associated latent heat, experience less frost penetration than do gravels under the same freezing conditions (discussed in Section 4.3.1.1). Because frost penetration into the ground and frost heaving occur in directions parallel to the direction of heat flow, the direction of heave may be other than vertical if the surface is not horizontal. Frost penetration and heaving can be in a horizontal direction, as in the case of a vertical wall, or at various angles, in the case of merging frost penetration from two directions. Nonuniform frost penetration also occurs at isolated and unheated foundation members exposed to the air above the ground surface, such as steel or concrete transmission tower foundations.

Foundations constructed on frost-susceptible frozen ground can experience severe settlement and distress when the ground thaws in the spring. Foundations subjected to frost action during winter construction, when the structure is unheated and not backfilled to final grade, can experience severe differential heaving. Similarly, unheated full basements, with only the first floor capped, may suffer severe wall and floor heaving during the winter. Thus foundation designs based on the normal operating conditions of a heated basement must include a consideration of the possible consequences of frost heaving that may occur before such conditions are actually achieved.

*Normal and Tangential Frost-Heave Forces.* The forces on foundations created by the heaving of frost-susceptible soils can be subdivided into two groups: those normal and those tangential (Phukan et al., 1978; Linell et al., 1980). The normal forces are those at right angles to the plane of freezing. A continuous strip footing, or isolated footing, placed on the surface of frost-susceptible material is therefore subject to these normal forces (Fig. 4–3) and consequent heaving during freezing of the soil directly beneath the footings. When the footings are placed a foot or more below the ground surface, the normal forces are not activated until the frost has penetrated below the base of the footing. If the footing base is placed below the maximum frost depth, the footing base will not experience any normal heaving forces unless the presence of the foundation member locally accelerates the frost penetration. If a deep footing, such as an unheated basement wall or a retaining wall, is exposed to below-freezing temperatures on one side, there may be unevenly distributed normal forces on the base of the footing, as well as horizontally oriented normal forces on the back of the wall (Fig. 4–3b). These normal forces are generally assumed to be only as great as the resisting forces, such as the weight of the footing, the imposed load, and the surcharge weight of the overlying soil. While the amount of displacement caused by frost can be reduced by applied pressures, the normal forces generated by ground freezing usually exceed the conventional loads imposed on footings. To suppress all frost heaving the footing loads would have to be in excess of the thawed bearing capacity of fine-grained soils, since the maximum potential force generated by such frost-heaving stresses can approach that generated by the confined freezing of water, that is, greater than 10 tons/ft$^2$ (9.7 kg/cm$^2$) (Crory et al., 1982). Thus there are limited options in suppressing the normal forces and heaving associated with the freezing of frost-susceptible soils. Such forces and displacements can be avoided, however, by placing the footing at depths that exceed the maximum frost penetration, or backfilling around the footings using non-frost-susceptible soils.

Tangential frost-heaving forces are the result of the action of soils that freeze to the side faces of foundations, the forces being tangent to such surfaces (Kinoshita, 1962). Tangential forces generated by frost heaving can best be illustrated by a timber pile that extends well below the maximum depth of frost, such that it is not subjected to the normal heaving forces described above (Fig. 4–3e). Assuming that the pile does not thermally influence the frost penetration, the soil surrounding the pile will heave upward as it freezes. The soil will also freeze to the pile. Shearing and tension stresses will be generated in the frozen soil layer immediately surrounding the pile since the normal vertical forces at the bottom of the surrounding frost front will be thrusting upward, while the frozen soils adhering to the pile are restrained by the imposed loads on the pile and the friction along the surface of the pile at greater depths. The tangential stresses on the frozen layer at the pile surface are greatest during periods of rapid frost penetration into the surrounding soil. During these periods the already frozen soils are coldest and have their greatest strength, as does the associated adfreeze bonding to the pile. During warm periods but not necessarily above freezing, the frost layer is weaker and can bend or otherwise relax. The adfreeze bond to the pile will also be weaker in warmer periods, with the creep rate increasing to the extent that the bond may be broken. When colder weather returns, the tangential forces increase as soon as the frost begins to penetrate further. In interior Alaska, the peak tangential heave forces on 8-in.-steel-pipe piles have been recorded at more than 55,000 lbf when the frost depth was only 2.5 to 3.0 ft (0.7 to 0.9 m) below the ground surface (Crory and Reed, 1965).

Foundation walls are also subjected to tangential heaving forces (Penner and Gold, 1971). When the insides of basement walls are heated, the adfreeze bond stresses on the outside of the wall will normally be very small and intermittent. If the wall is cold, such as in an unheated crawl space or in an unheated garage, the adfreeze bond and tangential forces may be very high. Distress to foundations can also occur by a combination of tangential and normal forces acting in unison, particularly in the latter part of the winter.

When piles, poles, or other foundation members are incapable of resisting such upward heaving forces, the foundations are progressively heaved upward each year. The cumulative heaving over many years may then be several inches to several feet. The frost-heaving forces on poles are discussed further in Chapter 6.

#### *4.4.3.1 Foundation Design for Frost-Heave Conditions*

Building codes commonly specify that the bottom surfaces of footings, grade beams, pile caps, or other foundation construction shall be below the frost line for the locality, as established by local records or experience (BOCA, 1976; National Building Code, 1976). The intent of such codes is to avoid the normal heave forces described below. Tangential heave forces are addressed in recent codes in Canada and Norway, which emphasize an engineering approach to frost-heave design rather than the simple rule of thumb based solely on frost depth (NRC, 1980; Saetersdal, 1976).

The building codes of most cities and towns specify minimum footing depths which are usually less than the maximum frost depths reported by the National Weather Service (as discussed in Section 4.3.1.1). Frost penetration depths based on theoretical or empirical equations utilizing long-term freezing indices and site-specific thermal properties should be used for foundation designs, ice skating arenas, cold storage facilities, and snow-free pavements. While the minimum depths of footing placement for frost considerations in the northern states of the United States and in Canada usually range from 0.9 to 1.5 m, the depth does not increase as one goes north. The building code for the city of Fairbanks, Alaska, specifies a minimum foundation depth of only 1 m for frost considerations, yet the frost penetration may be more than 3 m in the Chena River sands and gravels which are commonly found throughout the city.

Foundation distress from frost heaving in the case of fully heated basements is rarely encountered, provided that the basements are protected during construction and kept heated or semiheated after construction. The greatest distress near heated basements occurs at outside steps, loading aprons, and isolated footings for carports or porches, which receive little or no heat from the basement. The shallow footings of some split-level houses or buildings erected with unheated crawl spaces are susceptible to frost-heave damage when founded on or backfilled with frost-susceptible soils. Insulating the interior of foundation walls to conserve heat can also aggravate frost heaving.

Accessory buildings, usually limited to one-story structures with less than 36 $m^2$ (400 $ft^2$) of floor space, are commonly exempt from the minimum foundation depths specified by national or local codes. When such buildings are not continuously heated in the winter, the foundations can suffer severe differential frost heaving if built on frost-susceptible soils. Structural damage of such structures may range from minor for wood frame to major in the case of masonry block.

In the case of piles, poles, foundations of unheated structures, and isolated footings, provisions must be included in the design to avoid or resist the displacements associated with the tangential forces of frost heaving. In such cases, heave may be avoided by backfilling around the foundation member to sufficient depth and distance with non-frost-susceptible material, by backfilling with a nonheaving low-shear-strength soil–oil–wax mixture, or by using an oil-and wax-filled casing which protects the member such as those used for frost-free benchmarks in cold regions. The tangential forces on piles or poles may also be reduced by the use of coatings, such as coal tar, plastic, Teflon, or similar nonadhering surface materials. Tapering footings through the seasonally frozen layer and installing piles butt down are construction techniques commonly used in cold regions to reduce and counteract tangential frost-heaving forces. Foundation elements may be designed to resist heave by such techniques as providing adequate depth of embedment in soil or permafrost below the annual frost zone, anchoring foundation elements into bedrock if circumstances permit, or using footing enlargements to develop passive soil resistance. In permafrost areas, thermal piles may be used to increase pile stability, including resistance to heave.

The design of foundations in areas of deep seasonal permafrost has undergone dramatic changes in the past decade. The foundations must be designed to prevent damage from frost action on a performance concept. Designs for basements, crawl spaces, and slab-on-grade construction in Norway are based on anticipated heat-flow patterns and frost susceptibility, taking full advantage of the use of insulation under and around foundations (Adamson, 1972; Herje, 1972; Klove, 1972; Svendsen, 1972). Canadian building codes and standards (NRC, 1980) also stress the importance of engineering the design of foundations in frost areas, rather than placing the burden of design on minimum depths of embedment.

### 4.4.4 Remedial Measures

When evaluation of frost uplift or thrust forces indicate potential problems, these forces must be taken fully into account in the design, as must the necessary measures required to avoid detrimental effects either from the structural stresses developed by the frost forces or from the resulting soil deformations.

Some of the active and passive methods of frost-heave force mitigative or remedial measures are:

1. Replace existing ground material by excavating and backfilling with non-frost-susceptible (NFS) granular soils.
2. Eliminate total bond contact area between foundation and soil (isolation).
3. Reduce frost penetration.
4. Use thermal heat probes to eliminate the formation of ice segregation.
5. Provide sufficient embedment or anchorage to resist movement under the heaving forces (sufficient integral strength must be provided in foundation members to ensure such forces).

6. Provide sufficient loading on the foundation to counterbalance heaving forces.
7. Use exterior heat tracing.
8. Cut off the availability of water to the freezing plane by well points, impermeable barriers, or other means.
9. Trade off between a basic nonheaving design and a heaving design with monitoring and heave correction during operation.

The design of foundations against frost-heave forces may consist of: designing for maximum anticipated heave force, modifying or improving soil conditions by excavation and backfill, and eliminating frost-heave force by employing, singly or in combination, any of the various design measures indicated above. Design and construction of foundations against frost heave are discussed further in Chapters 5, 6, and 7.

## 4.5 THAW SETTLEMENT AND CONSOLIDATION

From a design and construction point of view, thaw settlement and consolidation characteristics of frozen ground are important phenomena and must be considered in the design of foundations of various structures, such as heated structures, warm oil pipelines, roads and airports, water-retaining structures, and cut slopes in frozen ground. In permafrost regions, extensive damage may be done to the terrain as a result of surface disturbances. These disturbances may in turn produce thaw settlement and consolidation of frozen soils.

As shown in Fig. 4–8, when frozen ground thaws, it generates thaw strain

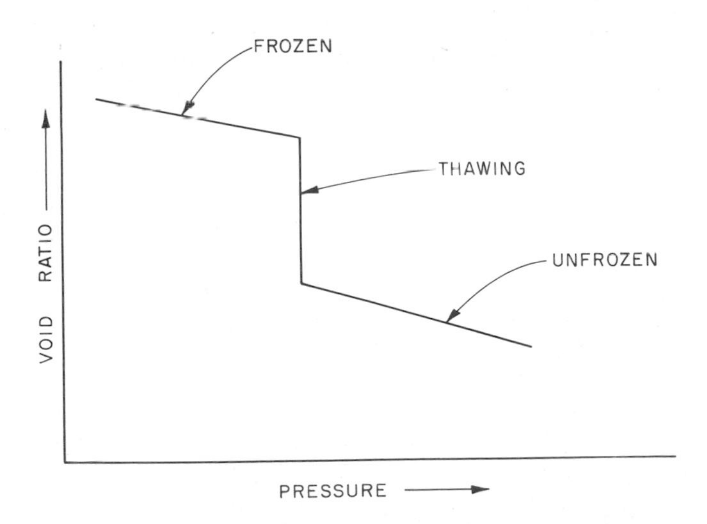


**Figure 4–8** Thaw-settlement process.

and volume changes of soil structure as a result of phase changes of ice to water and the dissipation of melted water. The magnitude of settlement and consolidation produced by the thawing of frozen ground depends on the type and nature of soil, ice content, density, excess pore pressure, and compressibility of soils. The development of potential methods for determining thaw settlement of frozen soils and for predicting pore pressures generated in thawing soils has been a prerequisite to improved engineering design in frozen ground. Laboratory tests are generally carried out on frozen soils to determine the thaw strain parameter and the compressibility characteristics of soils. The relationship between the dry density of soil and thaw strain may be obtained from laboratory tests (Figs. 4–9 and 4–10) which are performed either with a one-dimensional consolidation device or isotropically in a triaxial pressure cell. Luscher and Afifi (1973) and Speer et al. (1973) obtained a positive correlation between the thaw strain *(A)* and the dry density of silts, clayey silts, sands, and gravels (Figs. 4–9 and 4–10). Using a least-squares technique, the thaw strain *(A)* may be calculated from the following relationship (Nixon and Ladanyi, 1978):

$$A = 0.76 - 1.018 \ln \rho_f \pm 0.07 \tag{4–26}$$

Crory (1973a) developed a relationship by which the thaw strain *(A)* can be calculated directly from the physical properties of frozen and thawed soils as

$$A = \frac{e_f - e_t}{1 + e_f} \tag{4–27}$$

where $e_f$ = initial void ratio of frozen sample
$e_t$ = final void ratio of thawed sample

Thus the thaw settlement of *n* layers can be determined by the following equation:

$$\frac{\Delta H}{H} = \text{sum of thaw strain + volumetric strain}$$

or

$$\frac{\Delta H}{H} = \sum_{i=1}^{i=n} A_i + \sum_{i-1}^{i=n} \frac{\Delta V_i}{V_i}$$

or

$$\frac{\Delta H}{H} = \sum_{i=1}^{i=n} A_i + \sum_{i=1}^{i=n} m_{vi}\sigma'_i$$

or

$$\Delta H = \sum_{i=1}^{i=n} A_i H_i + \sum_{i=1}^{i=n} m_{vi}\sigma'_i H_i \tag{4–28}$$

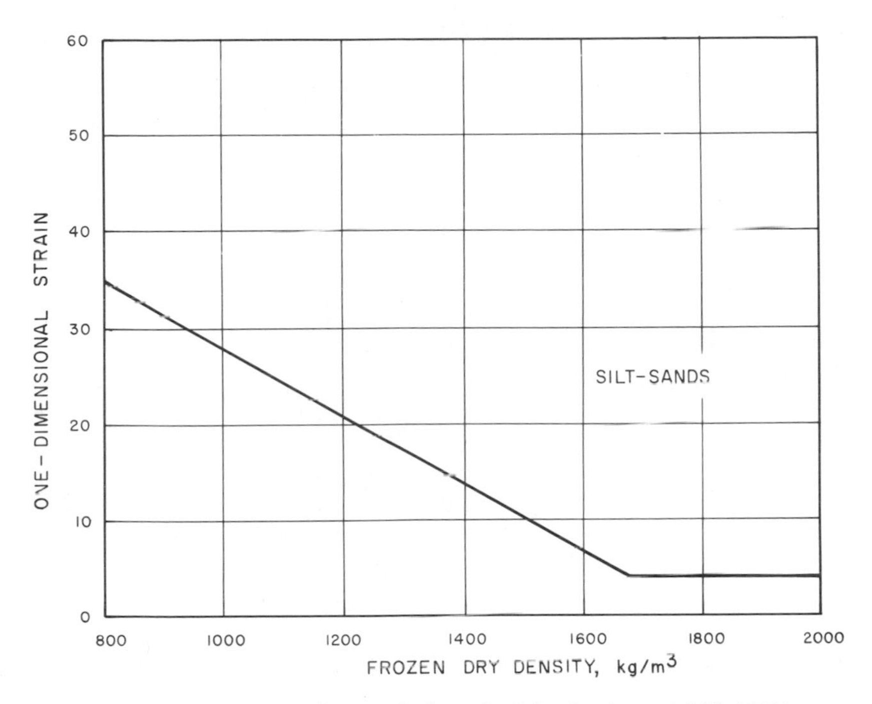


**Figure 4–9** Thaw consolidation of silt-sands. (After Luscher and Afifi, 1973.)

where $\Delta H$ = total thaw settlement
$A$ = thaw strain of each layer
$m_v$ = coefficient of volume compressibility of each layer
$\sigma'$ = effective stress at midheight of each layer

Typical values of $m_v$ are presented in Table 3–5 and for engineering purposes the value of $m_v$ is taken to be 10% of thaw strain, $A$.

Crory (1973) and Nixon and Morgenstern (1973) reported the effect of swelling characteristics and residual stress (the initial effective stress in soil thawed under undrained conditions is referred to as the *residual stress*) on the thaw settlement of frozen soils, respectively. It is a reasonable assumption to assume that the effective stress is equal to zero if the soil is rich in ice or has a high void ratio in the thawed undrained state. However, it should be noted that in cases where the stress and the thermal histories associated with the formation of a specimen of permafrost are such as to reduce the void ratio of the soil, a significant residual stress may be present in the soil upon thawing. Speer et al. (1973) tried to determine the thaw settlement at a field condition by the determination of ground-ice content. USSR (1960) recommended the calculation of thaw settlement due to ice inclusion as follows:

$$S = \sum_{i=1}^{i=n} h_i r_i \tag{4–29}$$

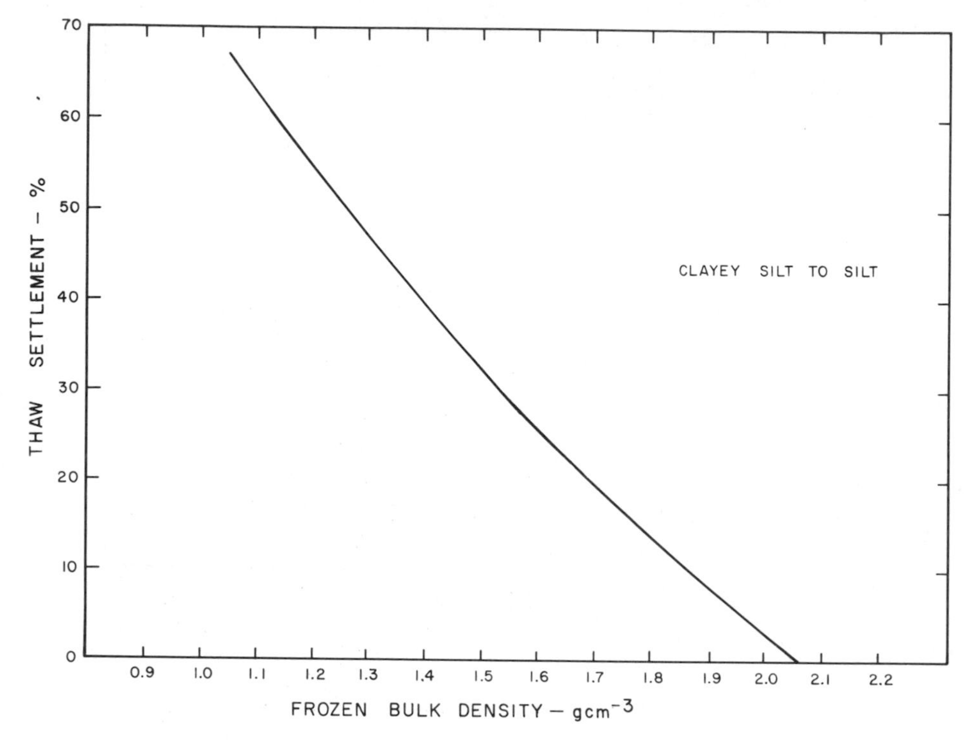


**Figure 4–10** Thaw settlement versus frozen bulk density relationship. (After Speer et al., 1973.)

where

$$
\begin{aligned}
h_i &= \text{thickness of individual ice inclusion} \\
r_i &= \text{reduction coefficient} \\
&= 0.4 \quad \text{for } h_i < 30 \text{ mm} \\
&= 0.6 \quad \text{for } 30 < h_i < 100 \text{ mm} \\
&= 0.8 \quad \text{for } h_i > 100 \text{ mm}
\end{aligned}
$$

### 4.5.1 Thaw Consolidation

In cases where the changes in the thermal regimes cause the frozen ground to melt at a faster rate than the dissipation of pore water, excess pore pressures can be generated; thereby, the thaw consolidation phenomena may cause serious total settlement or differential settlement. Morgenstern and Nixon (1971) developed a one-dimensional linear theory of thaw consolidation, and this theory could be used for predicting settlement. As shown in Fig. 4–11, a one-dimensional configuration is considered where a "step" increase in temperature is imposed at the surface of a semi-infinite homogeneous mass of frozen soil. The solution to the heat-conduction problem is given by Carslaw and Jaeger (1959). The advancement of the thaw plane is given by

$$x(t) = \alpha\sqrt{t} \tag{4–30}$$

where $x$ denotes the distance to the thaw plane from the soil surface and $\alpha$ denotes a constant determined in the solution of the heat-conduction problem. It is assumed that the soil is compressible in the thawed region and that the theory of consolidation (Terzaghi, 1943) is applicable. Hence it may be shown that

$$\frac{\partial u}{\partial t} = C_v \frac{\partial^2 u}{\partial x^2} \tag{4–31}$$

where $u$ = excess pore pressure
$C_v$ = coefficient of consolidation
$t$ = time period

The boundary conditions for Eq. (4–31) are

1. $0 < x < X$
2. $0 < t < \infty$

At the thaw line, water is liberated and it flows upward if there is any excess pore pressure. For a saturated soil, the boundary condition is any flow from the thaw line that is accompanied by a change in volume of the soil. Darcy's law may be used to calculate the volume of pore fluid expelled from a small layer $\Delta x$ as

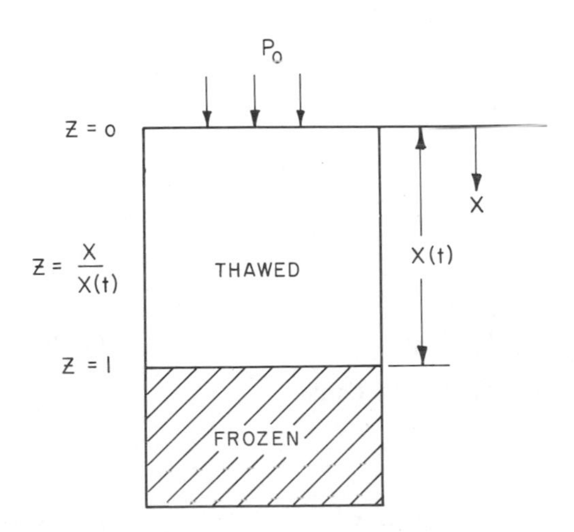


**Figure 4–11** One dimensional thaw consolidation. (After Morgenstern and Nixon, 1971.)

the thaw line advances through an increment in time $\Delta t$. We then have the volume of pore water expelled:

$$\Delta V = \frac{A'k\,[\partial u/\partial X](X, t)}{\rho_w}\Delta t \qquad (4\text{–}32)$$

where $A'$ = cross-sectional area of a soil element
$k$ = coefficient of permeability of soil
$\partial u/\partial X(X, t)$ = excess pore pressure gradient at the thaw surface
$\rho_w$ = unit weight of water

The discharge $V$ is equal to the change in volume of a layer of thickness $\Delta X$. The volumetric strain is given by

$$\frac{\Delta V}{V} = \frac{\Delta V}{A'\Delta X} = \frac{A'k\,[\partial u/\partial X]\,(x, t)}{A'\Delta x\,\rho_w}\Delta t$$

or

$$\frac{\Delta \text{V}}{\text{V}} = \frac{k\,[\partial u/\partial x](x, t)}{\rho_w(dx/dt)} \qquad (4\text{–}33)$$

For a compressible soil,

$$\frac{\Delta V}{V} = -m_v\,\Delta\sigma' \qquad (4\text{–}34)$$

where $m_v$ denotes the coefficient of volume compressibility and $\Delta\sigma'$ denotes the change in effective stress. From Eqs. (4–33) and 4–34), we have

$$\frac{k[\partial u/\partial x](x, t)}{\rho_w(dx/dt)} = -m_v\,\Delta\sigma$$

or

$$\Delta\sigma = \frac{C_v[\partial u/\partial x](x, t)}{dx/dt} \qquad (4\text{–}35)$$

(Note that $C_v = \dfrac{K}{m_v\rho_\omega}$

The total stress at $x = X$ is given by

$$\sigma(X, t) = P_o + \rho X \qquad (4\text{–}36)$$

where $P_o$ = stress applied to the surface by the overburden
$\rho$ = bulk density of the soil

At $x = X$, the pore pressure is

$$P_w(X, t) = U(X, t) + \rho_w X \qquad (4\text{–}37)$$

From Equations (4–36) and (4–37), we obtain the effective stress as

$$\sigma'(X, t) = P_o + (\rho - \rho_w)X - U(X, t)$$

or

$$\sigma'(X, t) = P_o + \rho_b X - U(X, t) \tag{4–38}$$

where $\rho_b$ denotes the submerged or buoyant weight of the soil. Now, the change in effective stress is given by

$$\Delta\sigma' = \sigma_0'(X, t) - \sigma_0' \tag{4–39}$$

where $\sigma_0'$ denotes the initial effective stress in the soil if no volume change were permitted on thawing. Assuming that $\sigma_0' = 0$, we have from Eqs. (4–38) and (4–39),

$$\Delta\sigma' = P_o + \rho_b X - U(X, t) \tag{4–40}$$

Substituting Eq. (4–40) into Eq. (4–31) gives the following condition at the thaw line:

$$P_o + \rho_b X - U(X, t) = \frac{C_v[\partial u/\partial x](X, t)}{dx/dt} \tag{4–41}$$

at $x = X(t)$ for $t > 0$.

Morgenstern and Nixon (1971) gave the complete solution of Eq. (4–41) as

$$U(X, t) = \frac{P_o}{\text{erf}\,(R) + e^{-R^{2}/}\sqrt{\pi}R}\,\text{erf}\left(\frac{x}{2\sqrt{C_v t}}\right) + \frac{\rho_b X}{1 + 1/2R^2} \tag{4–42}$$

where $R = \alpha/2C_v^{1/2}$ and is termed the thaw-consolidation ratio.

The error function erf (·) has been tabulated by Carslaw and Jaeger (1959) and the approximate solution is

$$\text{erf}\,(R) = \frac{2}{\sqrt{\pi}}R \tag{4–43}$$

Equation (4–42) represents the pore pressure generated under an applied loading $P_o$. The second term in the equation provides the pore pressures maintained in a soil thawing and settling under the action of its own weight. We have

$$\frac{U(Z, t)}{\rho_b X} = \frac{Z}{1 + 1/2R^2} \tag{4–44}$$

using $Z = x/X(t)$. The average degree of consolidation of the thawed soil is given by the ratio of the consolidation settlement that has occurred at $t$ to the total consolidation settlement that would occur if thawing were stopped at $t$. The ratio is denoted by $S$ and may be given as

$$S = \frac{S_t}{S_{max}} = \frac{\int_0^x [P_o + \rho_b X - U(X, t)]dx}{\int_0^x (P_o + \rho_b X)^{dx}} \tag{4–45}$$

Using Eq. (4–42), and the load ratio, $\text{Wr} = \rho_b X(t)/P_o$, Eq. (4–45) becomes

$$S = 1 - \frac{\text{erf}(R) + (e^{-R^2} - 1)/\sqrt{\pi}\, R}{[\text{erf}\,(R) + e^{-R^2}/\sqrt{\pi}\, R](1 + \text{Wr}/2)} - \frac{1}{(1 + 1/2R^2)(1 + 2/\text{Wr})} \tag{4–46}$$

For Wr = 0,

$$S = 1 - \frac{\text{erf}\,(R) + (e^{-R^2} - 1)/\sqrt{\pi}\, R}{\text{erf}\,(R) + e^{-R^2}/\sqrt{\pi}\, R}$$

and for Wr = ∞,

$$S = 1 - \frac{1}{1 + 1/2R^2}$$

The maximum settlement is given by

$$S_{max} = m_v\left(P_o X + \frac{\rho_b X^2}{2}\right) \tag{4–47}$$

Using Eqs. (4–46) and (4–47), the settlement at any time $(S_t)$ may be readily calculated.

As a first approximation, a value of $R$ greater than unity would appear to predict the substantial pore pressure at the thaw plane which may result in instability. The magnitude of $R$ is given by $\alpha/(2\sqrt{c_v})$. The value of $\alpha$ varies from 0.2 to 1.0 $\text{mm/s}^{1/2}$ and $C_v$ varies from 10 (sandy silt) to 0.01 $\text{mm}^2/\text{s}$ (clays). A knowledge of the $R$ value is therefore important to determine the thaw-consolidation settlement. Excess pore pressures and the degree of consolidation in thawing soils depends primarily on the thaw-consolidation ratio $R$.

The theory described above has been extended to nonlinear thaw consolidation (Nixon and Morgenstern, 1973). The stress–strain relationships of many frozen soils may be markedly nonlinear. Nixon (1973) has also extended this linear thaw consolidation theory to nonhomogeneous or layered soils. Both laboratory and field experimental confirmations of the theory of thaw consolidation are given by Morgenstern and Smith (1973), Nixon and Morgenstern (1974), and Watson et al. (1973b).

**Example 4–4**

A dike of 15 ft is constructed over frozen silty soils with the following conditions:

- Average rate of thaw penetration = 0.025 cm/s$^{1/2}$
- Dry density of thawed soils = 90 lb/ft$^3$
- Water content = 22%
- Coefficient of consolidation $C_v$ = 220 ft$^2$/yr

Estimate the total settlement of thawing soils that may be anticipated after 10 years.

## Solution

1. Determine the anticipated thaw depth below the dike after 10 years.

$$X = \alpha\sqrt{t}$$
$$\alpha = 0.025 \text{ cm/s}^{1/2}$$
$$= \frac{0.025}{(2.54)(12)}\sqrt{60(60)(24)(365)}\text{ft/yr}^{1/2}$$
$$= 4.61 \text{ ft/yr}^{1/2}$$

We have $X = 4.61\sqrt{10} = 14.58$ ft; say, 15 ft.

2. Total settlement = settlement due to thaw strain $S_{TS}$, compressibility $S_C$, and consolidation $S_{TC}$; or

$$S = S_{TS} + S_C + S_{TC}$$
$$S_{TS} = (A)\text{ (thaw depth)}$$
$$= (0.15)(15) \qquad \text{(assumed that } A = 15\%)$$
$$= 2.25 \text{ ft}$$
$$S_C = 10\% \text{ of } S_{TS}$$
$$= (0.10)(2.25)$$
$$= 0.23 \text{ ft}$$
$$S_{TC} = \Delta S \times S_{max}$$

Assuming that Wr = 0,

$$S = 1 - \frac{\text{erf }(R) + (e^{-R^2} - 1)/\sqrt{\pi}\,R}{\text{erf }(R) + e^{-R^2}/\sqrt{\pi}\,R} \qquad \text{(from Eq. 4–46)}$$

$$R = \frac{\alpha}{2\sqrt{C_v}} = \frac{4.61}{2\sqrt{200}} = 0.16$$

$$\text{erf }(R) = \frac{2}{\sqrt{\pi}}(R)$$

or

$$\text{erf }(0.16) = \frac{2}{\sqrt{\pi}}(0.16) = 0.18$$

we have

$$S = 1 - \frac{0.18 + (e^{-16^2} - 1)/(\sqrt{\pi})(0.16)}{0.18 + e^{-16^2}/(\sqrt{\pi})(0.16)} = 0.97$$

$$S_{\max} = m_v\left(P_o X + \rho_b \frac{x^2}{2}\right) \qquad \text{(from Eq. 4–47)}$$

Assuming that $m_v = 10^{-4}$ ft²/lb and that the density of the dike material $\rho = 130$ lb/ft³,

$$\begin{aligned} P_o &= (\rho) \times (\text{height of dike}) \\ &= (130)(15) \\ &= 1950 \text{ lb/ft}^2 \end{aligned}$$

The submerged weight

$$\rho_b = [90(1 + 0.22) - 62.4] = 47.4 \text{ lb/ft}^3$$

so we have

$$\begin{aligned} S_{\max} &= 10^{-4}\left[1950(15) + (47.4)\left(\frac{15^2}{2}\right)\right] \\ &= 3.46 \text{ ft} \\ S_{TC} &= (0.97)(3.46) \\ &= 3.28 \text{ ft} \end{aligned}$$

The total settlement

$$S = 2.25 + 0.23 + 3.28 = 5.76 \text{ ft}$$

## PROBLEMS

**4–1.** Determine the depth of frost penetration into a homogeneous silt for the following conditions:

- Mean annual air temperature (M.A.T.) = 35.5°F
- Surface freezing index = 2200 degree-days
- Length of freezing season = 140 days
- Dry unit weight of silt = 90 lb/ft³
- Water content = 20%

**4–2.** Determine the depth of thaw penetration into a homogeneous silt for the following conditions:

- Mean annual air temperature (M.A.T.) = −5°C
- Air thawing index = 900 degree-days
- Length of thaw season = 100 days
- Silt: unit weight = 1280 kg/m³; water content = 25%

Assume that the ground surface is covered with weeds.

**4–3.** Estimate the depth of thaw penetration at the site of a proposed building project where the mean annual air temperature is −3°C, the design air thaw index equals to 3300 degree F-days, and the thaw period lasts for 160 days. Assume that $n$ = 1.0 (elevated building with an air space between the building floor and the ground surface). Soil data for the site are as follows:

| Depth (m) | Group Symbol | Soil Description | Data |
|---|---|---|---|
| 0.6 | OL | Organic sandy silt, unfrozen | — |
| 0.6–3 | $ML\text{-}Nb_e$ | Black, slightly organic sandy silt, frozen; no visible segregation but bonded | $\rho_d$ = 1360 kg/m$^3$<br>$\omega$ = 33% |
| 3–16 | $SP\text{-}Nb_n$ | Brown, uniform medium sand; no visible segregation but bonded | $\rho_d$ = 1680 kg/m$^3$<br>$\omega$ = 20% |

Assume thermal properties of soils and that the organic layer is replaced with silt.

**4–4.** **(a)** Determine the depth of thaw penetration into a homogeneous deposit for the following conditions:

- Mean annual air temperature (M.A.T.) = 2°C
- Surface gravelly soils
- Air thawing index = 3500 degree-days
- Thawing season = 200 days
- Soil conditions: silty gravel deposit; unit weight = 2080 kg/m$^3$; water content = 6%

**(b)** If the surficial soil consist of a 0.3-m vegetative mat underlain by gravelly soils as above, what will be the depth of thaw penetration?

**4–5.** Determine the frost penetration below the bituminous concrete pavement described below.

| Layer | Thickness (ft) | Thermal Conductivity (Btu/hr ft °F) | Volumetric Heat Capacity (Btu/ft$^3$ °F) | Latent Heat of Fusion (Btu/ft$^3$) |
|---|---|---|---|---|
| Bituminous concrete | 0.333 | 0.8 | 28 | 0 |
| Base course | 0.5 | 1.00 | 23 | 850 |
| Subbase | 1.5 | 1.30 | 25 | 1200 |
| Subgrade | — | 1.70 | 27 | 2900 |

The pavement is kept free of snow. The mean annual temperature is 46°F, the surface freezing index is 1473 degree F-days, and the freezing period is 98 days.

**4–6.** Estimate the total settlement of a pipe in 30 years under the conditions shown in Fig. P4–6.

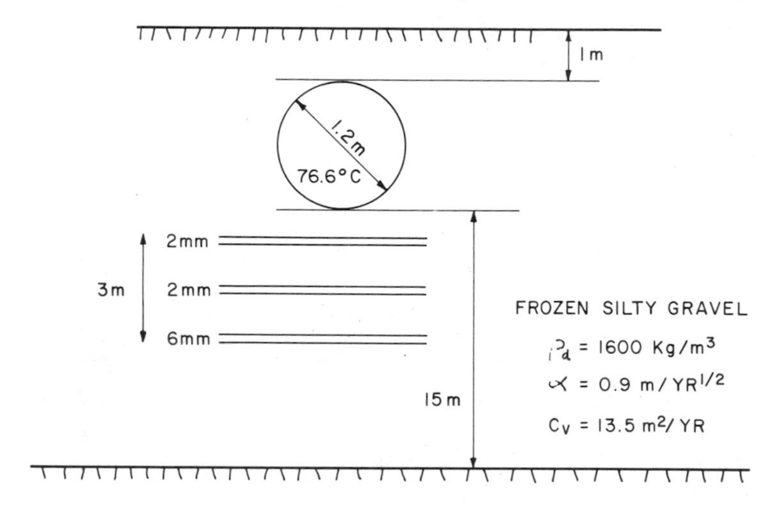


**Figure P4–6**

**4–7.** A dike 2 m in height is constructed over frozen silty soils with the following conditions:

- Average rate of thaw penetration = 0.05 cm/s$^{1/2}$
- Dry density of thawed soils = 1050 kg/m$^3$
- Water content = 40%
- Coefficient of consolidation $C_v$ = 220 ft$^2$/yr

Estimate the total settlement of thawing soils that may be anticipated after 10 years.

CHAPTER 5

# Shallow Foundations

## 5.1 INTRODUCTION

In Chapter 4 the basic approach and considerations of foundation design in frozen ground were discussed. It is the goal of engineering that foundations be designed in such a way that they not only meet the bearing capacity of the supporting materials but also the long-term settlement or deformation anticipated within the designed period of the structure. Figures 5–1 and 5–2 show the structural damage due to settlement. Anticipated settlement or deformation must be defined within

**Figure 5–1** Frame house on permafrost, Fairbanks, Alaska.

**Figure 5–2** Commercial building on permafrost, Fairbanks, Alaska.

tolerable design limits. Either shallow or deep foundations may be used depending on various factors, such as the nature of the structure and its purpose, the existence of competent load-bearing strata at known depths (noting the strata by its soil type, ice content, and water temperature), and the availability of suitable construction material. Shallow foundations are discussed in this chapter. Chapter 6 covers the design of deep foundations.

As shown in Fig. 5–3, shallow foundations may be defined as having a width $B$ equal or greater than the depth of foundation $D_f$, which is also called the

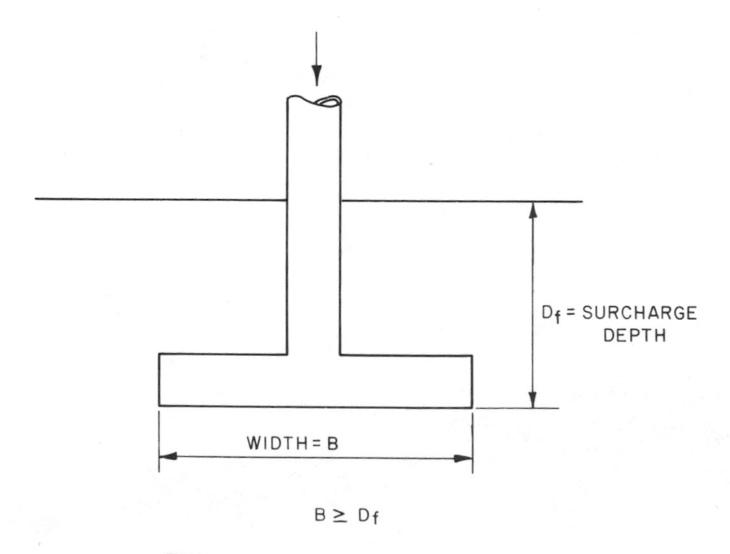


**Figure 5–3** Shallow foundation definition.

surcharge depth. Shallow foundations are usually selected when competent subsoil conditions are encountered at shallow depths or the design load of the superstructure can be economically satisfied without using deep foundations. Generally, the shallow foundations are placed in such a way that either thaw penetration or frost penetration does not affect the supporting material. Considerations with regard to shallow foundation design are discussed in the following sections.

Basic requirements for the satisfactory performance of shallow foundations are:

1. Adequate bearing capacity of the supporting soil strata
2. Tolerable short- and long-term settlements that do not damage the structure or destroy the structure's usefulness
3. Suitable foundation location with regard to any future factors that might cause serious adverse effects to the performance of the structure

The notation introduced in this chapter is given in Table 5–1.

**TABLE 5–1 Notation**

| Symbol | SI Unit | Definition |
|---|---|---|
| $A_d$ | $m^2$ | Duct area |
| $A_0$ | 0°C | Temperature amplitude (Eq. 5–1) |
| $B$ | m | Width of foundation |
| $C$ | $kN/m^2$ | Cohesion of soil (Eq. 5–2) |
| $C_f$ | $kJ/m^3$ K | Volumetric heat of air at constant pressure |
| $C_p$ | J/kg °C | Specific heat of air at constant pressure |
| $D$ | m | Duct diameter |
| $K_f$ | W/m K | Thermal conductivity of frozen soil |
| $L$ | m | Duct length |
| $N_c$, $N_r$, $N_q$ | — | Bearing capacity factors (Eq. 5–2) |
| $P$ | — | Period of 365 days |
| $Q$ | $J/m^2$ s | Total heat flow |
| $q_a$ | $kN/m^2$ | Bearing capacity (Eq. 5–2) |
| $q_{df}$ | $kN/m^2$ | Bearing capacity of frozen soil |
| $R_c$ | $m^2$ °C/W | Resistance of concrete |
| $R_f$ | $m^2$ °C/W | Resistance of floor system |
| $R_i$ | $m^2$ °C/W | Resistance of insulation |
| $S$ | m | Settlement of footing (Eq. 5–9) |
| $T_R$ | °C | Temperature rise in duct air |
| $T_{(d)}$ | °C | Temperature at depth $d$ (Eq. 5–1) |
| $V$ | m/s | Velocity of duct air |
| $\xi_c$, $\xi_r$, $\xi_q$ | — | Shape factors (Table 5–2) |
| $\alpha$ | $mm^2/S$ | Thermal diffusivity of soil |
| $\sigma_{ult}$ | $kN/m^2$ | Long-term strength |
| $\rho$ | $kg/m^3$ | Bulk density |
| $\rho_d$ | $kg/m^3$ | Dry density |
| $\rho_{df}$ | $kg/m^3$ | Dry density of frozen soil |
| $\omega$ | — | Water content |

## 5.2 TYPES OF SHALLOW FOUNDATIONS

The following types of shallow foundations may be used in continuous- and discontinuous-permafrost regions:

1. Footings or slab on gravel pad with or without insulation (Fig. 5–4)
2. Slab on self-refrigerated gravel pad (Figs. 5–5 and 5–6)
3. Footings or slab on ventilated gravel pad (Fig. 5–7)

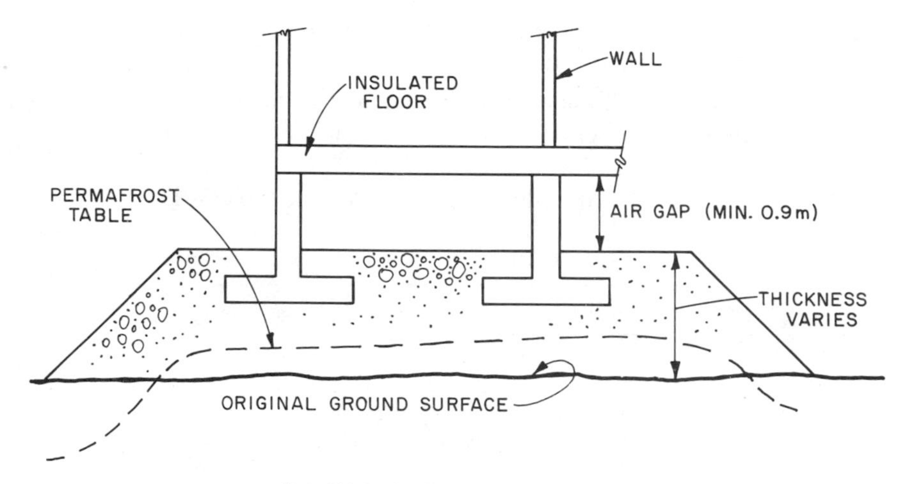


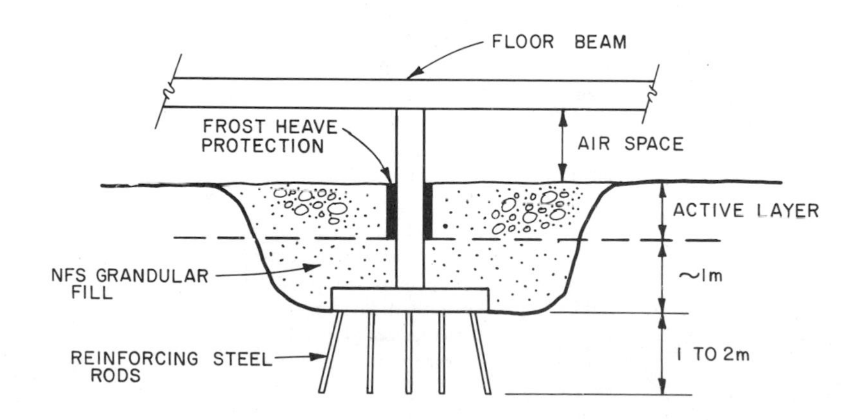


**Figure 5–4** Footings design.

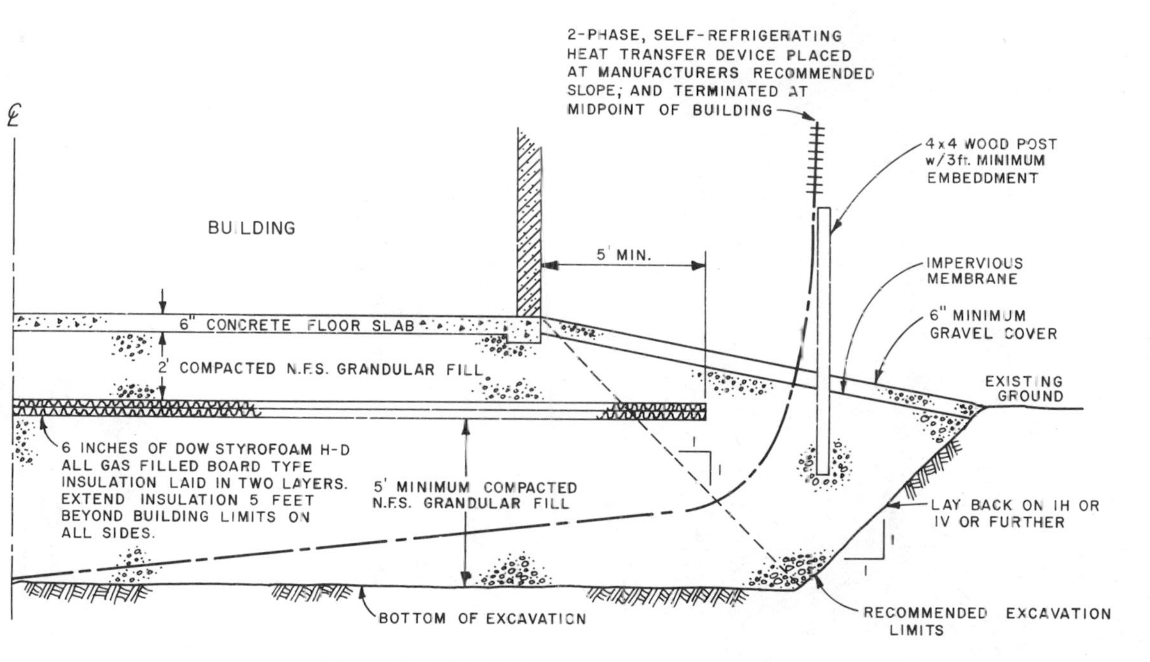


**Figure 5–5** Self-refrigerated insulated gravel pad.

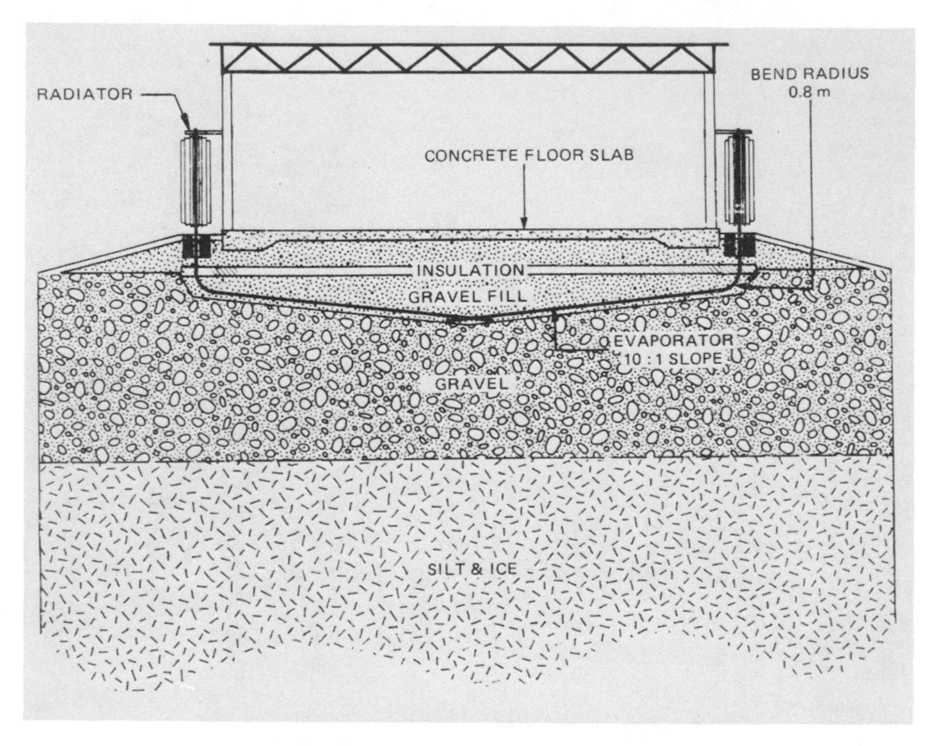


**Figure 5–6** Ross River School, Canada. (After Hayley, 1982.)

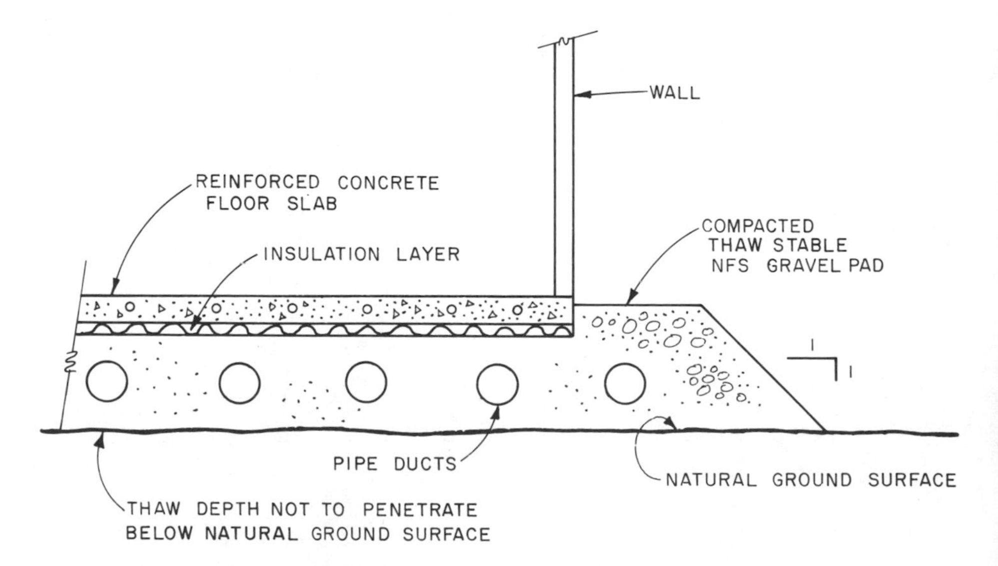


**Figure 5–7** Slab on ventilated gravel pad.

4. Footing or slab on mechanically refrigerated gravel pad (Fig. 5–8)
5. Post-and-pad type of foundation (Fig. 5–9)
6. Footings or shallow piers in natural soils (Fig. 5–10)
7. Ground sills on original ground or on gravel pad (Fig. 5–11)

Each foundation type has its particular advantages and disadvantages. Slab-on-grade construction with subgrade cooling systems are found to be very promising for structures with heavy floor loads (Phukan, 1981a; Phukan et al., 1978). Shallow foundations are usually placed on or just below the surface of compacted gravel pads. However, if the structure is temporary (design life period less than 5 years) or lightweight (single-story building), the shallow foundation may be placed on the original ground surface. Such foundations must have a clear air space between the floor of the structure and the ground level.

The gravel pad mentioned in the foregoing five typical foundation types (types 1 to 5 inclusive) should consist of thaw-stable, non-frost-susceptible (NFS) granular soils. The thickness of the gravel pad will depend on thermal input from the superstructure itself or the freezing and thawing index of the site. Also considered are the type of frozen soils and its ice content over which the gravel pad will be placed, the ground temperature profile, the availability of pad material, and the microclimate of the site.

The thaw-stable definition is applied to granular soils that are dense (required dry density of at least 100 lb/ft$^3$) and have adequate grain-to-grain contact,

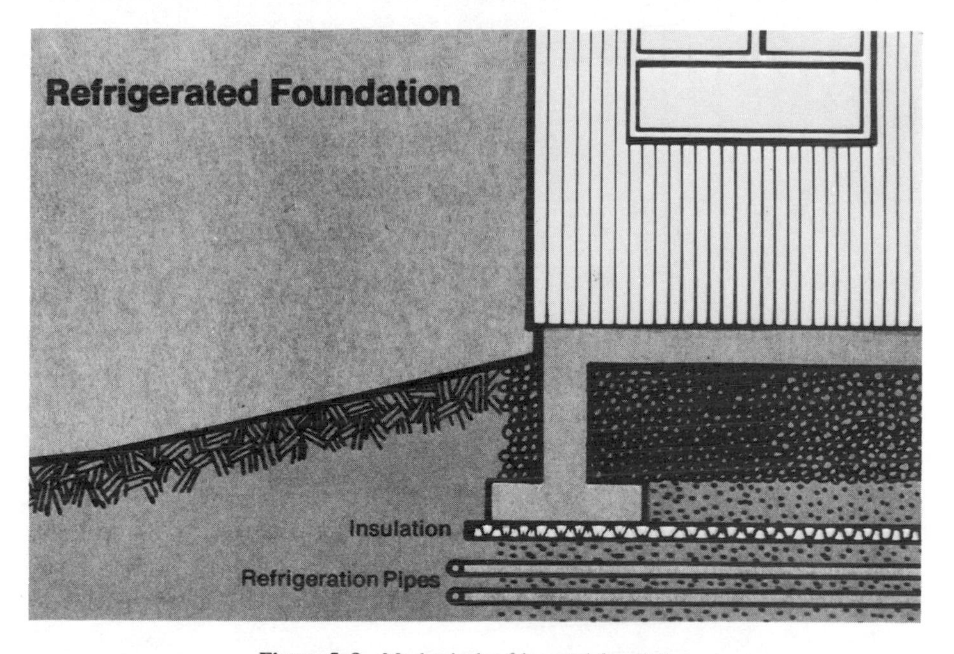


**Figure 5–8** Mechanical refrigerated foundation.

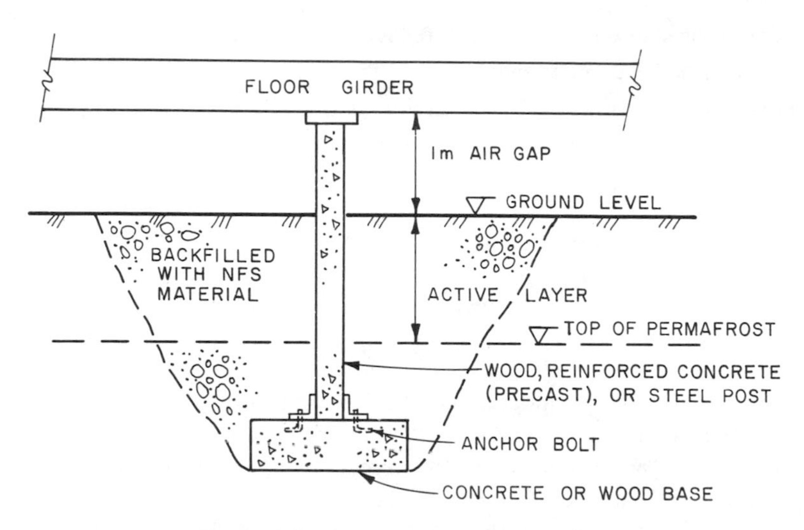


**Figure 5–9** Typical post and pad foundation.

which do not undergo significant deformation or strain after thawing (thaw strain less than 2%). The non-frost-susceptible definition discussed in Chapter 2 is varied in the literature according to the general purpose of a structure. For shallow foundation design, the granular material should have no more than 5% fines passing a No. 200 sieve. A typical gravel pad design is shown in Fig. 5–12. Generally, the

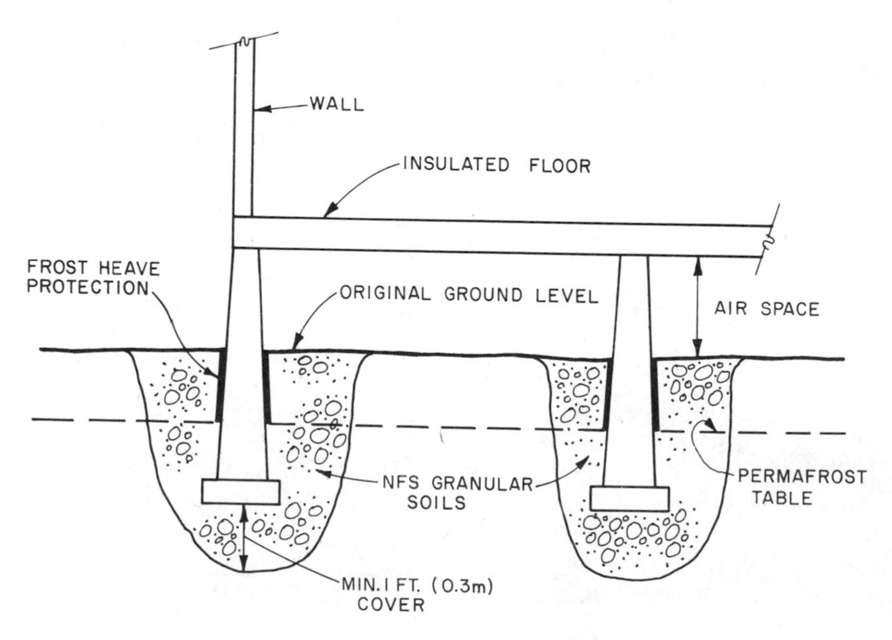


**Figure 5–10** Shallow piers in natural frozen ground.

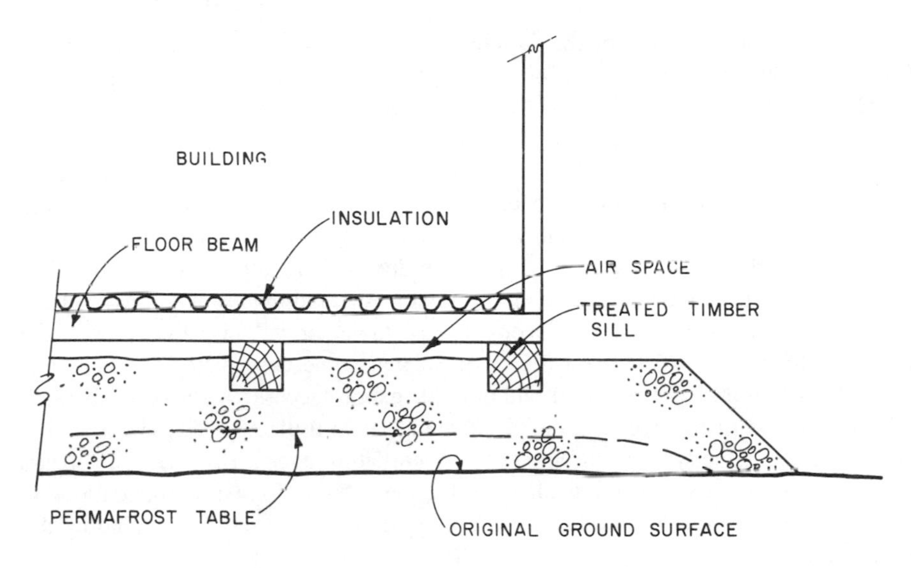


**Figure 5–11** Typical timber sills foundation.

anticipated thaw or frost penetration during the design period of the structure is arrested within the gravel pad, maintaining the existing "frozen" condition or thermal equilibrium.

A combination of gravel pad and insulation is generally used to reduce the quantity of granular soils required. As shown in Fig. 5–5, the self-refrigerated pad, consisting of gravel pad, insulation, and a two-phase self-refrigerating heat-transfer device (heat tube), may be used to maintain existing thermal conditions by extracting heat from the foundation ground. This subgrade cooling system is economically attractive. However, a major limitation of its use may be the

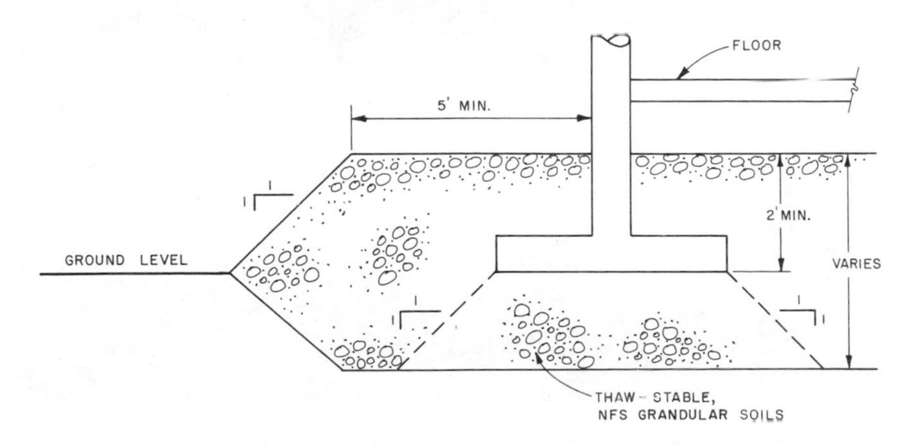


**Figure 5–12** Typical gravel pad design.

maximum allowable width of the structure (approximately 100 ft) for this type of foundation due to the limitation of the length of heat tubes which are presently manufactured. Figures 5–6 and 5–13 show some case histories where a slab on a self-refrigerated gravel pad has been utilized successfully.

Footings or ventilated slabs are widely used throughout the world. These include open-ended ducts in the pad, oriented in the direction of the prevailing winds (Auld et al., 1978). Another method of natural ventilation relies on the "chimney" effect to induce air motion in the ducted foundation (Fig. 5–14). This natural convection method, used in the design of shallow foundations, suffers from the disadvantage of reliance on unpredictable air velocity to remove heat from the foundation. Tobiasson (1973) presented some case histories where this type of foundation has been used without any major success. An inexpensive and reliable method of overcoming this disadvantage is to utilize forced-air circulation with a known velocity of air throughout the ventilating ducts. This method appears to be one of the most feasible techniques for the design of shallow foundations in frozen ground. Example 5–2 depicts a forced-air circulation design, including detailed calculations.

Principal components of the ventilated pad are (1) the pad thickness design to prevent thawing of the subgrade permafrost in summer, and (2) the ventilation requirements to remove heat from the gravel pad in winter. A simplifed method (Sanger, 1969) may be used to calculate the air velocity required to freeze the

**Figure 5–13** Shallow foundation on self-refrigerated insulated gravel pad, Fort Yukon, Alaska. (Courtesy of Kumin Associates.)

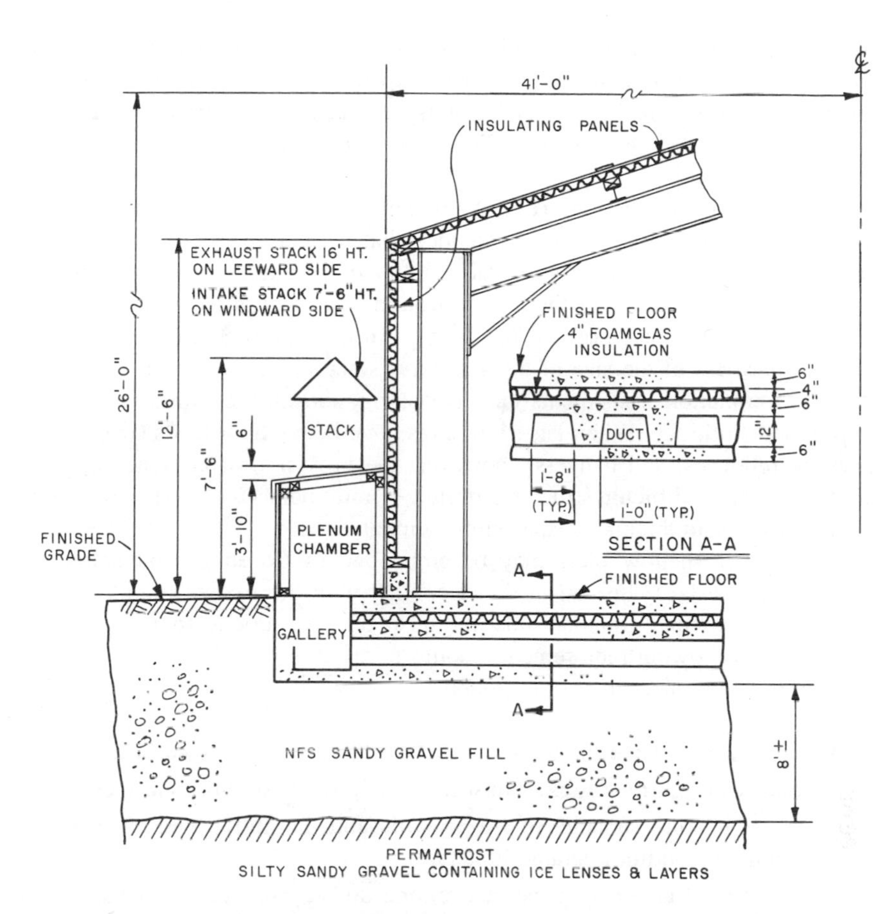


**Figure 5–14** Pan duct foundation, Thule, Greenland. (After NAVFAC DM-9, 1975.)

gravel pad. Nixon (1978) analyzed different case histories of ventilated pad foundations by determining the total heat flux to be removed from the system. Example 5–2 illustrates both of these methods to calculate the air velocity against the gravel pad thickness. A possible drawback of this system may be the high maintenance and operating costs. Poor maintenance of this system, caused by snowdrift, surface water infiltration, breakdown of air pumps, and other factors, may cause failure. A standby air-circulating pump or standby operating pump should be considered a necessary backup in this design.

Footings or a slab on a gravel pad with mechanical refrigeration using a brine cooling system may be used for shallow foundations. Such a system has been used under ice skating rinks. Several of the pumping stations along the Trans-Alaska Oil Pipeline system were built using mechanical refrigeration systems (Fig. 5–8). The performance of these systems has proven satisfactory. The cost of construction and maintenance of this type of system is generally high.

However, on large installations, footings or a slab on a gravel pad with mechanical refrigeration may be comparatively less expensive than the alternatives. Due to the performance reliability or end results, a mechanical refrigeration system may be utilized in the design of shallow foundations in soil conditions having poor thermal stability.

Post and pad foundations (Fig. 5–9) may be used for structures that can tolerate little or no more settlement, provided that the frozen ground remains in a "frozen" condition. Further, for lateral stability of structures, anchor bolts may be used to fasten the pad into the frozen ground, and such a technique has been used successfully in Spitzbergen (Fig. 5–15). An air gap of 3 ft (1 m) is generally used regardless of the size of the structure, since access to the air space may be used for foundation adjustments such as jacking, subsidence, shimming, repair, or inspection of utilities. When large buildings have heavy floor loads (workshops, garages, warehouses, and hangers), however, the provision of air space may necessitate extremely difficult structural design requirements with extensive formwork, increased slab thickness, and reinforcement.

Footings or shallow piers may be embedded in frozen ground below the original ground surface. Instead of direct contact with the frozen ground, a layer of non-frost-susceptible granular soils may also be used around footings or shallow piers. The following requirements should be used in the design and construction of footings and piers in frozen ground:

1. An air space between the floor and the natural ground surface.
2. Necessary design measures against frost jacking of footings and piers.
3. Adequate bearing capacity of frozen soils and the anticipated creep settlement must be within tolerable limits.
4. Disturbance of an existing thermal regime during excavations or backfilling of granular soils must be kept as low as possible to reduce the potential of thaw settlement.
5. The construction schedule should be such that the disturbed soils around footings and piers refreeze shortly after the construction activities.

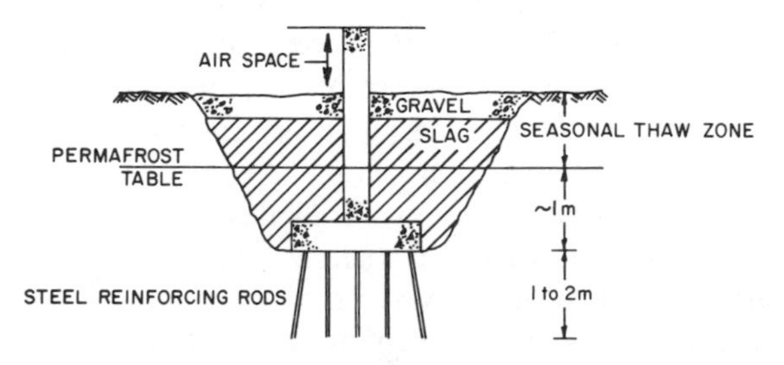


**Figure 5–15** Typical cast-in-place concrete footings supported on permafrost, Longyearbyen, Norway. (After Tobiasson, 1978.)

## 5.3 SHALLOW FOUNDATION DESIGN PROCEDURE

The following basic steps are generally used for the design of shallow foundations in frozen ground:

1. Based on various depth factors discussed in Section 5.3.1, a footing depth $D_f$ is selected using Fig. 5–3.
2. The warmest ground temperature below the shallow foundation is calculated using the standard solutions presented in Section 5.3.2.
3. Using bearing capacity theory and a suitable safety factor, a foundation size is selected.
4. A long-term settlement analysis (using the design period of the structure) is carried out based on the time–deformation relationships of the foundation soils.
5. If the settlement factor calculated is unacceptable, the foundation size as found in step 3 is modified.

### 5.3.1 Depth Factor

The depth of footings, piers, and post and pads used for shallow foundations should meet the following criteria:

1. The base of all foundations should be below the zone of high-volume change due to moisture fluctuations and the active layer.
2. Topsoil, peat, and organic materials should also be considered in the depth selection.
3. The location of foundations of adjacent structures should not interfere with the proposed foundation. As shown in Fig. 5–16, the new foundation must

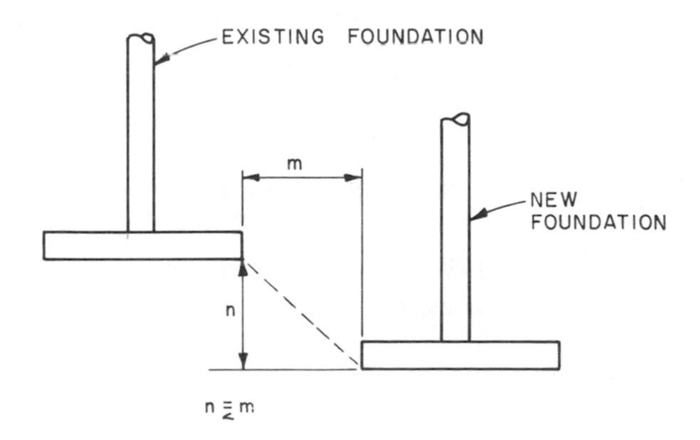


**Figure 5–16** Adjacent structure foundations.

be placed at a distance that will not cause any additional stress from the adjacent structural load.

### 5.3.2 Ground Temperature Profile

The following expression may be used to calculate the warmest frozen ground temperature below the footings:

$$T_{(d)} = 32 - A'$$

$$\text{where } A' = A_0 - A_0 e^{-d\sqrt{\pi/\alpha P}} \qquad (5\text{–}1)$$

$T_{(d)}$ = temperature at depth $d$
$A_0$ = amplitude of temperature difference between top of permafrost (0°C) and depth of no influence
$\alpha$ = thermal diffusity of frozen soils
$P$ = period of 365 days

The temperature calculated by Eq. (5–1) is conservative if maximum temperatures do not occur simultaneously at all depths. A positive approach is to install a series of thermistors at depths during the subsoil investigation of the foundation site and subsequent readings of thermistors for the time period (possibly to end of fall) when the ground temperature is maximal. Generally, a time lag of about 3 months exists between the air temperatures and the ground temperatures. Thermal properties of soils govern the distribution of temperatures throughout soil at depths. The thermal interaction between the structure and the ground was discussed in Section 4.3.1.

### 5.3.3 Bearing Capacity of Frozen Ground

The bearing capacity $q_d$ may be defined as the allowable pressure below the base of a foundation without causing any significant settlement of superstructure. Generally, a safety factor is applied to $q_d$ to determine the allowable soil pressure that governs the selection of foundation sizes to withstand the loads of the superstructure.

The bearing capacity in frozen soils may be determined in any of the following ways:

1. Bearing capacity theory and the formula for unfrozen soil are used. Time-dependent long-term strength is used in the final equation.
2. Guidelines from various published material are used.
3. Cavity expansion theory is used.

The bearing capacity of unfrozen soils may be determined by the following equation:

$$q_d = CN_c\xi_c + 0.5B\rho N_\gamma\xi_\gamma + \rho D_f(N_q - 1)\xi_q \qquad (5\text{-}2)$$

where $C$ = cohesion of soil
$N_c, N_\gamma, N_q$ = bearing capacity factors
$\xi_c, \xi_r, \xi_q$ = shape factors (given by Table 5–2)
$B$ = width of foundation
$\rho$ = unit weight of foundation soil
$D_f$ = depth of foundation

Equation (5–2) is based on the theory of plasticity and basic solutions (Prandle, 1921; Reissner, 1924). The pressure developed under the base of the footings is resisted by the shear strength of the underlying foundation soils and the surcharge pressure applied at the depth of the foundation. Solutions for the problem are given by many authors (e.g., Terzaghi, 1943).

The bearing-capacity factors are given by

$$N_q = \tan^2 (45° + \phi/2)e^{\pi \tan\phi} \qquad (5–3)$$

$$N_c = (N_q - 1) \cot \phi \qquad (5–4)$$

$$N_\gamma \simeq 2(N_q + 1) \tan \phi \qquad (5–5)$$

Values of $N_c$, $N_\gamma$, and $N_q$ for different values of the angle of internal friction of soils ($\phi$) are given in Table 5–3. Equation (5–2) may be used for determining the bearing capacity of a foundation in frozen ground by assuming that the shear strength of the foundation material is given by cohesion ($C$) only. The cohesion factor is obtained from long-term creep tests as discussed in Chapter 3. The frictional factors may be neglected for frozen ice-rich soils. So the bearing capacity $q_{df}$ in frozen soil may be written as

$$q_{df} = CN_c\xi_c \qquad (5–6)$$

where $C$ = $^{\sigma}$ult/2
$^{\sigma}$ult = long-term strength of foundation frozen soil given by Eq. (3–43)
$N_c$ = 5.14
$\xi_c$ depends on the geometry of foundations (Table 5–2)

**TABLE 5–2 Shape Factors**

| Foundation Type | $\xi_c$ | $\xi_\gamma$ | $\xi_q$ |
|---|---|---|---|
| Continuous | 1 | 1 | 1 |
| Circle and square | $1 + \frac{N_q}{N_c}$ | 0.6 | $1 + \tan\phi$ |
| Rectangular | $1 + \frac{BN_q}{LN_c}$ | $1 - 0.4\frac{B}{L}$ | $1 + \frac{B}{L}\tan\phi$ |

*Source:* After Vesic (1970).

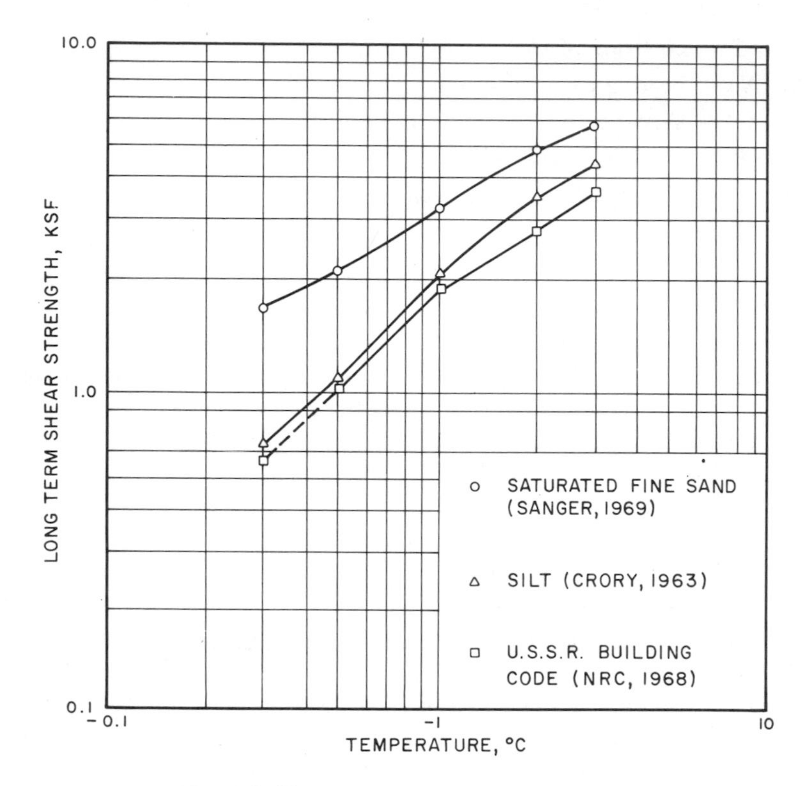


**Figure 5–17** Long-term strength of slurried pile.

The allowable soil pressure in frozen soil

$$q_{af} = \frac{q_{df}}{\text{safety factor}} \tag{5–7}$$

In ice-poor soils, the frictional component of the shear-strength parameter may be significant and the other factors on the right-hand side of Eq. (5–2) may also be used. Equation (5–6) is generally applicable to frozen fine-grained soils and ice-rich frozen soils where the cementing-bond effects of ice at negative temperatures make them behave like nonfrictional soil ($\phi = 0$).

A few published data sources are available for the long-term strength of ice-rich frozen soils at negative temperatures applicable to actual field conditions ($-5$ to 0°C or 23 to 32°F) where the determination of strength criteria is critical. Some of the published data are shown in Fig. 5–17. A tentative relationship (Johnston, 1981) is derived from

$$C = 35 + 28T \tag{5–8}$$

where $C$ is in kPa and $T$ is in °C below freezing (absolute value). This relationship

**TABLE 5–3 Bearing-Capacity Factors**

| $\phi$ | $N_c$ | $N_q$ | $N_\gamma$ | $\phi$ | $N_c$ | $N_q$ | $N_\gamma$ |
|---|---|---|---|---|---|---|---|
| 0 | 5.14 | 1.00 | 0.00 | | | | |
| 1 | 5.38 | 1.09 | 0.07 | 26 | 22.25 | 11.85 | 12.54 |
| 2 | 5.63 | 1.20 | 0.15 | 27 | 23.94 | 13.20 | 14.47 |
| 3 | 5.90 | 1.31 | 0.24 | 28 | 25.80 | 14.72 | 16.72 |
| 4 | 6.19 | 1.43 | 0.34 | 29 | 27.86 | 16.44 | 19.34 |
| 5 | 6.49 | 1.57 | 0.45 | 30 | 30.14 | 18.40 | 22.40 |
| 6 | 6.81 | 1.72 | 0.57 | 31 | 32.67 | 20.63 | 25.99 |
| 7 | 7.16 | 1.88 | 0.71 | 32 | 35.49 | 23.18 | 30.22 |
| 8 | 7.53 | 2.06 | 0.86 | 33 | 38.64 | 26.09 | 35.19 |
| 9 | 7.92 | 2.25 | 1.03 | 34 | 42.16 | 29.44 | 41.06 |
| 10 | 8.35 | 2.47 | 1.22 | 35 | 46.12 | 33.30 | 48.03 |
| 11 | 8.80 | 2.71 | 1.44 | 36 | 50.59 | 37.75 | 56.31 |
| 12 | 9.28 | 2.97 | 1.69 | 37 | 55.63 | 42.92 | 66.19 |
| 13 | 9.81 | 3.26 | 1.97 | 38 | 61.35 | 48.93 | 78.03 |
| 14 | 10.37 | 3.59 | 2.29 | 39 | 67.87 | 55.96 | 92.25 |
| 15 | 10.98 | 3.94 | 2.65 | 40 | 75.31 | 64.20 | 109.41 |
| 16 | 11.63 | 4.34 | 3.06 | 41 | 83.86 | 73.90 | 130.22 |
| 17 | 12.34 | 4.77 | 3.53 | 42 | 93.71 | 85.38 | 155.55 |
| 18 | 13.10 | 5.26 | 4.07 | 43 | 105.11 | 99.02 | 186.54 |
| 19 | 13.93 | 5.80 | 4.68 | 44 | 118.37 | 115.31 | 224.64 |
| 20 | 14.83 | 6.40 | 5.39 | 45 | 133.88 | 134.88 | 271.76 |
| 21 | 15.82 | 7.07 | 6.20 | 46 | 152.10 | 158.51 | 330.35 |
| 22 | 16.88 | 7.82 | 7.13 | 47 | 173.64 | 187.21 | 403.67 |
| 23 | 18.05 | 8.66 | 8.20 | 48 | 199.26 | 222.31 | 496.01 |
| 24 | 19.32 | 9.60 | 9.44 | 49 | 229.93 | 265.51 | 613.16 |
| 25 | 20.72 | 10.66 | 10.88 | 50 | 266.89 | 319.07 | 762.89 |

appears to be conservative in comparison to all available published data on the long-term strength of frozen soil at different negative temperatures.

Soviet workers (Vyalov, 1959; Poppe, 1976; Tsytovich, 1975) generally use empirical relationships for the determination of bearing capacity in frozen soils. The soil resistance at different negative temperatures and various factors depending on soil type and temperature are used in empirical relationships. Some of the recommended bearing capacity values for frozen soils given in Soviet publications are presented in Fig. 5–18. Special precautions must be used in using these values for the design of foundations in frozen ground, as they tend to be more general.

Ladanyi (1974) and Ladanyi and Johnston (1974) have used the cavity expansion theory for the determination of the bearing capacity of footings in frozen soils (for both cohesive and frictional conditions). They presented various bearing-capacity-factor values for use in estimating the depth factor and bearing capacity of footings of any shape. The use of cavity expansion theory should be viewed tentatively until more field and laboratory loading test results are made available.

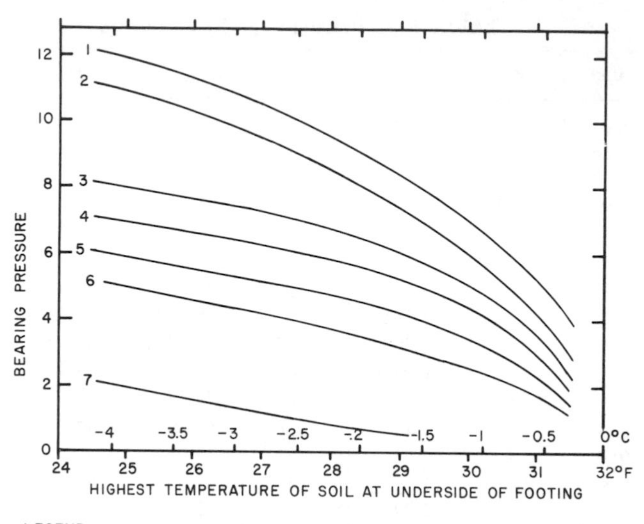


**Figure 5–18** Design bearing pressure. (After USSR, 1960.)

## 5.3.4 Long-Term Settlement Analysis

The bearing capacity discussion in the preceding section does not include the effect of long-term or creep settlement which may occur in ice-rich frozen soils or frozen soils at warm temperatures ($-2$ to $-0.02$°C). It is mentioned in the design procedures that long-term settlements must be calculated before finalization of foundation sizes. Generally, the long-term settlement values govern the design of foundations in frozen soils.

Long-term deformation due to creep in frozen soils is discussed in Chapter 3. Settlement due to compressibility of frozen soils, especially at warm temperatures, may also occur and calculations may be made to check whether they are of significant value in the total deformation calculations. Volume compressibility $m_v$ may be determined from Eq. (4–34). Table 3–5 may also be used for the calculation of $m_v$. Generally, the contribution of volume compressibility to the total settlement does not exceed more than one-third.

As presented in Chapter 3, various theories are available to determine the long-term or creep settlement of frozen soils at different phases. The settlement of footings is the sum of the strain multiplied by the respective layer thickness influenced by the applied load. Thus for $n$ layers,

$$S = \sum_{i=1}^{n} \varepsilon_i \, \Delta h_i \qquad (5\text{–}9)$$

where $\varepsilon_i$ = strain in layer $i$
$\Delta h_i$ = thickness of layer $i$

A knowledge of stress distribution under the foundation is required to calculate the settlement. Stresses induced by the foundation loads may be calculated by using standard elastic solutions available for a variety of problems (Poulos and Davis, 1984; Lambe and Whitman, 1969). The most common approach for determining stress under a concentrated load is the use of the Boussinesq equation, which can be integrated for other loading conditions. Methods such as the use of influence diagram charts (Fig. 5–19) and trapezoidal distribution (Fig. 5–20) may be utilized.

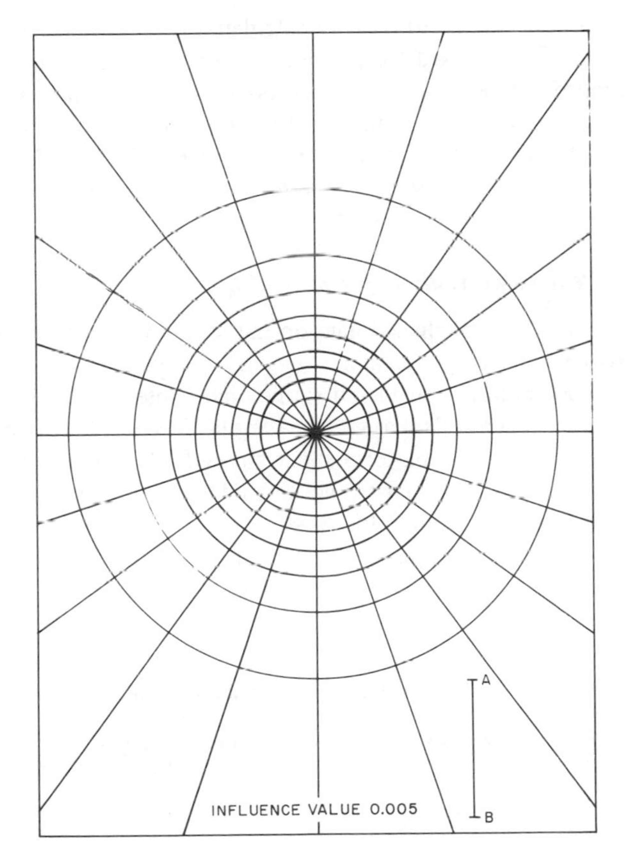


**Figure 5–19** Influence chart for computation of vertical pressure. (After Newmark, 1942.)

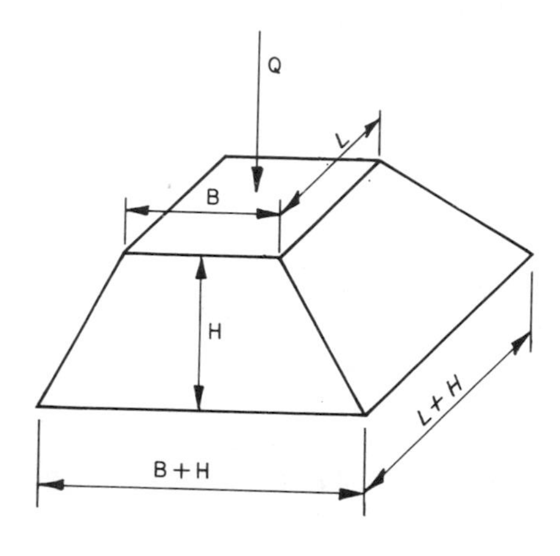


**Figure 5–20** Trapezoidal stress distribution.

The strain $\varepsilon_i$ in Eq. (5–9) may be calculated from Eqs. (3–34) and (3–42). Generally, Eq. (3–42) is used for ice-rich frozen soils which are dominated by secondary creep behavior, whereas the ice-poor frozen soils are analyzed by Eq. (3–34). It is most important to remember that the creep law and parameter values for undisturbed frozen soils at an applicable temperature range for the field condition under investigation must be accurately determined in order for the long-term settlement calculations to be made.

### *5.3.4.1 Settlement Analysis for Special Soil Conditions*

There are very few published data on the long-term shear strength of saline and peaty frozen soils at temperatures usually encountered in field situations. USSR (1976) workers have considered saline (salt content more than 0.25%) frozen soils as plastic-frozen soils that go into the plastic stage at a lower temperature than do salt-free soils. The long-term shear strength of such soils must be determined by special tests. If test data are not available, values for preliminary design purposes can be estimated from Table 5–4.

Equation (5–9) may be used to calculate settlement on saline and peaty frozen soils, but the analysis must be vigorous to calculate the secondary strain

**TABLE 5–4 Typical Bearing-Capacity Values for Saline Frozen Soils**

| Temperature (°C) | Bearing Capacity (KPa) |
|---|---|
| −1 | 30 |
| −2 | 65 |
| −3 | 100 |
| −4 | 140 |
| −5 | 180 |
| −6 | 230 |

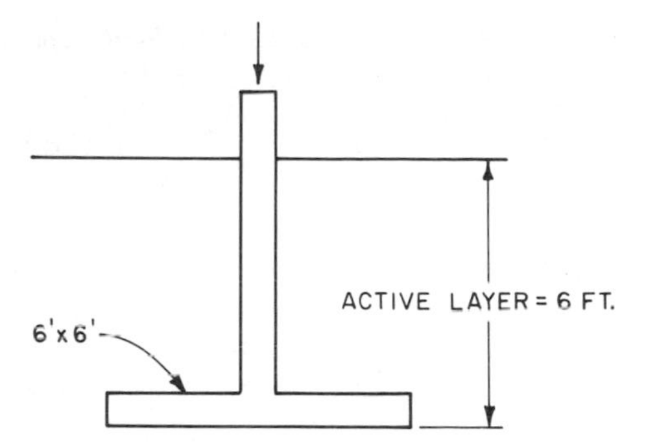


**Figure 5–21** Footing, 6 ft by 6 ft, resting at a 6-ft depth.

rate. Also, the safety factor should be at least 3, to limit the allowable soil pressure below the foundation.

**Example 5–1**

Determine the allowable load on a footing 6 ft by 6 ft, resting at a depth of 6 ft (below the active layer) on frozen sandy silt underlain by frozen silty sand as shown in Fig. 5–21. Assume that there is an air space between the floor and the natural ground which is provided to maintain the existing thermal equilibrium condition. The temperature at a depth of zero annual amplitude = 27°F.

For the sandy silt:

- Frozen dry density, $\rho_{df}$ = 85 lb/ft$^3$
- Water content $\omega$ = 33%
- Thermal conductivity $K_f$ = 1.2 Btu/ft hr °F
- Volumetric heat capacity $C_f$ = 28.5 Btu/ft$^3$

Long-term creep test results:

$$\sigma_{ult} = \frac{1816}{\ln t_f + 8.15}$$

For the silty sand:

- $\rho_{df}$ = 105 lb/ft$^3$
- $\omega$ = 20%
- $K_f$ = 2 Btu/ft hr °F
- $C_f$ = 28.3 Btu/ft$^3$

## Solution

1. *Depth factor:* the base of the footing is placed below the active layer = 6 ft.
2. *Calculation of warmest ground temperature below the footing:* Given $A_0$ = 5°F, the thermal diffusity of the frozen sandy silt

$$\alpha = \frac{K_f}{C_f} = \frac{(1.2)\,(24)}{28.5} = 1.01 \text{ ft}^2\text{/day}$$

From Eq. (5–1), the temperature at depth $d$

$$T_{(d)} = 32 - A'$$
$$A' = A_0 - A_0 e^{-d}\sqrt{\pi/\alpha p}$$
$$= 5 - 5e^{-d}\sqrt{\pi}/(1.01)(365)$$
$$= 5 - 5e^{-0.0923d}$$

so

$$T_{(d)} = 32 - 5 + 5e^{-0.0923d} \tag{5–10}$$

Following are the temperature calculations at different depths:

| Depth from Ground Level, $X$ (ft) | Depth from Base of Footing, $d$ (ft) | $0.0923d$ | $e^{-0.0923d}$ | $T_{(d)}$ (°F) |
|---|---|---|---|---|
| 6 | 0 | 0 | 1.0 | 32.0 |
| 7 | 1 | 0.092 | 0.91 | 31.6 |
| 8 | 2 | 0.184 | 0.83 | 31.2 |
| 10 | 4 | 0.369 | 0.69 | 30.4 |
| 13 | 7 | 0.646 | 0.52 | 29.6 |
| 18 | 12 | 1.107 | 0.33 | 28.6 |
| 25 | 19 | 1.754 | 0.25 | 28.2 |

The temperature in the silty sand layer is determined as follows:

$$\text{diffusivity of silty sand } \alpha = \frac{K_f}{C_f} = \frac{2\,(24)}{28.3} = 1.7\ \text{ft}^2/\text{day}$$

$$\text{diffusivity ratio } \frac{\sqrt{\alpha_{\text{silt}}}}{\sqrt{\alpha_{\text{sand}}}} = \frac{\sqrt{1.01}}{\sqrt{1.7}} = 0.77$$

Thus 1 ft of silty sand is equivalent to 0.77 ft of sandy silt with reference to temperatures under transient conditions.

Assume that all the soils were frozen sandy silt and the temperature distribution in the frozen silty sand is obtained by adjusting the depths beneath the silt–sand interface. In this case,

$$X' = \text{actual distance from ground level}$$
$$= \frac{d - 19}{0.77} + 25$$

| $X'$ (ft) | $d$ (ft) | $0.0923d$ | $e^{-0.023d}$ | $T_{(d)}$ (°F) |
|---|---|---|---|---|
| 32.8 | 25 | 2.31 | 0.099 | 27.4 |
| 39.3 | 30 | 2.76 | 0.062 | 27.3 |
| 44.5 | 34 | 3.14 | 0.043 | 27.2 |
| 46 | 35.2 | 3.24 | 0.039 | 27.1 |

3. *Allowable soil pressure:*
   Long-term strength:

$$\sigma_{ult} = \frac{1816}{\ln t_f + 8.15}$$

Assuming a design period $t_f$ = 100 years = $10^6$ hr, we have

$$\sigma_{ult} = \frac{1816}{\ln (10^6) + 8.16} = 82.7 \text{ psi}$$

$$C = \frac{\sigma_{ult}}{2} = \frac{82.7}{2} = 41.35 \text{ psi}$$

From Eq. (5–6), the bearing capacity

$$q_{df} = CN_c\xi_c = (41.35)(5.14)(1.2) = 255 \text{ psi}$$

The allowable soil pressure:

$$q_{af} = \frac{q_{df}}{\text{F.S.}} = \frac{255}{3} = 85 \text{ psi} = \frac{85\ (144)}{2000} = 6.12 \text{ tons/ft}^2$$

where F.S. is the safety factor. Thus the load capacity of a 6 ft by 6 ft footing = (6) (6) (6.12) = 220 tons

4. *Settlement Calculation:* The vertical stress below the footing may be obtained by using either the influence diagram (Fig. 5–19) or the trapezoidal distribution (Fig. 5–20). Then the settlement calculations may be performed using Eq. (3–34) or (3–42), depending on the creep behavior of the supporting soils. Generally, Eq. (3–34) is used to calculate the long-term settlement, and the settlement calculations for the case are given in Table 5–5.

   The assumed magnitudes of creep parameters in Eq. (3–34) are $m$ = 0.49, $K$ = 0.76, $\Lambda$ = 0.074, and $\omega$ = 570 [psi(hr)$^{\Lambda}$]/°F$^k$. Consider the design period $t$ = 20 years = $1.752 \times 10^5$ hr.

   The settlement calculated is equal to 1.2 in., which would be acceptable, and an allowable pressure of 85 psi under the footing could be tolerated.

**Example 5–2**

A building (15 m by 55 m) with an average floor temperature of 15.5°C is to be designed on a gravel pad with a forced-air duct system. Determine the necessary sizes and the air velocity required for the duct system. Assume:

- Freezing index at the building site = 3000 degree C-days
- Thawing season = 100 days
- Gravel: $\rho_d$ = 1920 kg/m$^3$; $\omega$ = 5%
- Average thermal conductivity of gravel = 1/2 (0.9 + 1.1) = 1.0 W/m °C
- Floor concrete thickness = 200 mm
- Thermal conductivity of concrete = 1.73 W/m °C
- Insulation thickness = 150 mm
- Thermal conductivity of insulation = 0.057 W/m °C

**TABLE 5–5**
**Settlement Calculation**

| Zone Thickness, $h$ (ft) | Temperature below Freezing, $\Theta$ (°F) | Vertical Stress, $\sigma$ (psi) | Strain, $\varepsilon = \left[\frac{\sigma t^{\lambda}}{\omega(\Theta + 1)^k}\right]^{1/m}$ | Deformation, $\Delta h = (\varepsilon)(h)$ (ft) |
|---|---|---|---|---|
| 2 | 0.4 | 62.5 | $\left[\frac{(62.5)(1.752 \times 105)^{0.074}}{570(0.4 + 1)^{0.76}}\right]^{1/0.49} = 0.04$ | 0.08 |
| 2 | 1.2 | 37.78 | $= 0.007$ | 0.014 |
| 3 | 2 | 23.78 | $= 0.0016$ | 0.0049 |
| 5 | 2.9 | 12.74 | $= 0.0003$ | 0.0016 |
| 7 | 2.6 | 6.62 | $= 0.00006$ | 0.0005 |
| | | | | $\Sigma = 0.101$ ft or 1.2 in. |

## Solution

1. *Gravel pad thickness determination:* Ducts are closed during the thawing season. The floor thawing index

$$I_{tf} = (15.5 - 0)\,(100) = 1550 \text{ degree-days}$$

$$\text{resistance of floor system } R_f = R_c + R_i \qquad (5\text{–}11)$$

where $R_c$ = resistance of concrete
$R_i$ = resistance of insulation

$$R_c = \frac{\text{thickness of concrete}}{\text{thermal conductivity of concrete}} = \frac{200}{1000\,(1.73)} = 0.116 \text{ m}^2\ ^\circ\text{C/W}$$

$$R_i = \frac{\text{thickness of insulation}}{\text{thermal conductivity of insulation}} = \frac{150}{1000\,(0.057)} = 2.632 \text{ m}^2\ ^\circ\text{C/W}$$

$$R_f = 0.116 + 2.632 = 2.748 \text{ m}^2\ ^\circ\text{C/W}$$

Latent heat of gravel:

$$L = (333.7)\,(1920)\,(5/100) = 32.87 \text{ MJ/m}^3$$

Depth of thaw given by the modified Berggren equation:

$$x = KR_f\left[\left(1 + \frac{7200\Lambda^2 I_{tf}}{KLR_f^2}\right)^{1/2} - 1\right]$$

$$= (1)\,(2.748)\left\{\left[1 + \frac{7200\,(24)\,(0.97)^2(1550)}{(1)(32.87 \times 10^6)(2.748)}\right]^{1/2} - 1\right\}$$

$$= \text{(say) } 3 \text{ m}$$

The required gravel pad thickness below the floor system = 3 m.

2. *Determination of the total amount of heat to be removed from the gravel pad:* The heat contained in the gravel pad is calculated as follows:

$$3\,(32.87) = 98.61 \text{ MJ/m}^2$$

$$+10\% \text{ sensible heat} = 0.986 \text{ MJ/m}^2$$

$$\text{total heat} = 99.60 \text{ MJ/m}^2$$

The heat must be removed during the freezing season (265 days), so the average rate of heat flow from the gravel pad is

$$\frac{99.60}{265\,(24)} = 1.566 \times 10^{-2} \text{ MJ/m}^2 \text{ N} = 4.13 \text{ W/m}^2$$

The average thawing index at the surface of the gravel of the pad is

$$I_{tp} = \frac{Lx^2}{7200\ \Lambda^2 K} = \frac{(32.87 \times 10^6)(3)^2}{7200\ (24)(0.97)^2(1)} = 1820 \text{ degree-days}$$

This thawing index must be compensated by an equal freezing index at the duct outlet to assure freeze-back.

Average pad surface temperature at the duct outlet:

$$\frac{\text{required freezing index}}{\text{length of freezing season}} = \frac{-1820}{265} = -6.87°\text{C}$$

Inlet duct air temperature:

$$\frac{\text{air freezing index}}{\text{length of freezing season}} = \frac{-3000}{265} = -11.32°\text{C}$$

Average possible temperature rise in the ducts:

$$-6.87 - (-11.32) = 4.45°\text{C}$$

Average heat flow between the floor and the inlet duct air:

$$\frac{15.5 - (-11.32)}{2.748} = 9.75 \text{ W/m}^2$$

Average heat flow between the floor and outlet duct air:

$$\frac{15.5 - (-6.87)}{2.748} = 8.13 \text{ W/m}^2$$

Average heat flow between floor and duct:

$$\frac{9.75 + 8.13}{2} = 8.94 \text{ W/m}^2$$

The total heat flow $= 8.94 + 4.13 = 13.07$ W/m$^2$. This heat flow to the duct air must be equal to the heat removed by the duct air.

$QLD = VA_d\rho C_p T_R$
$Q$ = total heat flow to duct, J/m$^2$ s
$L$ = duct length, m
$D$ = duct spacing, m
$V$ = velocity of duct air, m/s
$A_d$ = cross-sectional duct area, m$^2$
$\rho$ = air density, kg/m$^3$ (say, 1.33 kg/m$^3$)
$C_p$ = specific heat of air at constant pressure, J/kg °C (say, 1000 J/kg °C)
$T_R$ = temperature rise in duct air, °C

Assume that the diameter of the ducts = 203 mm and the spacing of the ducts = 1.5 m.

$$A_d = \frac{\pi\ (0.203)^2}{4} = 0.32 \text{ m}^2$$

The velocity of duct air required is

$$V = \frac{QLD}{A_d \rho C_d T_R}$$

$$= \frac{(13.07)(55)(1.5)}{(0.032)(1.33)(1000)(4.45)}$$

$$= 5.69 \text{ m/s}$$

An alternative method is given by Nixon (1978) to calculate the total heat flux to be removed by the ventilating ducts.

1. *Pad thickness determination:* The thickness of the pad may be calculated from the equation (Nixon and McRoberts, 1973)

$$X = \sqrt{\left(\frac{K_2}{K_1}H\right)^2 + \frac{2K_2T_st}{L}} - \frac{K_2}{K_1}H \tag{5–12}$$

where $K_1$ = thermal conductivity of insulation
$K_2$ = thermal conductivity of pad material
$H$ = insulation thickness
$T_s$ = surface temperature
$L$ = latent heat of the pad material
$t$ = time

The latent-heat term in Eq. (5–12) is given by

$$L = L_s + C_f|T_{av}| + \tfrac{1}{2}C_uT_s \tag{5–13}$$

where $L_s$ = fill latent heat
$C_f$ = volumetric heat capacity of frozen soil
$= \rho_{df}(0.2 + 0.5\omega)$
$C_u$ = volumetric heat capacity of unfrozen soil
$= \rho_d(0.2 + \omega)$
$|T_{av}|$ = mean ground temperature in degrees below freezing
$\omega$ = water content of fill material
$\rho_d$ = dry density of fill material

2. *Total heat flux calculation:* It is assumed that the heat is transferred to ducts from two sources: the heat gained from the structure above the ducts and the heat gained from the pad below the ducts. The ambient air temperature can usually be considered as being very sinusoidal, with a mean value $T_{av}$ and an amplitude $A_0$. Thus

$$T_{air} = T_{av} + A_0 \sin\frac{2\pi t}{365} \tag{5–14}$$

where $t$ is the time in days, measured from the time the air temperatures pass through their mean value (usually about May 1). The heat flux from the structure is given by

$$F_1 = \frac{K_1}{H_1}(T_s - T_{air}) \tag{5–15}$$

where $H_1$ is the thickness of insulation with the thermally equivalent thickness of fill between the structure and the duct centerline. The heat flux at the surface of this pad (i.e., at the duct centerline) can be calculated as

$$F_2 = K_2 A_0 \sqrt{\frac{\pi}{365a}} \sin\left(\frac{2\pi t}{365} + \cos\frac{2\pi t}{365}\right) \tag{5–16}$$

where $a$ is the thermal diffusivity of pad material, $m^2$/day. The total heat flux to be removed by the duct

$$F = F_1 + F_2 \tag{5–17}$$

It can be shown mathematically that the heat flux predicted by Eq. (5–17) reaches a maximum value when the time $t$ is given by

$$t_{max} = \frac{365}{2\pi} \tan^{-1}\left(1 + \frac{K_1}{H_1 K_2}\sqrt{\frac{365a}{\pi}}\right) + 182.5 \tag{5–18}$$

**Example 5–3**

Calculate the height of a chimney required to have adequate natural air draft for the subgrade cooling system in Example 5–2 if forced-air circulation is not used.

### Solution

The height of the chimney is determined by the relation (ASHRAE, 1977)

$$h_c = h_f + h_v \tag{5–19}$$

where $H_c$ = natural draft, mm $H_2O$
$h_f$ = friction head, mm $H_2O$
$h_v$ = velocity head, mm $H_2O$

The natural draft $H_c$ is given by (ASHRAE, 1977)

$$h_c = \frac{\rho \varepsilon H(T_c - T_o)}{T_c + 273} \quad \text{mm } H_2O \tag{5–20}$$

where $\rho$ = air density at average duct temperature, kg/$m^3$
$\varepsilon$ = efficiency of stack system $\simeq$ 0.80 (ASHRAE, 1977)
$H$ = height of chimney, m
$T_c$ = chimney air temperature, °C
$T_o$ = average outside air temperature, °C

The friction head ($h_f$) is given by (ASHRAE, 1977)

$$h_f = f' \frac{L_e}{D_e} h_v \quad \text{mm } H_2O \tag{5–21}$$

where $f'$ = dimensionless friction factor
$L_e$ = equivalent duct length, m
= 65 diameters for each right-angle bend plus 10 diameters for each entry and exit
$D_e$ = equivalent duct diameter (m)
= 4 times area ($m^2$) divided by perimeter (m)

$f'$ is a function of Reynolds number $N_R$ and the ratio $e/D_e$. The Reynolds number is given as

$$N_R = \frac{V(a' + 0.25D_e)}{\nu} \tag{5–22}$$

where $a'$ = shortest duct dimension, m
$\nu$ = kinematic viscosity of air at the average duct temperature, $m^2/s$

$$N_R = \frac{(5.69)[0.203 + 0.25(0.20)]}{1.27 \times 10^{-5}} = 113,000$$

Sanger (1969) gave the value of $e$ equal to 0.0003 m based on field observations.

$$f' = 0.0055\left[1 + \left(20,000\frac{e}{De} + \frac{10^6}{N_R}\right)^{1/3}\right]$$

$$= 0.0055\left[1 + (20,000)(0.0003) + \frac{10^6}{113,000}\right]^{1/3} \tag{5–23}$$

$$= 0.02$$

Assume an inlet length of duct = 2 m and an outlet length of duct = 4 m. The estimated length of the duct = 55 + 2 + 4 = 61 m, so

$$h_e = 61 + (0.2)[2(65 + 10)] = 91 \text{ m}$$

From Eq. (5–21) we get

$$h_f = \frac{(0.02)(91)}{0.20} h_v = 9.10h_v$$

so

$$h_c = 9.10h_v + h_v$$

$$= 10.10h_v = 10.10\left(\frac{5.69}{19.59}\right)^2 = 0.85 \text{ mm } H_2O$$

The height of the chimney ($H$) required to produce the draft of 0.85 mm $H_2O$ (from Eq. 5–20) is

$$H = \frac{h_c(T_c + 273)}{\rho\varepsilon(T_c - T_o)} = \frac{(0.85)(-6.87 + 273)}{(1.33)(0.80)(-6.87 - (-11.32)} = 47.82 \text{ m}$$

(*Note:* From Example 5–2, duct outlet temperature $T_c = -6.87°C$ and the duct inlet temperature $T_o = -11.32°C$.)

## PROBLEMS

**5–1.** Proportion a square footing to carry a column load of 100 tons at a safety factor of 2. The base of the footing will be 5 ft below the existing ground level. The frozen clay beneath the footing has a long-term strength of 2 tons/ft$^2$.

**5–2.** A building (20 m by 100 m) is to be supported on a slab-on-grade with an average floor temperature of 21°C. Design a duct system beneath the floor to maintain the existing frozen-ground condition at the site. Assume that the building site has the following soil properties and design data:

- Frozen silt: $\rho_d = 1450\ \text{kg/m}^3$; $\omega = 25\%$
- Freezing index at the site = 4000 degree-days
- Freezing season = 215 days

**5–3.** A hanger is to be supported on shallow foundations. Design an appropriate foundation system for the hanger using the following conditions:

- Mean annual air temperature = −3°C
- Freezing index = 5000 degree-days
- Average floor temperature = 18°C
- Thawing season = 100 days
- Temperature at depth of zero amplitude = −4°C
- Frozen silt: $\rho_{df} = 1400\ \text{kg/m}^3$; $\sigma = 30\%$; $\sigma_{\text{ult}} = 200$ kPa.

**5–4.** Estimate the settlement of the center of the tank shown in Fig. P5–4 due to creep settlement of a 10-m layer of frozen silt.

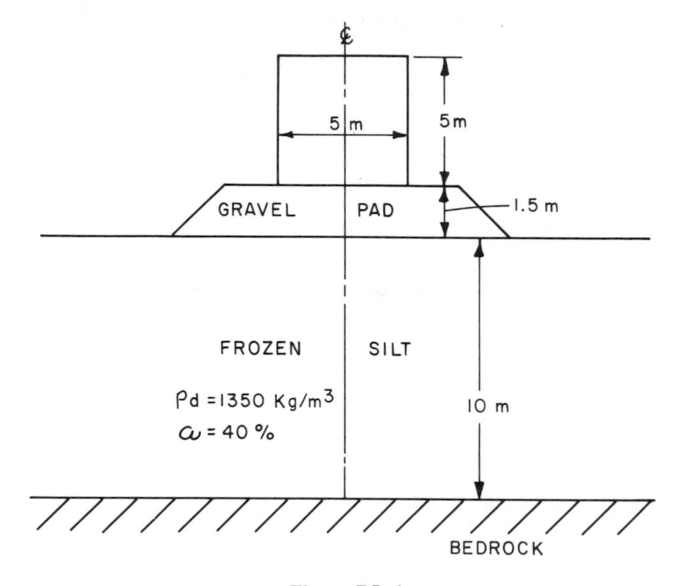


**Figure P5–4**

**5–5.** Discuss the merits and demerits of shallow foundations that can be used in continuous permafrost conditions.

**5–6.** Assume that a workshop (10 m by 30 m) with an average floor temperature of 20°C is to be designed on ice-rich frozen-silt ground conditions. Give recommendations on the most appropriate shallow foundations with a subgrade cooling system for the workshop.

**5–7.** If the footing shown in Fig. P5–7 is not to settle more than 1 in., what is the maximum load that it can carry? The long-term strength of supporting frozen silty sand is 3 tons/ft?

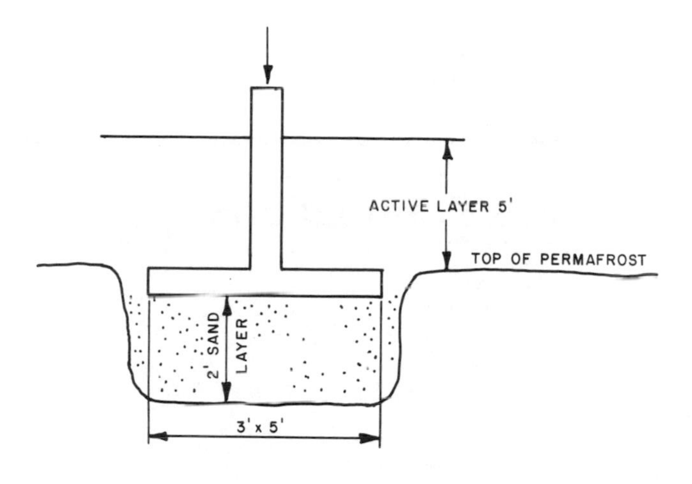


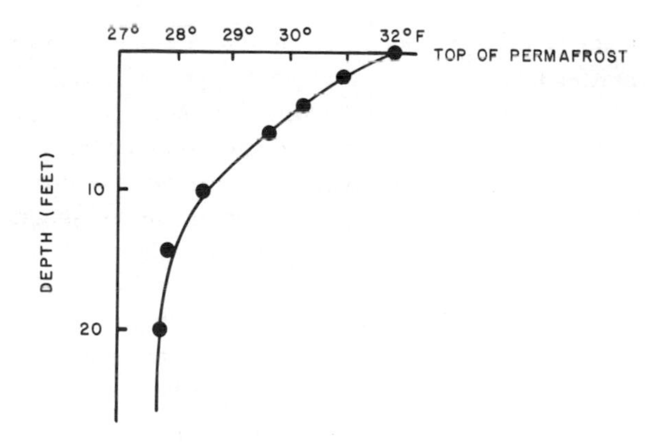


Figure P5–7

CHAPTER 6

# Pile Foundations

## 6.1 INTRODUCTION

Pile foundations are commonly used in frozen ground and provide structural support to transfer loads to a depth where competent subsoil conditions exist. Piles can provide an effective support to withstand heavy loads even in difficult frozen ground by maintaining existing thermal equilibrium conditions or preventing thermal degradation. At shallower depths where volume changes, compressibility, loss of shear strength, and change of temperatures of subgrade soils may occur, pile foundations offer an effective tool to isolate the structure from existing ground conditions. In comparison to shallow foundations, pile foundations are generally suitable in remote places as well as in ice-rich frozen soils. The basic design philosophy used in the design of pile foundations in unfrozen soils may be used in frozen ground. Unlike frozen soils, the behavior of frozen soils depends on some primary factors, such as ice matrix and negative temperature. These factors must be applied in the design of pile foundations in frozen soils (Phukan, 1977). The load-carrying capacity of piles in frozen ground is achieved by the adfreeze or shear strength developed between the pile surface and the existing natural soils or backfill material that may be placed between the pile surface and the natural soils. In addition, end-bearing support may also be obtained in suitable strata such as ice-free bedrock and thaw-stable dense granular soils. The fundamental load transfer mechanism of pile foundations in frozen soils is shown in Figs. 6–1 and 6–2. Both summer and winter conditions at the construction site must be considered in the design of pile foundations in frozen soils. The use of a thermal device generally called *heat tubes* associated with piles (named *thermal piles*) has signif-

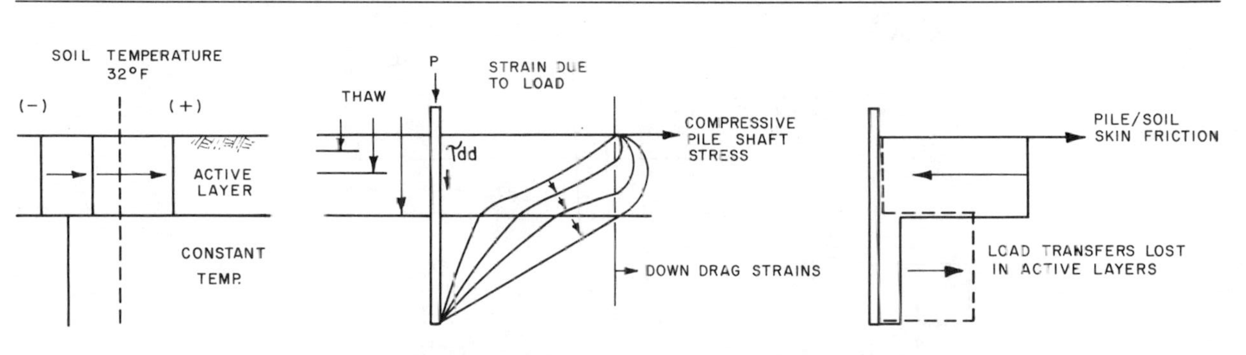


**Figure 6–1** Generalized load transfer condition: winter-to-summer transition.

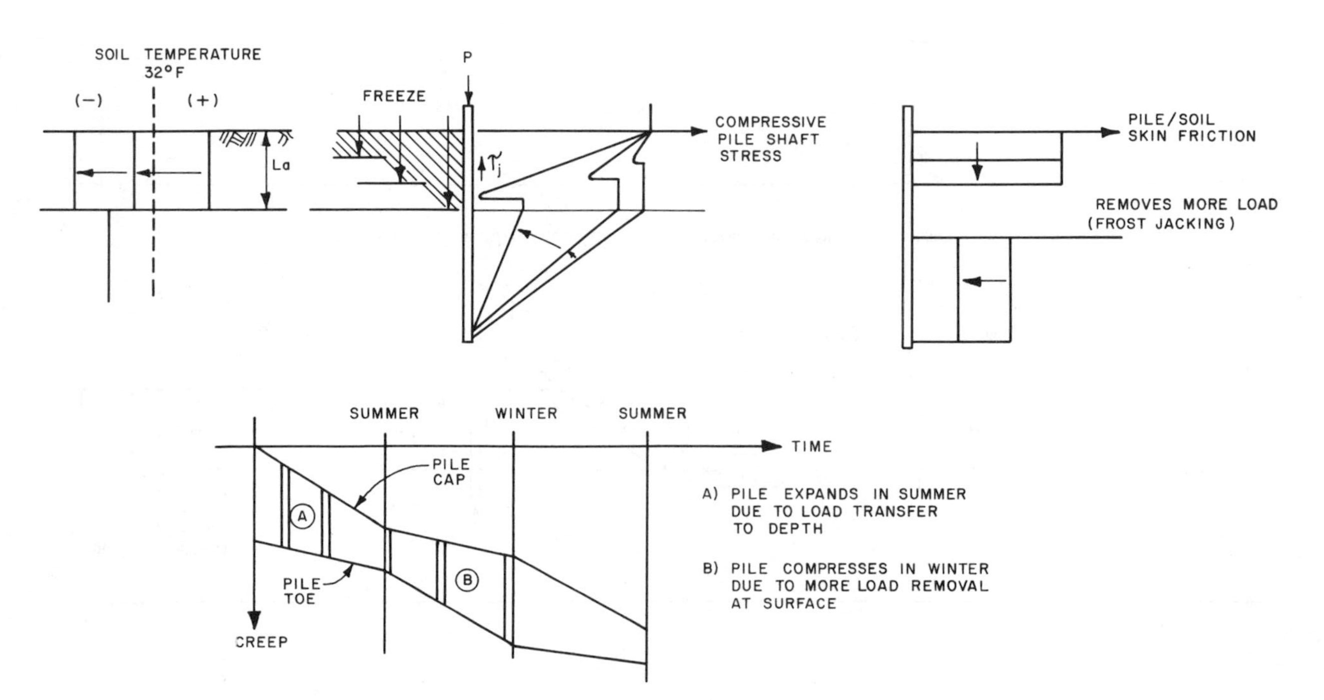


**Figure 6–2** Generalized load transfer condition: summer-to-winter transition.

icantly advanced the state of the art of pile design in the last decade. Many theories and concepts have been developed and may be considered when designing pile foundations in frozen soils. This chapter presents some of the more appropriate concepts and practice case histories to be considered when evaluating suitable designs options.

The notation used in this chapter is given in Table 6-1.

**TABLE 6–1 Notation**

| Symbol | SI Unit | Definition |
|---|---|---|
| $a$ | m | Pile radius (Eq. 6–9) |
| $C$ | kJ/m$^3$ | Volumetric heat capacity of frozen ground (Eq. 6–2) |
| $d$ | m | Diameter of pile (Eq. 6–4) |
| $d_1$ | m | Diameter of pile (Eq. 6–1) |
| $d_2$ | m | Diameter of augered hole (Eq. 6–1) |
| $L$ | J/m$^3$ | Volumetric latent heat of slurry (Eq. 6–1) |
| $L_A$ | m | Active-layer thickness (Eq. 6–5) |
| $L_e$ | m | Effective pile embedment layer (Eq. 6–5) |
| $P$ | kN | Structural load on pile (Eq. 6–3) |
| $P_d$ | kN | Downdrag force (Eq. 6–3) |
| $P_s$ | kN | Shaft resistance (Eq. 6–3) |
| $P_u$ | kN | Uplift force (Eq. 6–6) |
| $Q$ | J/M | Latent heat of slurry per linear meter |
| $S$ | m | Grid pile spacing (Eq. 6–2) |
| $\sigma_a$ | kN/m$^2$ | Allowable soil pressure at pile tip (Eq. 6–5) |
| $\dot{\varepsilon}$ | — | Secondary creep rate (Eq. 6–8) |
| $\tau_A$ | kN/m$^2$ | Shear strength of soil in active layer (Eq. 6–4) |
| $\tau_s$ | kN/m$^2$ | Shaft adfreeze strength (Eq. 6–5) |
| $\dot{U}_a$ | — | Pile velocity (Eq. 6–9) |

## 6.2 PILE TYPES USED IN FROZEN GROUND

Types of piles used in frozen ground generally include timber, steel, concrete, and composite materials of steel and concrete. The type of pile to be used at a particular site will depend on several factors, such as soil type and composition, temperature profile, loads to be supported, length of embedment, availability of material and construction equipment available, initial installation cost considerations, and construction schedule.

### 6.2.1 Timber Piles

Timber piles are generally the least expensive pile type and have been used extensively in the permafrost areas of the world. Spruce, Douglas fir, and pine are the most common source of timber materials. Standard, readily available timber stock come in lengths from 6 to 15 m and diameters ranging from 0.03 to 0.35 m butt and 0.15 to 0.25 m tip. Timber piles must be protected in the active layer against deterioration and decay. Either pressure treating with various wood preservatives, creosoting, or brushed-on preservative applications may be used to protect timber piles from deterioration. However, the use of such preservatives may reduce the adfreeze bond between the timber pile surface and frozen soils.

Due to favorable insulation properties, higher adfreeze bond strengths may develop between timber piles and frozen soils than with other pile types. Timber piles have a comparatively low structural capacity, however, and have further limitations in that they cannot be driven mechanically into the ground. Thus the limited installation procedures (cannot be driven) have narrowed their use for frozen-ground applications.

### 6.2.2 Steel Piles

Steel pipe piles and steel H-section piles are the main types of pile used in frozen ground. Oxidation and corrosion of steel must be protected in the active layer. Closed-ended or capped pipe piles are generally placed in preaugered holes. The annulus between the hole and the pile surface is filled with sand–water slurry or may be filled with concrete or sand to provide higher load capacities. Due to uniform cross sections, pipe piles are superior to H-section piles when designing against lateral loads. Heat tubes may readily be placed inside the pipe pile if required to maintain existing thermal regime. In recent years, both open-ended pipe and H-section steel piles have been successfully driven in warm permafrost (Davison et al., 1978; DiPasquale et al., 1983). Mechanically driven, H-section steel piles can penetrate better in frozen ground and heat tubes may also be used with the addition of channel sections (Fig. 6–3a). A photographic view of a water tank and a pump house founded on driven H piles is shown in Fig. 6–3b.

### 6.2.3 Concrete Piles

In North America, economic considerations have dictated the preference for steel piles over concrete piles. However, precast, reinforced circular and square piles are commonly used in the USSR, where steel is a relatively expensive material. Then piles can be driven and can also support heavy loads. The major concern with concrete piles are their low tensile strength, which precludes their use in sites where significant frost-heave forces are anticipated. Concrete piles are particularly vulnerable to frost-heave damage. Cast-in-place concrete piles have been used only occasionally in frozen ground due to problems of thermal degradation strength and freezing of concrete.

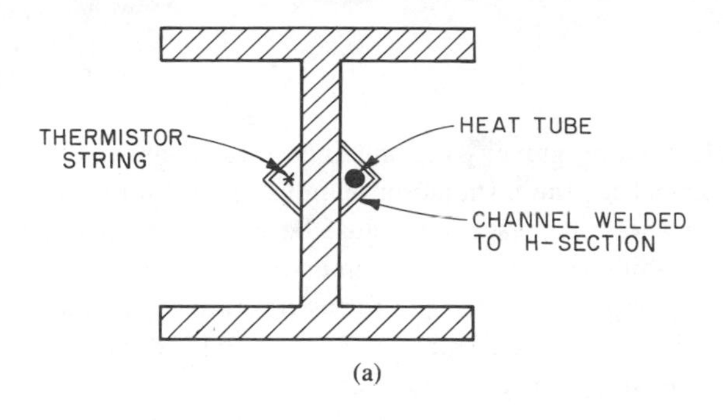


(a)

**Figure 6–3** (a) Typical H section with heat tube; (b) water tank and pump house on driven H piles, Kipnuk, Alaska. [(b) courtesy of Alaska Area Native Health Services.]

## 6.3 INSTALLATION OF PILES

The following methods are used to place piles in frozen ground:

1. Placement in a bored hole of diameter greater than the pile diameter
2. Placement in a prethawed hole of diameter greater than the pile diameter
3. Direct driving into natural frozen ground
4. Driving into undersized holes or steamed holes of diameter less than the pile diameter
5. Vibratory pile driving
6. Various combinations or modification of techniques 1 through 5

Generally, working gravel pads and access roads are required on the site to use heavy equipment for the installation of piles. The disturbances of natural frozen ground during the pile placement must be minimum. If adequate gravel fill material is not readily available for summer construction, temporary measures, such as installing temporary timber platforms or corduroy or metal mats to facilitate employing pile-driving equipment, may be used. A snow work pad may be used if a winter construction schedule is undertaken.

### 6.3.1 Pile Placement in Bored Oversize Holes

The method of placing a pile in a bored oversize hole is widely used in cold permafrost regions. In this method the pile is placed in a prebored oversize hole (generally greater than 6 in. or 152 mm of the pile diameter) and the annulus around the pile is backfilled with a sand–water slurry. Holes may be drilled by dry augering techniques using standard equipment that can penetrate most frozen soils. Hard-frozen soils with little ice and granular material may not be drilled easily. Due to wear and tear on the equipment, the dry augering technique may not be desirable in hard-frozen soil conditions, so that equipment such as a rotary or churn drill with air or refrigerated fluid to remove cuttings may be required. After completion of hole drilling, slurries are normally prepared in a portable concrete mixer to a consistency of about 0.15 m or 6 in. slump concrete and placed in approximately 1-m or 3-ft lifts around the pile. A small spud vibrator may be used to compact the placed slurry and ensure against air voids. Generally, the slurry temperature should not exceed 5.4°C or 40°F. A typical sand–slurry gradation is presented in Table 6–2. The cuttings from the drilled hole may also

**TABLE 6–2 Typical Gradation of Sand Slurry**

| U.S. Standard Sieve Size | Percent finer by Weight |
|---|---|
| 9.52 mm | 100 |
| No. 4 | 93–100 |
| No. 10 | 70–100 |
| No. 40 | 15–57 |
| No. 200 | 0–17 |

be used as slurry material. The cutting materials should be thawed and dried before placement, and careful attention must be given on the control of water content (not to exceed 15%). Excess water tends to decrease the adfreeze strength while increasing the refreezing time (Section 6.5). Moreover, timber piles may float when the holes are backfilled with a slurry and then the piles must be weighted down or restrained until the freeze-back has taken place. Figures 6–4 and 6–5 illustrate the placement of a timber pile with a heat tube in a bore hole.

**Figure 6–4** Timber pile with heat tube.

**Figure 6–5** Slurry placement around timber pile with heat tube. (Courtesy of Cold Regions Consulting Engineers.)

### 6.3.2 Placement in Prethawed Holes

Before the availability of modern drilling and driving techniques, piles were traditionally installed in steam- or water-thawed holes (Sanger, 1969; Pihlainen, 1959). The method is quick and fairly inexpensive. However, precautions must be taken not to generate any metastable condition in the thawed soil. Sanger (1969) documented several case histories where piles have failed as a result of poorly controlled steaming. Further, injection of steam into frozen ground introduces large quantities of heat and water, which can increase considerably the freeze-back time of piles. Also, holes are frequently made too large in ice-rich and stony frozen soils by prethawing methods.

This method is generally used in frozen ground with temperatures greater than −3°C or 27°F and the holes are limited to 7 m or 20 ft. To reduce the amount of heat input into the frozen ground, Goncharov (1964) proposed drilling a small pilot hole in advance of steaming. This method has been used successfully in the USSR in frozen clays, silts, and sands. However, this method cannot be used in coarse-grained frozen soils. A recent method developed in the USSR (Porkhaev et al., 1977) uses a "steam vibroleader," which consists of a hollow tube with a cutting bit attached to its lower end to which steam is supplied. The steam thaws frozen soil around the leader perimeter and then escapes to the surface through the hollow tube. A vibratory hammer is used to drive the leader into the thawed soil, and upon removal of the leader, the pile is driven into the thawed column of soil. The pile displaces the thawed material, which acts as a slurry around the pile.

### 6.3.3 Direct Driving

Recently, steel H-section and hollow-pipe piles have been successfully driven in frozen ground (DiPasquale et al., 1983; Davison et al., 1978). Generally, H-section piles are modified with angle pockets on either side of the web and driven directly in frozen ground using a wide range of impact driving equipment with highly rated energy (15,000 to 40,000 ft-lb or 20,000 to 54,000 J).

Vibratory hammers have also been used for driving H-section or circular hollow piles (Fig. 6–6). Bendz (1977) reported the use of high-frequency (sonic) driving equipment. Vibratory hammers and high-frequency equipment may also be used for extraction of piles installed in frozen ground.

Some of the main advantages of driving techniques are minimal disturbance of the physical and thermal regime of the soil during installation, quick freeze-back and immediate load application potential, placement under severe environmental conditions, and construction speed. However, the damage to the driven pile must be avoided by careful observations of the penetration rate of the installed pile. Generally, the direct driving technique is most efficient in frozen fine-grained soils.

**Figure 6–6** Vibratory hammer for pile placement in frozen ground. (Courtesy of Norcon.)

### 6.3.4 Driving into Undersize Holes

In this method a small pilot hole is either drilled or steamed; then the pile is driven into the undersized hole. This method expedites pile driving in difficult ground conditions. Also, with this method, vertical alignment of pile is correctly maintained. Nottingham and Christopherson (1983) have reported the successful use of this technique in the North Slope area of Alaska. Figure 6–7 presents the relationship between the distance from the pile hole edge and the elapsed time.

## 6.4 PILE CATEGORY

Given the various techniques used for the installation of piles and the use of thermal devices or heat tubes, piles in frozen ground can be categorized into five main groups.

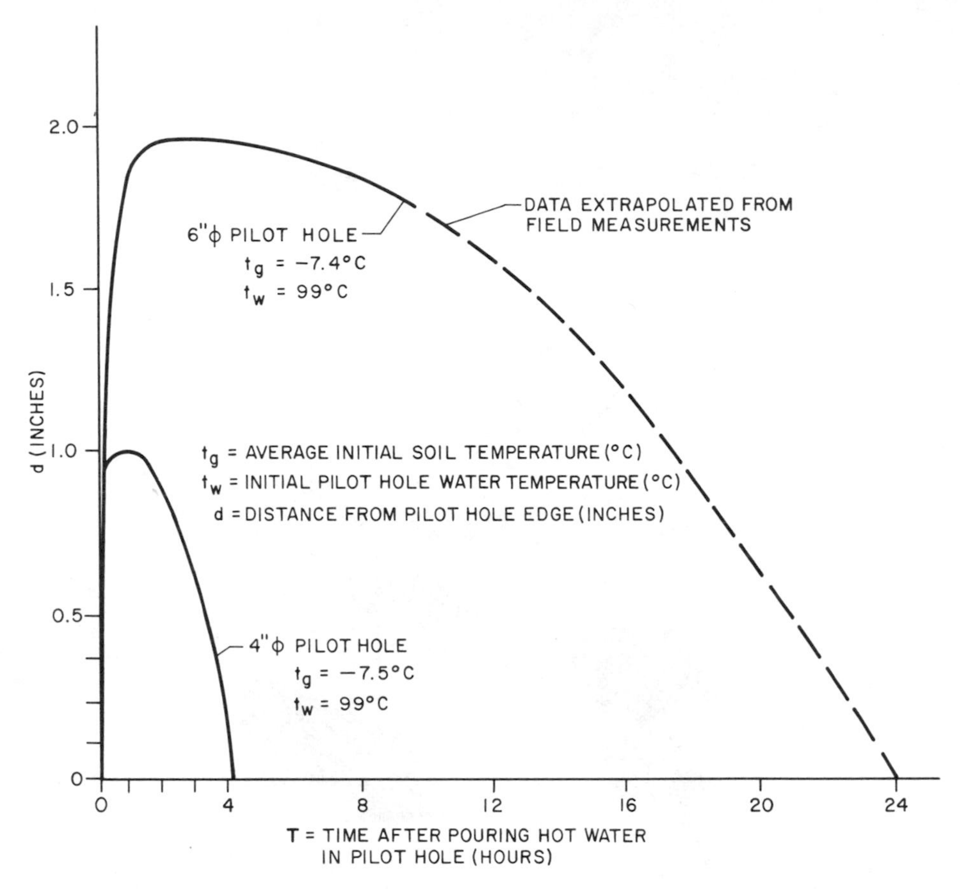


**Figure 6–7** Thermal growth/collapse of −5°C isotherm in silt. (After Nottingham and Christopherson, 1983.)

1. *Slurried pile:* This type of pile system is one of the most economical and feasible foundation supports for any type of construction in cold frozen ground. As shown in Fig. 6–8, the slurried pile is installed by predrilling a hole larger than the diameter of the pile, inserting the pile, and then backfilling the annular space around the pile with a consolidated sand–water slurry. Occasionally, the cuttings or the natural materials encountered during the drilling of holes may be used in place of sand–water slurry. But such construction techniques give a much lower load capacity of installed piles. Criteria to be used for the selection of slurried piles include:

(a) The average ground temperature must be cold (1.1°C or 30°F) and the depth of the active layer is relatively shallow (less than 1 m).
(b) The ground must be relatively flat or in sloping situations where there is no potential of ground motion problems such as slope stability, thaw plug stability, and liquefication during an earthquake.
(c) No end-bearing stratum is encountered at shallow depths.

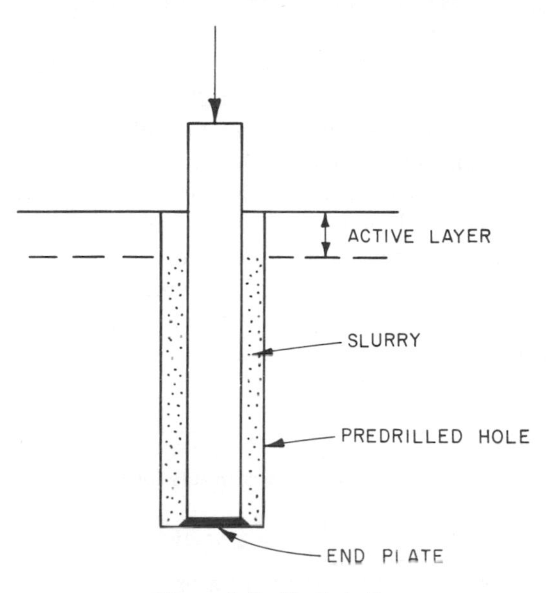


**Figure 6–8** Slurried pile.

2. *Thermal pile:* The slurried pile may be converted to a thermal pile by the addition of one or more heat tubes which extract heat from the supporting subsoils during the winter period. A typical thermal pile used in the construction of the Trans-Alaska Oil Pipeline is shown in Fig. 6–9. During the summer months, the thermal devices are dormant and the thermal pile installation is similar to the installation of the slurried pile except that thermal devices are included inside the pile and the pile inside is generally filled with unsaturated sand–water slurry. In

**Figure 6–9** Aboveground warm oil pipeline supported on thermal pile.

cases where the slurrying is not used around the pile, the thermal devices are placed near the pile in the native soils. As the heat pipes withdraw heat from the soil surrounding the pile in winter, there is a buildup of the reduced-temperature volume of the material around the pile that is not returned to its original temperature by the summer thermal regime. As a result, no thermal degradation will take place in the frozen soils during the life of the project and in many cases will actually upgrade. On the basis of supporting soils, the thermal pile may be classified into the three following categories: friction, end-bearing, and special. Ad-

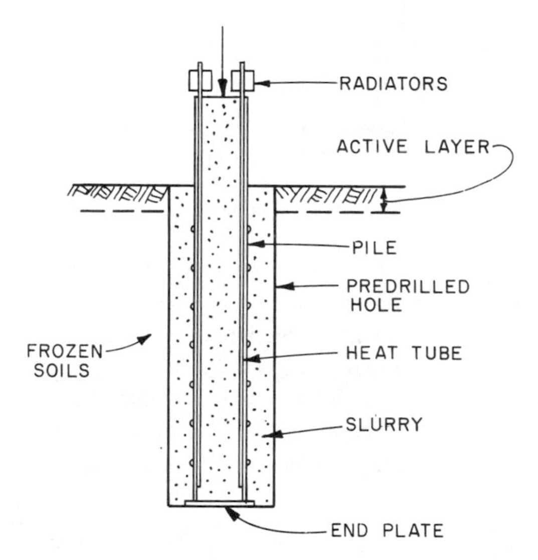


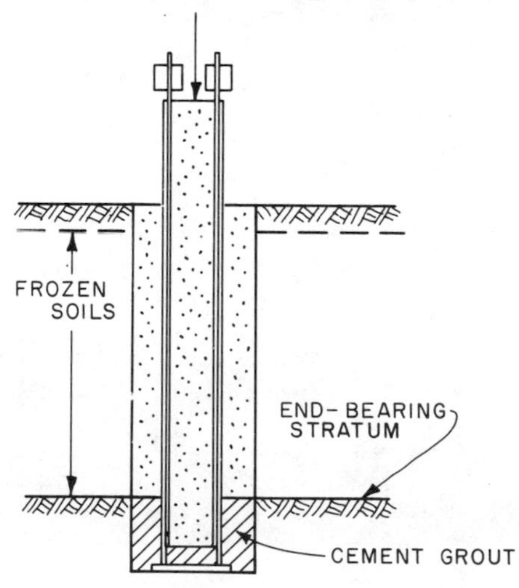


**Figure 6–10** (a) Adfreeze thermal pile; (b) thermal end-bearing pile.

freeze and end-bearing thermal pile are shown in Fig. 6–10. The friction thermal piles are similar to the design of slurried piles where the load capacity relies on the adfreeze strength between the soil and the slurry and the pile wall. The end-bearing thermal pile is used where the bearing stratum is encountered at a relatively shallow depth and the thermal devices stabilize the overburdened soils. The load capacity of this pile type is provided only by the end-bearing stratum. In cases where the thermal devices may not significantly aggrandize the thermal regime of the supporting subsoils, a special thermal pile may be used to support the design loads. Such special thermal piles are to be designed based on the reduced adfreeze bond strength. They may be used in frozen granular soils with traces of fine-grained soils (5 to 12% fines passing a No. 200 sieve).

Criteria to be used for the use of thermal piles are as follows:

(a) Where the slurried pile is not suitable in cold frozen soils (also in frozen soils where the thermal degradation has to be prevented)
(b) In areas where potential for liquefication, thaw plug stability, or thaw settlement would be a problem after the development of a deep thaw bulb
(c) In problem-thawed areas that should be kept frozen
(d) In areas where frost heaves govern the design
(e) Where massive ice is encountered in the soil profile

3. *End-bearing pile:* This pile type (Fig. 6–11a) is similar to a conventional pile type in unfrozen soil. The only difference is the presence of thaw-stable frozen soils in the overburden. The use of this pile type may be limited by the stability of the overburdened soils. Criteria to be used for the selection of this type are as follows:

(a) A suitable end-bearing stratum (bedrock, thaw-stable sands and gravels) must be encountered at a shallow depth
(b) The overburdened soil should be thaw-stable material (it should not contribute excessive loads to the pile)
(c) The ground may be flat or in sloping ground where there is no existing ground-motion problem

4. *Special friction pile:* A special friction pile may be used where frozen soils overlay thawed soils. It is similar to a thermal pile, the only difference being the use of concrete as the slurry-backfill material instead of sand–water slurry. Criteria to be utilized for the selection of this pile type are:

(a) A profile where the frozen soils consisting of fine-grained soils overlay thawed refreezable soils
(b) No end-bearing stratum is encountered at shallow depths
(c) The soil profile does not contain any massive ice
(d) The thawed soils in the profile should not cause any convective heat-flow problems

5. *Drilled-in-place pile:* As shown in Fig. 6–11b, this pile type is generally used in cases where a thawed soil profile is underlain by frozen soils. The thawed soil profile with a high water table generally causes the drill hole to collapse as the drilling advances. Also, under these thawed conditions an effective soil-to-pile

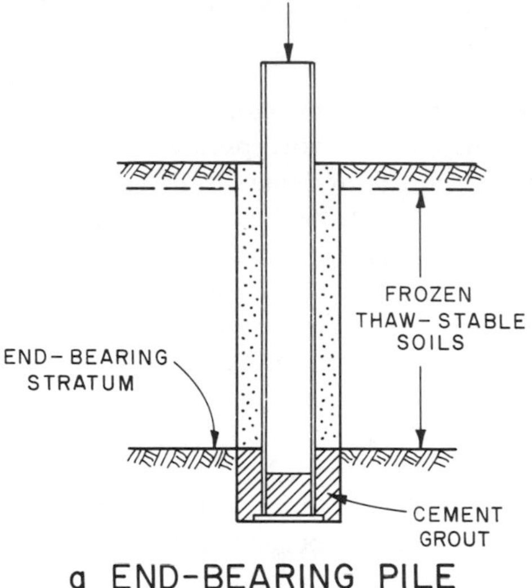


a END-BEARING PILE

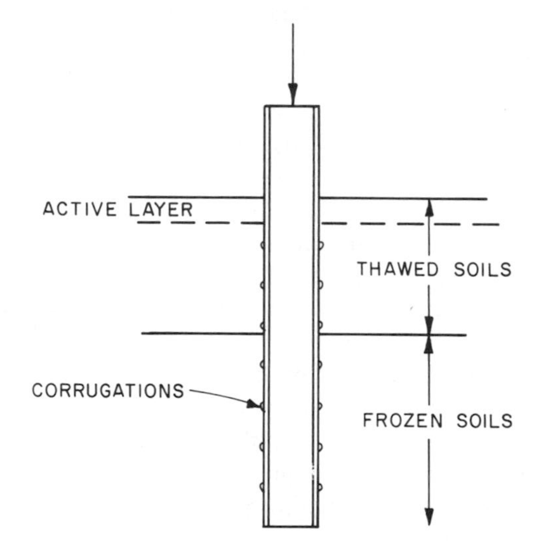


b DRILLED-IN-PLACE PILE

**Figure 6–11** (a) End-bearing pile; (b) drilled-in-place pile.

adfreeze strength may not be possible. In such cases, a drilled-in-place pile is placed where the pile acts as the casing to stabilize the collapsing walls. The drilled-in-place pile may be classified into two categories, depending on the use of heat tubes. One is standard friction and the other is thermal. End bearing may also be used if an end-bearing stratum is encountered. Criteria to be used for the selection of a drilled-in-place pile are:

(a) Where sidewall caving or unstable drill-hole conditions make the installation of piles difficult

(b) Where the soil may be blown inside the pile during the installation process
(c) Where no low-density ice-rich soils are encountered
(d) When the designed adfreeze bond may be equal to 75% of the frozen shear strength

## 6.5 PILE FREEZE-BACK

It is necessary that enough time be allowed for the installed pile to freeze back before any load is applied on the pile. Natural freeze-back may be allowed when there is a sufficient reserve of cold in the ground or sufficient time is available for the freeze-back. Artificial refrigeration methods may be used to freeze back quickly and they are useful when the permafrost temperatures and the high and/or the load application time is limited. The natural freeze-back method is generally preferable when the ground temperature is at the lowest level (usually late fall or spring). Sometimes, local ground conditions dictate the application of artificial freeze-back methods.

As discussed in Section 6.3, it appears that the slurried pile is the widely used technique for the installation of different types of piles. Factors such as the volume of slurry per unit length, the latent heat of fusion of the slurry, the spacing of piles, and the ground thermal regime must be considered for estimating the freeze-back time. Both latent heat and sensible heat are conducted into the frozen ground surrounding a slurried pile. Under given ground thermal conditions, the freeze-back time will be minimum when the diameter of the hole as well as the water content of slurry is minimum (thereby, latent heat and sensible heat) and the spacing of the piles is at maximum. For each unit length of pile embedment, the latent heat of slurry is given by

$$Q = L \frac{\pi}{4} (d_2^2 - d_1^2)\, \rho_d \omega \qquad (6\text{–}1)$$

where $L$ = latent heat of slurry, J/m$^3$ or Btu/ft$^3$
$d_2$ = diameter of hole
$d_1$ = diameter of pile
$\rho_d$ = dry density of slurry
$\omega$ = water content of slurry

The following two methods may be used to calculate the freeze-back time:

METHOD 1. A general solution given by U.S. Army/Air Force (1966)

METHOD 2. Step-by-step computation of the total heat dissipation or heat-removal requirements

Based on the work of various authors (Crory, 1966; Sanger, 1969; U.S. Army/Air Force, 1966), method 1 can be used to estimate the freeze-back time for a slurried pile. The relationships shown in Fig. 6–12 may be used and a detailed procedure for the use of method 1 is given in Example 6–1. Method 2 involves the following steps:

1. Determine the volumetric latent heat of backfill.
2. Determine the heat capacity of frozen backfill.
3. Determine the heat capacity of thawed backfill.
4. Calculate the heat required to depress the backfill temperature to the freezing point.
5. Calculate the heat required to depress the backfill temperature from the freezing point to the required below-freezing temperature.
6. Add the total heat to be abstracted from the slurry.

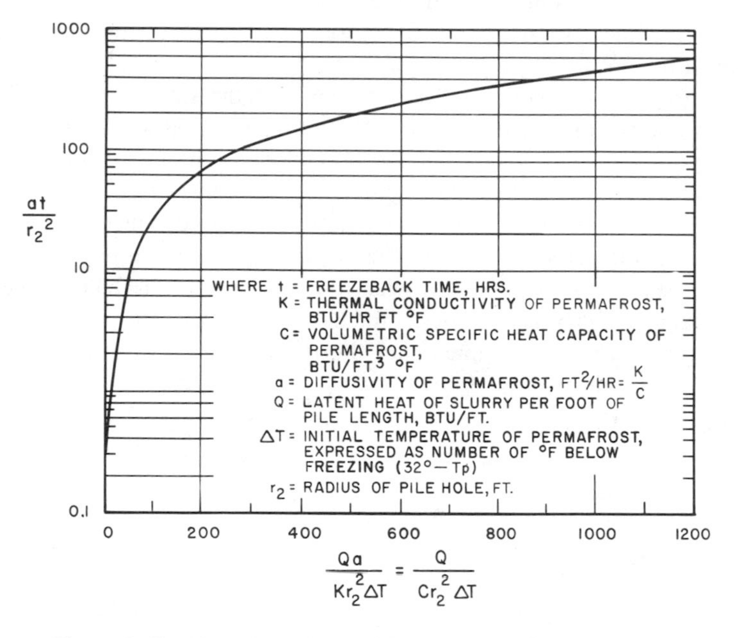


**Figure 6–12** Slurry freeze-back. (After U.S. Army/Air Force, 1966.)

It should be noted that natural heat exchange cannot take place when the backfill and the surrounding ground are at the same temperature. The temperature of the frozen ground between the piles increases as the latent heat is absorbed and the temperature increases slowly, dissipating over a long period to the surrounding undisturbed frozen soils. During late winter, freeze-back time is quickest nearest the coldest surface as heat dissipates upward and outward (Fig. 6–13a). During

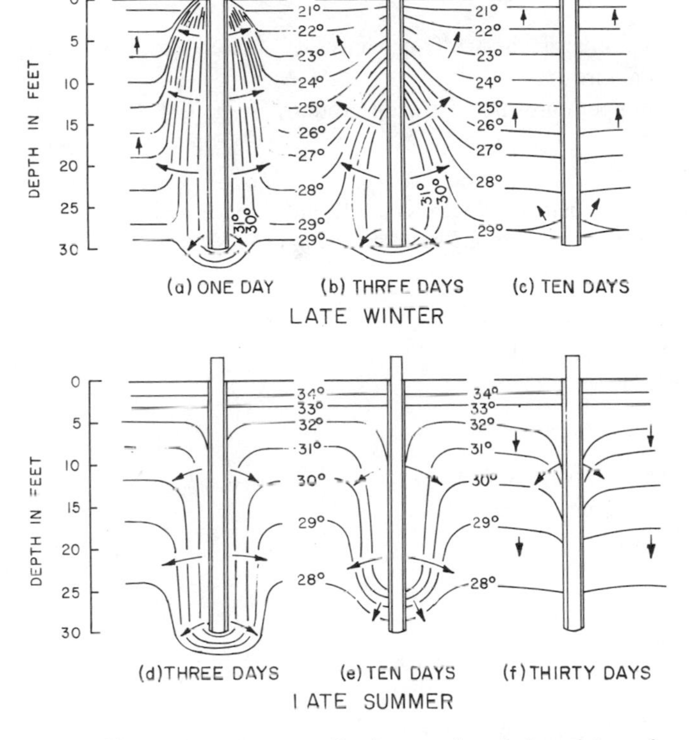


**Figure 6–13** Natural freeze-back on piles in permafrost during winter and summer. (After Crory, 1966.)

the summer, heat dissipates downward and outward, as shown in Fig. 6–13b, and freezeback depends on the cold reserve of the frozen ground below the active layer.

Another important consideration is the pile spacing, which affects the freeze-back time of the backfill material. The pile spacing can be calculated, as heat dissipated by the backfill is equal to the heat gained by the frozen ground between the piles:

$$\frac{\pi}{4} d_2^2 C \Delta t + Q = C \Delta t S^2$$

or

$$S = \sqrt{\frac{\pi d_2^2}{4} + \frac{Q}{C \Delta t}} \tag{6–2}$$

where $S$ = grid pile spacing, ft or m
$d_2$ = diameter of augered hole, ft or m
$Q$ = latent heat of slurry per lineal foot or meter, Btu/ft or J/m

$C$ = volumetric heat of capacity of frozen ground, in Btu/ft °F or cal/cm$^3$ °C

$\Delta t$ = temperature rise of frozen ground, °F or °C

**Example 6–1**

Calculate the time required to freeze back a 12-in.-diameter timber pile placed in an 18-in. hole preaugered in permafrost and backfilled with a slurry for the following conditions.

Permafrost: silt

- Initial temperature = 25°F
- Dry unit weight = 80 lb/ft$^3$
- Water content = 25%

Slurry backfill: sand–water

- Placement temperature 40°F
- Dry unit weight = 100 lb/ft$^3$
- Water content = 12%

### Solution

Thermal conductivity of the permafrost:

$$K = 0.8 \text{ Btu/ft hr °F} \qquad \text{(Fig. 3–23)}$$

Volumetric heat capacity of permafrost:

$$\begin{aligned} C &= 80[0.17 + (0.5)(0.25)] \\ &= 23.60 \text{ Btu/ft}^3 \text{ °F} \end{aligned}$$

Thermal diffusivity of permafrost:

$$\begin{aligned} \alpha &= \frac{K}{C} = \frac{0.8}{23.6} \\ &= 0.03 \text{ ft}^2\text{/hr} \end{aligned}$$

Volumetric heat capacity of the slurry (neglecting sensible heat):

$$\begin{aligned} L &= 144(100)(0.12) \\ &= 1728 \text{ Btu/ft}^3 \text{ °F} \end{aligned}$$

Latent heat per linear foot of slurry:

$$\frac{\pi}{4}(1.5^2 - 1^2)1728 = 1695.6 \text{ Btu}$$

$$\frac{Q}{Cr_2^2\,\Delta T} = \frac{1695}{(23.6)(0.75)^2(7)} = 18.24$$

Using Fig. 6–12, we found

$$\frac{\alpha t}{r_2^2} = 2$$

or

$$t = \frac{(2)(0.75)^2}{0.03}\ \text{hr}$$
$$= 37.5\ \text{hr}$$

**Example 6–2**

The average volume of slurry backfill for a group of pile is 1 m$^3$ each. The slurry is placed at an average temperature 7°C and must be frozen to −1°C. The sand–water slurry with a dry density of 1000 kg/m$^3$ and 15% water content is used as backfill. Calculate the freeze-back time required for a group of 30 piles if an artificial refrigeration device with a capacity of 50 kJ/s is available.

## Solution

1. Compute the volumetric heat capacity of slurry. Unfrozen heat capacity of slurry:

$$C_u = \rho_d\left[0.71 + (2.1)\left(\frac{15}{100}\right)\right]$$
$$= 1450(0.71 + 0.315)$$
$$- 1486.25\ \text{kJ/m}^3\ {}^\circ\text{C}$$

Frozen heat capacity of slurry:

$$C_f = 1450\left[0.71 + (4.19)\left(\frac{15}{100}\right)\right]$$
$$= 1940.83\ \text{kJ/m}^3\ {}^\circ\text{C}$$

2. Compute the latent heat of the slurry, $L$.

$$L = 333.7\rho_d \frac{\omega}{100}$$
$$= (333.7)(1450)\left(\frac{15}{100}\right) = 72{,}579.75\ \text{kJ/m}^3$$

3. Compute the heat required to depress the slurry temperature to the freezing point.

$$(C_u)(\text{volume of slurry})(\text{temperature above freezing})$$
$$= (1486.25)(1)(7)$$
$$= 10{,}403.75\ \text{kJ/pile}$$

4. Compute the heat required to freeze the slurry.

$$(L)(\text{volume of slurry}) = (72{,}579.75)(1) = 72{,}579.75\ \text{kJ/pile}$$

5. Compute the heat required to depress the slurry temperature from the freezing point to −1°C.

$$(C_f)(\text{volume of slurry})(\text{required temperature below freezing})$$
$$= (1940)(1)(1) = 1940\ \text{kJ/pile}$$

6. Compute the total heat to be abstracted from the slurry.

$$(30)(10{,}403.75 + 72{,}579.75 + 1940) = 2{,}547{,}705.00\ \text{kJ}$$

7. Calculate the freeze-back time.

$$\frac{2{,}547{,}705}{50 \times 60 \times 60} \text{ hr} = 14.15 \text{ hr}$$

**Example 6–3**

Determine the minimal allowable pile spacing in Example 6–1 that will permit natural freeze-back without allowing the permafrost temperature to rise above 31°F.

Solution

From Equation (6–3), the pile spacing may be calculated as

$$S = \sqrt{\frac{\pi d_2^2}{4} + \frac{Q}{C_f \, \Delta t}}$$

$$= \sqrt{\frac{(3.14)(1.5)^2}{4} + \frac{1695.60}{(23.6)(32 - 25)}}$$

$$= 3.47 \text{ ft}$$

## 6.6 LOAD CAPACITY OF PILES

Pile design in frozen ground must satisfy both the rheological and thermal considerations discussed in Section 4.3. In addition, there must be an adequate margin of safety against gross failure. The rheological aspect of pile design assures against gross failure and unacceptable deformation or settlement (both short term and long term; see Section 6.3). The thermal aspect of pile design considers the following factors:

1. Unacceptable changes in the ground thermal regime imposed by the aboveground structure
2. Unacceptable changes in the ground thermal regime transmitted by heat conduction down the pile
3. Freeze-back time of slurried piles
4. Heat flow around a thermal pile

Pile design in frozen ground is generally based on the long-term adfreeze strength between the pile surface and the adjacent frozen backfill and/or natural soil. The end bearing is considered when the pile tip is embedded in dense granular soils or bedrock. Generally, the shaft resistance due to adfreeze strength and end-bearing capacity should not be combined, as some movement in shaft will be required to mobilize the end-bearing capacity.

The load capacity of a single pile against a statical vertical load must consider the following:

1. The warmest or average frozen-ground temperature profile at the site.
2. The adfreeze strength as a function of temperature.
3. Downward structural loads, down-drag force, and frost-heaving forces along the active layer.
4. Pile type, size, pile surface characteristics, and construction procedures.
5. Complete freeze-back of slurry before the application of loads.

Field and laboratory tests are generally used to establish the pile load capacity and associated long-term settlements.

As shown in Fig. 6–14, both summer and winter loading conditions should be considered for determining the pile load capacity. In summer conditions, the equation of vertical equilibrium gives (Fig. 6–14)

$$P + W_p + P_d - P_s - P_E = 0 \tag{6–3}$$

or

$$P + W_p + \pi d L_A \tau_A - \pi d L_e \tau_s - \frac{\pi d^2}{4} \sigma_a = 0$$

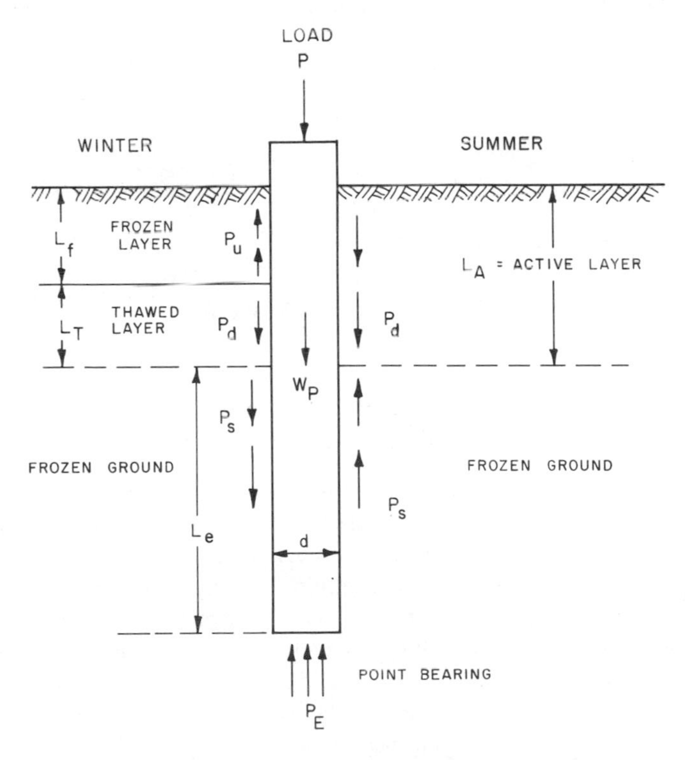


**Figure 6–14** Pile design: summer and winter conditions.

or

$$P + W_p = \pi d L_e \tau_s + \frac{\pi d^2}{4} \sigma_a - \pi \mathrm{d} \mathrm{L}_A \tau_A \tag{6–4}$$

where $P$ = structural load on pile
$W_p$ = weight of pile
$P_d$ = down-drag force
$P_s$ = shaft resistance
$P_E$ = end-bearing capacity
$d$ = diameter of pile
$\tau_s$ = shaft adfreeze strength
$L_e$ = effective pile embedment length
$\sigma_a$ = allowable soil pressure at pile tip
$\tau_A$ = shear strength of soil in active layer
$L_A$ = active-layer thickness

If the end bearing is neglected, Eq. (6–4) becomes

$$P + W_p = \pi d L_e \tau_s - \pi d L_A \tau_A \tag{6–5}$$

In winter conditions, the equation of vertical equilibrium becomes (Fig. 6–14)

$$P_u - P - W_p - P_d - P_s + P_E = 0 \tag{6–6}$$

$$P_u - P - W_p - \pi d L_T \tau_A - \pi d L_e \tau_s + \frac{\pi d^2}{4} \sigma_a = 0$$

or

$$P_u = P + W_p + \pi d L_e \tau_s - \frac{\pi d^2}{4} \sigma_a + \pi d L_T \tau_A \tag{6–7}$$

where $P_u$ = uplift force or frost heave force
= $\pi d L_f \tau_u$
$L_T$ = thawed layer thickness
$L_f$ = frozen layer thickness
$\tau_u$ = adfreeze strength in the frozen active layer

During the winter, the uplift or frost heave developed in the active layer must be resisted by the adfreeze shaft resistance (neglecting $P$, $W_p$, and $P_d$). Temperatures in the frozen active layer are much colder than those in the frozen ground; consequently, the adfreeze strength in the active layer ($\tau_u$) is much greater than in the frozen ground ($\tau_s$). Therefore, a design against frost-heave force based on embedment area alone is usually inadequate and designs must employ passive or active measures to mitigate frost-heave force. Otherwise, greater embedment length will be used to counteract the frost-heave force. One or more of the following techniques are used to mitigate frost heave:

1. The adfreeze bond may be reduced in the active layer by isolating the pile surface from the active layer from the soil grip by applying oil–wax grease to the pile surface.
2. The overall stability of the pile may be increased either by using a corrugated pile or by grouting the pile into a firm bearing stratum.
3. The frost-susceptible soil around the pile in the active layer may be replaced with NSF material.
4. A dead load may be applied before the generation of frost-heave force.
5. Heat tubes may be used in the active layer to reduce the frost-heave force.

If the frost-heave force is eliminated, the approach presented in Example 6–4 will govern the pile design. It is seen from Eqs. (6–5) and (6–7) that the adfreeze bond strength must be selected to determine the effective pile length. Adfreeze strengths for rough-surfaced piles (i.e., wooden or corrugated steel piles) approach the shear strengths of the frozen soils. However, the adfreeze strength of smooth piles is sharply influenced by the presence of ice at the pile interface. The factors influencing adfreeze bond strength (Kaplar, 1971) are:

1. Type of soil
2. Amount of water and ice in soil
3. Degree of saturation of soil
4. Dry unit weight of soil
5. Viscoplastic properties of soil
6. Temperature
7. Type of material in contact with soil
8. Size and shape of contact surface
9. Surface roughness of material in contact with soil
10. Porosity of contact material
11. Wetness or dryness of material
12. Young's modulus of material
13. Poisson's ratio of material
14. Type of test
15. Rate of loading
16. Duration and magnitude of loading

Another approach regarding the tangential adfreeze bond strength states that adhesion between two materials involves chemical bonding and an associated electrical effect, resulting in the formation of an electrical double layer in which one of the materials becomes a donor and the other an acceptor (Parameswaran, 1981). By this process an electrical potential is generated during the freezing of frozen soils to piles, especially metallic piles. One can calculate the charge concentration at the soil–pile interface and the force of adhesion as a function of change of concentration, based on the electrical bond theory of solids. Figures 6–15 to 6–17 present data on the adfreeze strength of soil and soil–pile interface at various temperatures.

**Example 6–4**

An 8-in.-diameter timber pile is placed in a frozen silt deposit which is frozen well below the depth explored by drilling. Assume that the pile was placed to a depth of 35 ft in a hole of 14 in. diameter and that the annular space between the pile and the hole was slurried back with sand at a temperature of 40°F. The following information is given for the site:

- Average warmest temperature of frozen ground = 31°F
- Uplift force due to frost heave = 15 tons/ft of perimeter of pile
- Depth of active layer = 5 ft

Determine the maximum load capacity of the pile.

## Solution

1. Consider the summer conditions.

$$\text{down-drag force } P_d = (\pi)\left(\frac{8}{12}\right)(5)(500) = 5233.3 \text{ lb}$$

(assuming shear strength of soil in the active layer = 500 psf)

$$\text{shaft resistance } P_s = (\pi)\left(\frac{8}{12}\right)(35 - 5)(5 \times 144) = 45{,}216 \text{ lb}$$

(From Fig. 6–15, the adfreeze strength = 5 psi at 31°F.) Neglecting the weight of pile and the end-bearing capacity, we have, from Eq. (6–4),

$$\text{maximum load capacity of pile } P = \frac{45{,}216 - 5233.3}{2000} = 20 \text{ tons}$$

2. Consider the winter conditions.

$$\text{uplift force } P_u = (15)(\pi)\left(\frac{8}{12}\right) = 31.4 \text{ tons}$$

Neglecting the weight of the pile and the end-bearing capacity, we have, from Eq. (6–7),

$$P = 31.4 - \frac{45{,}216}{2{,}000} = 8.8 \text{ tons}$$

Therefore, the maximum load capacity of the pile = 20 tons.

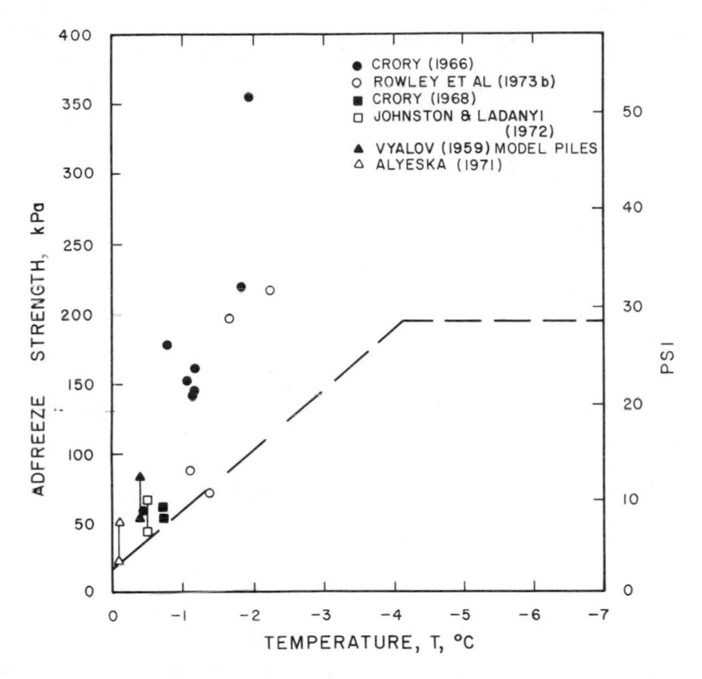


**Figure 6–15** Recommended adfreeze strength of soil interface. (Modified from Johnston, 1981.)

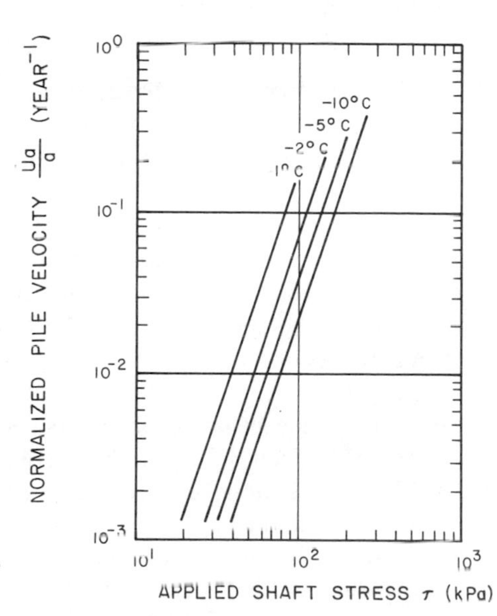


**Figure 6–16** Pile creep in ice. (After Morgenstern et al., 1980.)

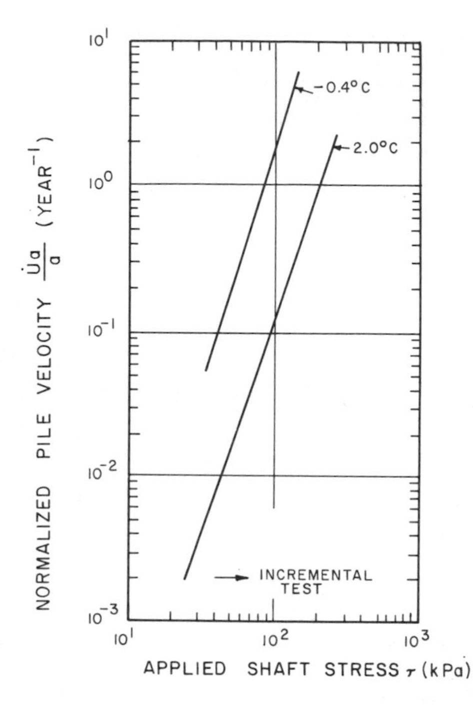


**Figure 6–17** Pile creep in ice-rich soil. (After Morgenstern et al., 1980.)

As the long-term deformation characteristics in ice-rich frozen soils are different from those in ice-poor frozen soils, the design approaches to such frozen soils should be different. These design differences are discussed next.

## 6.6.1 Pile Design in Ice-Poor Frozen Soils

The basic pile design philosophy in frozen ground as discussed above applies to ice-poor frozen soils. Design methods in both North American and Soviet literature are usually concerned with assessing the tangential adfreeze strength at the soil–pile interface. Once the distribution of allowable adfreeze strength along the pile shaft (see Fig. 6–15) is known, the depth of embedment in frozen ground required to carry the applied load can be calculated. On the basis of engineering judgment and experience, a suitable safety factor may be employed to limit settlement (Fig. 6–18).

In general, the behavior of ice-poor soils at low stresses is dominated by primary creep (Section 3.3.5.1). These soils, defined as "structured" soils (Vyalov, 1962), have grain-to-grain contact, which influences the primary creep behavior. A deformation analysis may not be required in the pile load capacity under the sustained loading for these structured soils. Sanger (1969) has developed a design procedure to limit settlement. This empirical method is based on site pile load tests where a series of short-duration load increments are applied and the rate

**Figure 6–18** Pile settlement.

of settlement is recorded. Then a plot of the log of settlement rate against load (giving a linear relationship) can be extrapolated to obtain the design load that would result in an acceptable rate of pile settlement.

## 6.6.2 Piles in Ice-Rich Frozen Soils

The load–deformation behavior of ice-rich frozen soils is largely dominated by the ice that separates the soil particles or group of particles. The properties of ice are responsible for the viscoplastic and deformation properties, which are unique characteristics of frozen soils. The design methods presented in the preceding sections are concerned with assessing the tangential adfreeze strength, which corresponds to the stress that causes rupture of the adfreeze bond. This design approach does not include any rationale for the settlement that may be encountered over the design life of the structure, although it is possible to apply, on the basis of geotechnical judgment, a suitable safety factor in order to limit settlement (Nixon and McRoberts, 1976). As deformation behavior of ice-rich frozen soils is dominated by secondary creep behavior, it is appropriate to introduce a creep-flow law into a mathematical model of pile behavior, and thereby a relationship between load and settlement or settlement rate may be obtained from the investigations. Thus pile design in ice and ice-rich frozen soils must limit the pile shaft stresses to keep the pile displacements within allowable limits.

Many investigators (Glen, 1955; Butkovich and Landauer, 1959; Mellor and Testa, 1969) have reported on the creep behavior of ice (Figs. 6–16 and 6–17).

By reviewing these creep data, a creep equation may be written (Johnston and Landanyi, 1972; Nixon and McRoberts, 1976; Morgenstern et al., 1980).

Power flow laws have been proposed to describe steady-state creep in incompressible materials at constant temperature as a function of multiaxial stress. Applied to ice-rich frozen soils, this relationship (Vyalov, 1966) may be written as

$$\dot{\varepsilon} = B\sigma^{n} \tag{6–8}$$

where $\dot{\varepsilon}$ = secondary creep rate
$\sigma$ = level of a general state of stress
$B, n$ = temperature-dependent constants unique to frozen soil

Typical values of creep parameters ($B$ and $n$) for ice are summarized in Section 3.3.1. DiPasquale et al. (1983) reported the creep parameter for ice-rich silt as follows:

$$n = 2.67$$
$$B = 1.02 \times 10^{-2}\ \mathrm{MN/m^{2-n}\ Year^{-1}} \text{ at } -1^\circ\mathrm{C}$$
$$B = 1.55 \times 10^{-3}\ \mathrm{MN/m^{2-n}\ Year^{-1}} \text{ at } -2^\circ\mathrm{C}$$

As discussed in detail by others (Vyalov, 1959; Johnston and Ladanyi, 1972; Nixon and McRoberts, 1976), the deformation of frozen ground around a pile shaft may be idealized as the shearing of concentric cylinders. If the constitutive behavior of material is characterized by Eq. (6–8), the pile velocity $\dot{U}_a$ under constant tangential shear stress $\tau$ and uniform constant ground temperature is given by (Morgenstern et al., 1980)

$$\frac{\dot{U}_a}{a} = \frac{3^{(n+1)/2}\, B\tau^{n}}{n-1} \tag{6–9}$$

where $a$ = pile radius
$n$ = temperature–dependent stress exponent approximately given by $(T + 1)^{0.61}$
$T$ = temperature expressed in celsius degrees below freezing
$B$ = creep parameter (see Table 6–1)

Figures 6–16 and 6–17 summarized the long-term pile creep in ice and ice-rich frozen soils. It should be noted that the flow law for ice at temperatures warmer than $-1^\circ$C is poorly defined and the design of piles in such conditions must be accomplished, providing special design provisions.

Equation (6–9) is insensitive to changes in normal stress on the lateral surfaces of the pile and to pile compressibility. It assumes that soil properties are homogeneous and isotropic with depth and that there is no slip at the pile–soil interface (i.e., tangential stress less than adfreeze strength). Equation (6–9) may be algebraically rearranged and applied incrementally to account for variations in soil-creep properties with depth. This allows investigation of a pile's velocity–load relationship under different ground thermal regimes.

Equation (6–9) has been further modified to account for pile compressibility (Nixon and McRoberts, 1976):

$$\dot{U}_a = D(Z, t)\left(\frac{aE}{2}\right)^n\left(\frac{\partial^2 Ua}{\partial Z^2}\right)^n \tag{6–10}$$

The fundamental relationship between pile strain and shaft stress over time may be explored in terms of pile-cap settlement. This information is most helpful in interpreting short-term load test data. It is generally assumed that lacking creep data for a particular ice-rich soil, the use of pure ice constants instead will yield a safe design. This is helpful, as the preponderance of such data exists for ice (see Fig. 6–16).

**Example 6–5**

Determine the anticipated pile settlement after 30 years for the calculated load of 20 tons in Example 6–4. Assume the frozen silt to be ice-rich.

### Solution

Assuming the creep parameter to be

$$n = 2.67$$

$$B = 1.02 \times 10^{-2}\ \text{MN/m}^2$$

$$\text{tangential shear stress} = \frac{(20)(0.00996)}{(\pi)(8/12)\ (0.305)}$$

$$= 0.275\ \text{MN/m}^2\ (1\ \text{ton}_f = 0.00996\ \text{MN})\ \text{and}\ 1\ \text{ft} = 0.305\ \text{m}$$

From Eq. (6–9), we have

$$\frac{\dot{U}_a}{a} = \frac{3^{(2.67+1)/2}(1.02 \times 10^{-2})(0.275)^{2.67}}{2.67 - 1}$$

or

$$\dot{U}_a = (1.46 \times 10^{-4})\left(\frac{4}{12}\right)(0.305)$$

or

$$U_a = (1.46 \times 10^{-4})\left(\frac{4}{12}\right)(0.305)(30)$$

or

$$\text{pile settlement} = 4.5 \times 10^{-4}\ \text{m}$$

which is negligible.

## 6.7 LATERALLY LOADED PILES

The capacity of vertical piles to resist lateral loads due to earth pressure, wind, and earthquake effects may be limited by the following factors:

1. Large horizontal movements of the piles and failure of the foundation.
2. Structural failure of the pile because bending moments have generated excessive bending stress in the pile material.
3. Significant movements of the superstructure may take place due to very large deflections of the pile heads.

The capacity of a laterally loaded vertical pile in frozen ground is dependent on the rate of loading and the duration of the applied load. Therefore, different lateral loading conditions must be distinguished. Some of the lateral loading conditions are: (1) short-term loadings such as those imposed during construction by seismic forces or wind, (2) seasonal loadings such as those caused by thermal expansion and contraction, and (3) long-term loadings such as those imposed on aboveground pipeline (Fig. 6–19) at intermediate points and bends in the line. These loads would normally apply for the life of the structure.

**Figure 6–19** Lateral load on aboveground pipeline.

Two design approaches may be used to determine the lateral load capacity of piles in frozen ground. The ultimate lateral soil resistance per linear meter of pile can be established from Brom's model (1965) formulated as $9dS_u$, where $d$ is the diameter of the pile and $S_u$ is the long-term strength of frozen soils (for unfrozen soils, $S_u$ is the undrained shear strength). The other method is called *beam on elastic foundation,* reported by Matlock and Reese (1962). To use this method, the load–deflection *(P–Y)* relationship of the pile and the supporting soil must be known.

Very limited information is available on the capacity of laterally loaded vertical piles in frozen ground. Based on load tests carried out on 30-cm-diameter steel pipe and timber piles, Johnston (1981) reported that the short-term lateral load on any pile not less than 30 cm (12 in.) in diameter and embedded not less than 6 m (20 ft) in frozen ground must be restricted to 8 tons for steel piles and 6 tons for timber piles. Loading criteria for vertical piles subjected to seasonal or

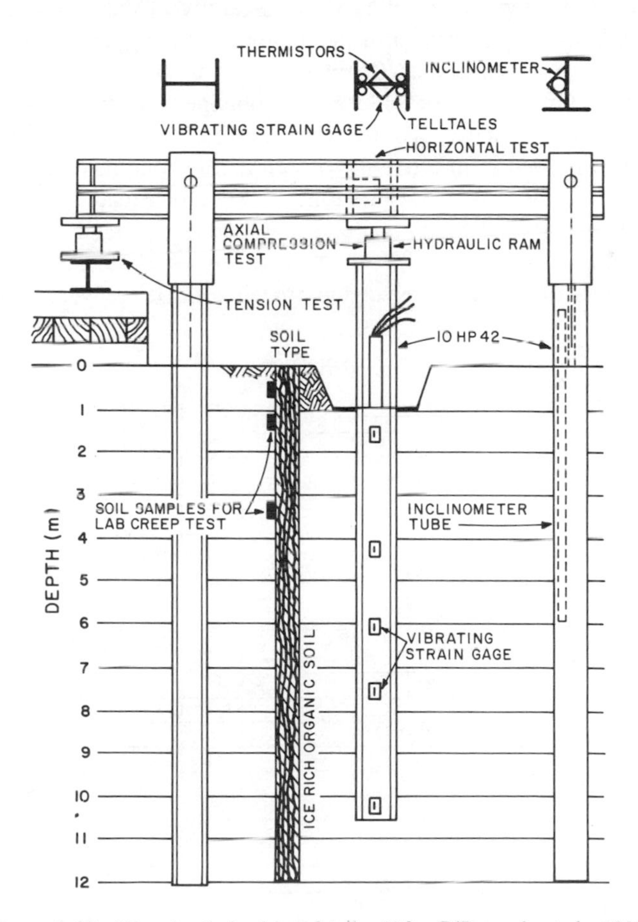


**Figure 6–20** Kipnuk pile load test details. (After DiPasquale et al., 1983.)

long-term lateral loads are generally governed by creep of the frozen soil supporting the pile section. If the sustained loadings exceed the lateral capacities, it will be necessary to install a group of battered piles to withstand the lateral capacity. Drilling and backfilling or driving battered piles in frozen ground will be more difficult and expensive than for equivalent vertical piles. It may be possible to drive piles at batter angles of 5 to 20° to the vertical. Although increasing the batter angle increases the lateral resistance of the pile group, it also increases the problem of slurring the annulus between the pile and the drilled hole if slurried pile is used. This aspect should be borne in mind when selecting the batter angle.

## 6.8 PILE LOAD TESTS

Few pile load tests have been conducted in frozen ground, especially ice-rich warm permafrost. Factors contributing to this scarcity of data (and to the usefulness of such data) include costs, instrumentation problems, and lack of a recommended protocol. As discussed in Chapter 3, frozen soil tends to creep under a fraction of the ultimate load or failure load obtained under short-term loading tests. Pile load tests are necessary to verify design parameters used on specific jobs and are also needed to improve the design methods in general. A typical pile loading test arrangement is shown in Fig. 6–20. DiPasquale et al. (1983) and Crory (1966) have published the details of pile load test methods to determine the adfreeze strength.

Two aspects of pile load testing in frozen ground are very important: (1)

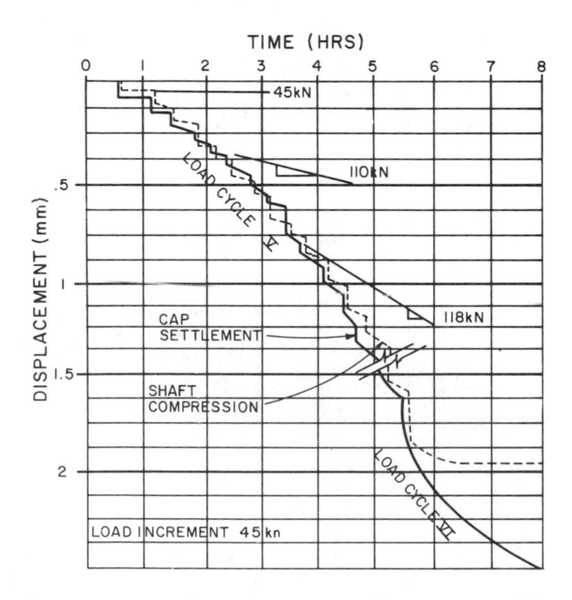


**Figure 6–21** Load settlement with time, Kipnuk pile load test.

load application and (2) instrumentation of the tested pile. It is preferable that tests on piles in frozen ground be carried out using sustained dead loads left in position for as long a period as possible. Loading may be increased after the creep rate is established before the next load increment is applied. The test pile should be instrumented with vibrating strain gauges and thermistors. As shown in Fig. 6–20, sensitive dial gauges (reading up to 0.001 in.) may be used to monitor the pile-head movements. The test facility should be covered and protected from external disturbances, including sunlight, precipitation, and frost-heave forces.

A typical load–settlement curve obtained from pile loading tests on driven H piles in ice-rich frozen soils is presented in Fig. 6–21. Figure 6–22 illustrates the stress redistribution near pile–soil elements during various time periods after the application of the load.

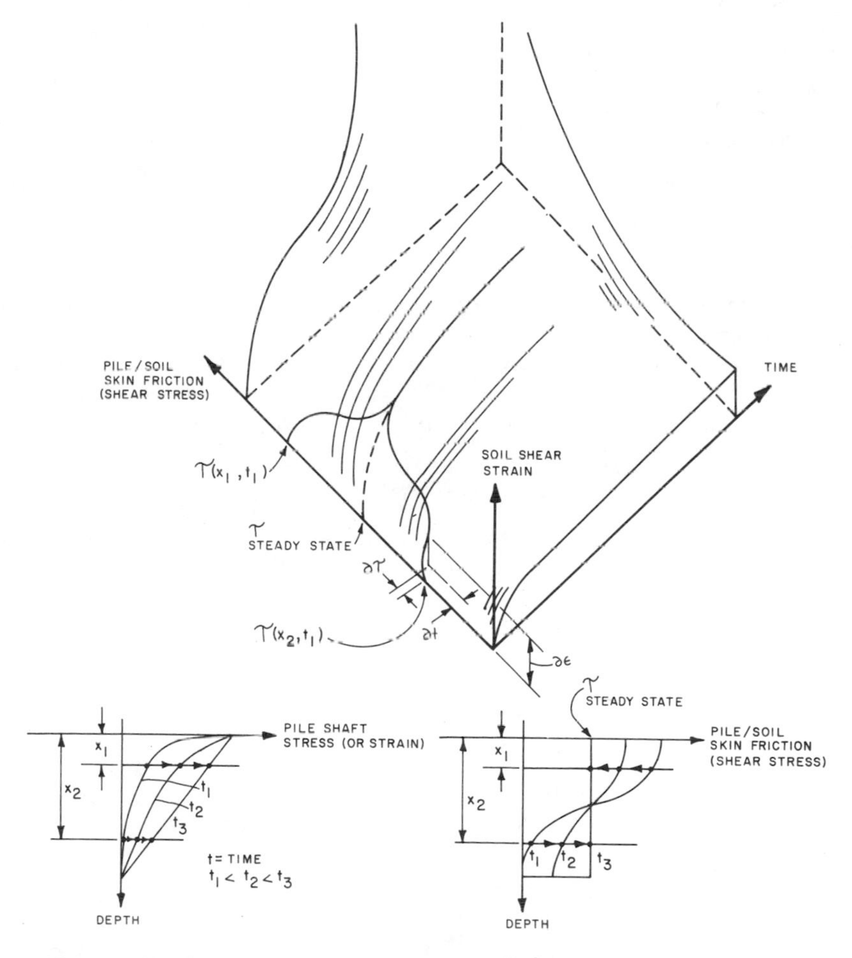


**Figure 6–22** Stress redistribution of near pile–soil elements (constant load, constant soil temperature).

## PROBLEMS

**Problems**

**6–1.** What is the most preferable method of pile installation in continuous and discontinuous permafrost? Discuss the problem considering a wide range of permafrost soils.

**6–2.** Estimate the time required to freeze back a 0.305-m-diameter steel pile placed in a 0.457-m hole preaugered in frozen ground and backfilled with a slurry for the following condition:

Frozen soils: silty sand

- Initial temperature = −5°C
- Dry unit weight = 1450 $kg/m^3$
- Water content = 35%

Slurry backfill: sand–water

- Placement temperature = 4°C
- Dry unit weight = 1600 $kg/m^3$
- Water content = 15%

**6–3.** Redo Problem 6–2 with a frozen soil temperature of −3°C and a slurry temperature of 3°C.

**6–4.** Calculate the length of time required to freeze back a cluster of 50 piles. The average volume of slurry backfill for a group of piles is 0.8 $m^3$ each. The slurry is placed at an average temperature of 4°C and must be frozen to −2°C. A sand–water slurry of 1600 $kg/m^3$ dry weight and 12% water content is used as backfill material and an available refrigeration unit is capable of receiving 45 kJ/s.

**6–5.** Determine the minimum allowable pile spacing in Problem 6–4 so that the permafrost temperature will not rise above −1°C.

**6–6.** A 0.3-m-diameter steel pile is placed in an ice-rich sandy silt deposit which is frozen well below the depth explored by drilling. The hole is 0.46 m in diameter and the annular space between the pile and the hole was slurried back with sands at a temperature of 4°C. The following information is given for the site:

- Freezing index = 5000 degrcc-days
- Thawing index = 2500 degree-days
- Length of freezing season = 200 days
- Mean annual air temperature = −3°C
- Structural load = 250 kN/pile

| Sand Silt | Sand Slurry |
|---|---|
| $\rho_d$ = 1360 $kg/m^3$ | $\rho_d$ = 1700 $kg/m^3$ |
| $\omega$ = 40% | $\omega$ = 15% |

- Average warmest temperature of permafrost = −2°C

- Slurry must freeze back to $-2°C$
- Uplifting force due to frost heave = 10 tons/ft of perimeter of pile

Determine:

**(a)** The total length of pile required to withstand the structural load.
**(b)** The freeze-back time for the slurry to the required temperature.
**(c)** State design measures to eliminate frost jacking forces.

**6–7.** An 0.2-m-diameter timber pile is placed in an ice-rich silt deposit which is frozen well below the depth explored by drilling. Assume that the pile was placed to a depth of 10 in. in a hole of 0.36 m diameter and that the annular space between the pile and the hole was slurried back with sands at a temperature of 4°C. The following information is given for the site:

- Thawing index = 2500 degree-days
- Length of thawing season = 150 days

| Sand Silt | Sand Slurry |
|---|---|
| $\rho_d$ = 1450 kg/m$^3$ | $\rho_d$ = 1600 kg/m$^3$ |
| $\omega$ = 40% | $\omega$ = 15% |

- Average warmest temperature of permafrost = $-1°C$
- Slurry must freeze back to $-1°C$
- Uplifting force due to frost heave = 15 tons/ft of perimeter

Determine:

**(a)** The maximum load capacity of the pile.
**(b)** The freeze-back time to the required temperature for the slurry.

**6–8.** Estimate the pile velocity in Problem 6–7.

**6–9.** Estimate the pile velocity in Problem 6–7 considering only the creep parameter of ice.

**6–10.** A steel pipe pile of 0.305 m is loaded with a 500-kN structural load. Determine the required effective embedment length of pile for both summer and winter conditions. The soil may be assumed to be permafrost with the following characteristics: sand and gravel; dry density = 1700 kg/m$^3$; water content = 10%. The highest-temperature profile is measured as follows:

| Depth (m) | 0 | 0.15 | 0.3 | 1 | 3 | 8 | 13 | 20 |
|---|---|---|---|---|---|---|---|---|
| Temperature (°C) | 4 | 3 | 2 | 0 | −1 | −2 | −3 | −4 |

Assume that the uplift force due to the frost heave is 150 kN/m$^2$.

**6–11.** Discuss the different design measures to eliminate frost jacking forces on piles.

**6–12.** An H-pile (10 HP42) is driven to a depth of 10 m in a warm permafrost (silty soils) of average temperature $-1°C$. Calculate the pile capacity in both summer and winter conditions. The thickness of the active layer is 1 m and the uplift force is 100 kN/m$^2$.

CHAPTER 7

# Roadway and Airfield Design in Frozen Ground

## 7.1 INTRODUCTION

Roadways and airfields have been built on frozen ground to serve the transportation needs of populations in northern regions of the world. The demand in the future for expansion in the North will increase as the demand for energy and resource development projects increase and become economically feasible.

The performance of the existing roadways on permafrost has been unsatisfactory; and with the poor performance, maintenance costs have been high. This poor performance is due primarily to various factors having to do with inadequate understanding of how to design for frost action and thaw settlement. During the freezing and thawing cycles, the pavement serviceability is seriously affected by various modes of pavement distress unless proper long-term design and construction techniques are used. The requirement for pavement serviceability throughout the year in permafrost regions imposes special consideration of roadway design and maintenance. Modes of distress in pavements are presented in Table 7–1. Figures 7–1 to 7–3 illustrate some of the problems faced by civil engineers.

Design considerations for roadways and airfields in frozen ground are very similar. The essential differences between the two types of pavements, flexible and rigid, are well known and their design concepts are well developed (Transportation Research Board, 1974; Yoder and Witcvzak, 1975; Federal Aviation Administration, 1978). This chapter is concerned primarily with the design of roadway and airfield sections consisting of base, subbase layers, and the fill materials below the surface course of pavement. The ''roadway'' term refers to the materials below the road surface, as indicated in Fig. 7–4. Similarly, the airfield

**TABLE 7–1 Modes of Distress in Pavement**

| Distress Mode | General Cause | Specific Cause |
|---|---|---|
| Cracking | Traffic load | Repeated loading (fatigue)<br>Slippage (resulting from braking stresses) |
| | Other | Thermal changes<br>Moisture changes<br>Shrinkage of underlying materials |
| Distortion (may also lead to cracking) | Traffic load | Rutting or pumping and faulting (from repetitive loading)<br>Plastic flow or creep |
| | Other | Differential heave<br>Frost action in subgrades or bases<br><br>Swelling of clays in subgrade<br>Differential settlement<br>Permanent from long-term effects; transient from reconsolidation after heave (may be accelerated by traffic)<br>Curling of rigid slabs from moisture and temperature differential |
| Disintegration | May be advanced stage of cracking mode of distress or may result from differential effect of certain materials contained in the layered system or from abrasion by traffic; may also be triggered by freeze–thaw effects | |

*Source:* After Transportation Research Board (1974).

**Figure 7–1** Alligator cracking and surface distress on pavement surface. (Courtesy of B. Connor.)

**Figure 7–2** Culvert icing. (Courtesy of D. Esch.)

**Figure 7–3** Surface erosion resulting from roadways culvert, Longyearbyen, Norway.

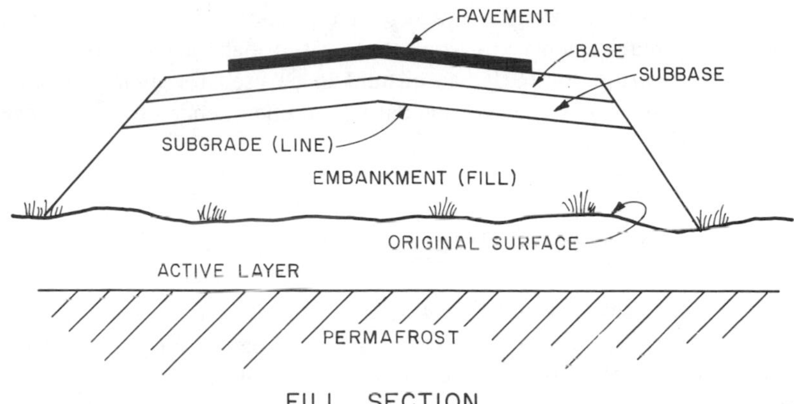


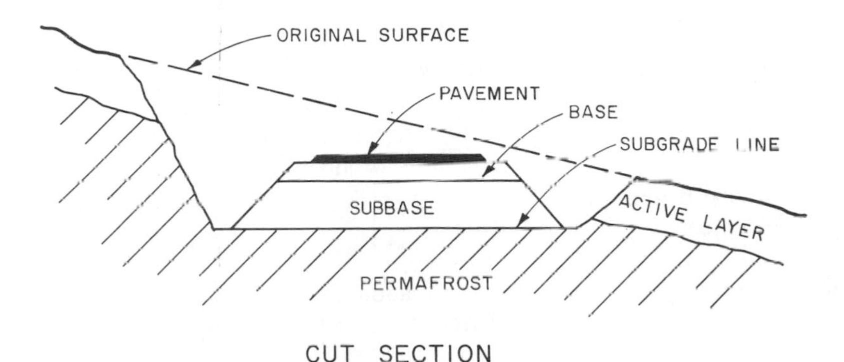


**Figure 7–4** Roadway sections.

term refers to the embankment section, as shown in Fig. 7–4. As flexible pavements are the most common in northern environments, Chapter 7 will address the design of roadways and airfields under a flexible pavement. Particular emphasis will be placed on designing the supporting fill thickness against the potential detrimental effects of frost action and to preserve the thermal equilibrium of frozen ground.

The design of roadway and airfield sections should consider the economics of alternative construction techniques, the use of existing native soils, the nature of foundation soils, drainage, thermal regime, topography, construction scheduling, and environmental impact. Construction methods may be geared toward minimal thermal degradation of existing permafrost conditions or toward removal of the seasonal frozen layer or thawing of permafrost. Design and construction techniques must also satisfy the level of service to be provided. For example, a lower standard may be used for an access road for exploration works compared to highways where the requirements to carry heavy traffic loads and volumes dictate a higher design standard. Similarly, an airfield to be used for light planes or low use rates may be built to a different standard from one intended for heavier planes or more frequent use.

## 7.2 ROADWAY DESIGN AND CONSTRUCTION

The general influences of environmental and physical factors on the design and construction of roadways on frozen ground are described in this section. The fundamental factors of permafrost and climatic factors, discussed in Chapter 1, are essential to determine the appropriate roadway fill thickness. In addition, material characterization and exploration approaches play an important role in the design and construction of roadways. Roadway terminology to be used is detailed in the glossary of terms (AASHTO, 1972) presented in Appendix A.

### 7.2.1 Characteristics of Materials

Various standard tests are performed to determine the properties and characteristics of roadway materials. These tests are categorized into the following groups for unfrozen and frozen soils.

Unfrozen soils:

1. Routine tests for strength and deformation
   (a) California bearing ratio (CBR)
   (b) Stabilometer (*R* value)
   (c) Triaxial
   (d) Pile loading
   (e) Indirect tensile test
2. Sophisticated tests for theoretical solutions
   (a) Resilient modulus
   (b) Dynamic modulus
3. Long-term tests to determine material distress conditions
   (a) Fatigue
   (b) Heave rate

Frozen soils:

1. Routine tests for strength and deformation
   (a) Unconfined compression
   (b) Triaxial
   (c) Indirect tensile

(d) Frost heave
(e) Freeze–thaw cycles

2. Sophisticated test for theoretical and long-term conditions
   (a) Creep test
   (b) Resilient modulus
   (c) Dynamic modulus
   (d) Thaw consolidation

Common standard tests that are used to determine physical and mechanical properties of soils include:

1. Grain size distribution by sieve and hydrometer tests
2. Specific gravity
3. Water content
4. Optimum moisture content
5. Shear strength
6. Permeability
7. Consolidation (fine-grained subgrade soils only)

Standard procedures (such as AASHTO) are available to perform these tests and should be followed to obtain a variety of data that will be useful in the design and construction of roadways.

### 7.2.2 Stress Distribution

The determination of stress distribution due to various traffic loading conditions is essential for design of roadway sections. A widely accepted standard vehicle type is the 18-kip single axle (generally, equivalent to a 32-kip dual tandem). Thus the effects of other vehicles are normally accounted for in the design by the equivalent 18,000-lb single-axle load (EAL).

Theoretical solutions (one-layer, two-layer, and three-layer systems) are available to determine the stress distribution at various design depths. Boussinesq's formula is commonly used to obtain the vehicle stress ($\sigma_z$) under a point load as follows:

$$\sigma_Z = \frac{P}{Z^2} \frac{3}{2\pi} \frac{1}{[1 + (r/z)^2]^{5/2}} \tag{7–1}$$

where $\sigma_Z$ = depth
$r$ = radial distance from point load ($P$)

Boussinesq's formula may also be used to calculate vertical stress under the distributed load.

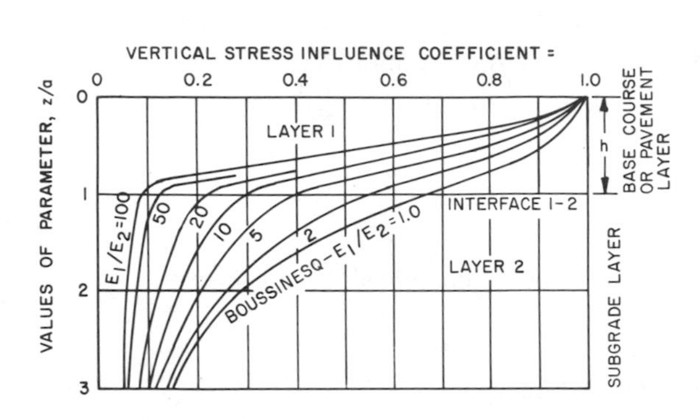


Figure 7–5 Vertical stress distribution: two-layered solution. (After Burmister, 1943.)

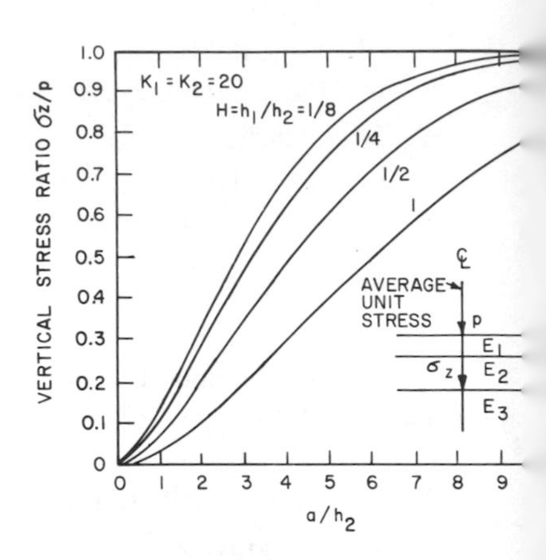


Figure 7–6 Vertical stress distribution: three-layered solution.

Other methods, such as the two-layered and three-layered theories given by Burmister (1945, 1958) and Acum and Fox (1951), are often used to calculate stress and displacements in a layered soil profile. Burmister's two-layered solution under the center of a circular plate is given in Fig. 7–5.

Figure 7–6 illustrates the influence of changing the pavement layer thicknesses based on the vertical compressive stress factor for a three-layered pavement system. It can be seen from the figure that the subgrade stress decreases greatly with a decrease in the ratio of the $a/h_2$ parameter. In the three-layered system, the base-course modulus ($E_2$) has a pronounced effect on stress reduction, while the pavement layer modulus ($E_1$) controls the subgrade stress for two-layered systems.

### 7.2.3 Site and Route Exploration

Site and route exploration are required for the final design and construction of roadways. Reconnaissance surveys, detailed site and route studies, environmental assessments, and the availability of construction materials must be taken into account when analyzing site and route selection. Lack of information on permafrost conditions and terrain factors may lead to unsatisfactory performance of roadways on permafrost. Maintenance costs can also be anticipated to be excessive. The scope of site and route exploration depends on the amount of available information, geographical location, and the type of roadway proposed. The site and route information required is presented in Table 7–2.

**TABLE 7–2 Site and Route Information**

| | |
|---|---|
| A. Terrain factors | |
| (1) Relief terrain | (a) Topography |
| | (b) Slope, including degree and orientation |
| | (c) Surficial information (vegetation and features) |
| (2) Geology and soils | (a) Surface and subsurface geology |
| | (b) Temperature profile |
| | (c) Ground ice conditions |
| | (d) Engineering properties of soils and rocks |
| | (e) Borrow areas |
| (3) Hydrology | (a) Surface and subsurface |
| | (b) Groundwater |
| | (c) Flooding |
| | (d) Icing |
| | (e) Ice jams |
| | (f) Snow melts |
| B. Climate | |
| (1) Temperature | |
| (2) Precipitation (rain and snow) | |
| (3) Wind | |
| C. Environmental factors | |
| (1) Water pollution | |
| (2) Land abuse (specifically, wetlands) | |
| (3) Wildlife (animal, fish, birds) | |
| D. Logistics | |
| (1) Winter and summer construction | |
| (2) Access to and from the site | |
| (3) Local materials and equipment | |
| (4) Transportation facilities | |

## 7.2.4 Frost Action

Frost action is one of the major factors affecting roadway performance. The term *frost action* includes both frost heave and loss of subgrade support during the thawing period. The problem of frost-action damage can be widespread. Two destructive effects pertinent to the frost action of soils within or beneath a roadway section are the expansion and lifting of the soils (frost heaving) in winter and the loss of soil bearing strength by thaw weakening in the spring. Soils that show one or both of these characteristics are referred to as being frost susceptible.

### *7.2.4.1 Frost-Heaving Process*

As discussed in Chapter 4, three conditions must exist simultaneously for frost heaving to occur in soils. These conditions are a soil-moisture supply, a sufficiently cold temperature to cause soil freezing, and a frost-susceptible soil.

The heaving of frost-susceptible soil can be attributed to two processes: freezing of in situ pore water during frost penetration (ice-lens formation), or ice segregation, which occurs in soils due to migration of soil moisture toward the freezing plane. Thus heaving can occur mainly with the growth of ice lenses in frost-susceptible soils. The freezing of the pore-water process contributes only a 9% volume change in the soil-water volume unless pore-water expulsion occurs. Ice lenses usually form parallel to the isothermal freezing plane, and heaving is always in the direction of heat flow. Heave may be uniform or nonuniform, depending on variations in the characteristics of the soils and the groundwater conditions.

Frost heave in roads is indicated by the raising of the pavement. When a nonuniform heave occurs in pavement, there are appreciable differences in the heave of adjacent areas, resulting in objectionable unevenness or abrupt changes in grade along the pavement surface. Conditions that contribute to irregular heave generally occur in areas where subgrades vary between clean non-frost-susceptible soils and silty frost-susceptible soils or in places where abrupt transitions between two different subgrade materials occur. Nonuniform heaves also occur in places where abrupt transitions from cut to fill sections are made with the groundwater table close to the surface and where excavation cuts are made into water-bearing strata.

Uniform heaving also occurs when soil and moisture conditions conducive to ice segregation exist under a pavement, but are uniform in longitudinal and traverse directions. A uniform heave that raises the pavement uniformly may not be noticeable to motorists even though the vertical displacement may amount to several inches. Conditions where uniform heave has taken place may not have an effect on the serviceability of the pavement as long as the frozen and heaved condition lasts. The undesirable effects of heaving, whether uniform or nonuniform, may become noticeable during the spring when the thaw weakening and release of water into the base course may increase the rate of deterioration and thus affect the pavement's performance.

#### *7.2.4.2 Thaw Weakening*

Soil weakening from the action of thawing is particularly evident early in the spring when thawing is occurring at the top of the subgrade and the rate of melting is rapid compared to the rate of drainage. Melting of the ice from the surface downward releases water that cannot drain through the still-frozen soil below, nor redistribute itself readily. As a result, the base course becomes completely saturated, resulting in the reduction of the bearing capacity of the base.

Attention has been drawn to the detrimental effect of high traffic density and load during the thawing period. Such conditions may cause excessive pore pressures and greatly reduce the load-carrying capacity of the pavement. Ice segregation during freezing may not always be a necessary precursor for the thaw weakening of soils. The supporting capacity in a clayey soil subgrade may be reduced even though significant ice lenses or heave have not occurred. Reduced subgrade soil strength may result with freeze–thaw shrinkage and remolding processes. The reduction of soil strength during the frost melting periods and the length of time during which the strength of soils is reduced depends on the type of soil; the temperature conditions during the freezing and thawing periods; the amount and type of traffic during frost-melting periods; the moisture supply during fall, winter, and spring; and existing drainage conditions.

#### *7.2.4.3 General Soil and Frost-Susceptibility Classification*

For design and construction purposes, the use of the soil classification systems are necessary. As discussed in Chapter 2, the USCS has been extended for use with frozen soils. The frozen soil classification is useful for the design of roadways in permafrost regions.

The U.S. Army Corps of Enginners (1965) has developed criteria to delineate the degree of frost susceptibility of various soils and these criteria are the most commonly used in North America as a basis for frost design (Table 2–6). In this system frost-susceptible soils (with 3% or more, by weight, finer than 0.02 mm) are classified into one of the four groups F1, F2, F3, and F4. Soil types are listed in Table 2–6 in approximate order of increasing susceptibility to frost heaving and/or weakening as a result of frost melting, although the order of subgroups under F3 and F4 does not necessarily indicate the order of susceptibility to frost heaving of these subgroups. The basis for distinguishing between the F1 and F2 groups is that F1 material may be expected to show higher bearing capacity than F2 material during thaw, even though both may experience equal ice segregation.

## 7.3 DESIGN METHODS

The ridged and flexible methods of pavement design in permafrost regions have not been found to differ significantly in seasonal frost areas. In fact, the structural zone of the roadway embankment that must support the imposed vehicle loadings and prevent fatigue cracking of the pavement surface exists essentially in the seasonal frost or active-layer region of the roadway embankment. For this reason, flexural design methods for seasonal frost can be applied directly to permafrost regions. Further discussion of pavement flexural design methods will not be attempted herein. The current Alaska Department of Transportation (1982) pavement design method for roadway pavement may be referred to for the flexural design methods.

Roadway design in seasonal frost as well as in permafrost areas depends on the type of subgrade soil as well as the fill materials used for the roadway section. The main concerns are the limitation of surface deformation resulting from seasonal thaw-related settlement, the prevention of long-term thawing of the roadbed soils, and the adequacy of the bearing capacity during the seasonal thawing period. The first factor is more severe in seasonal frost areas, whereas the second factor is more predominant in permafrost regions. Additional consideration must be given to the influence of construction on the existing ground thermal balance. Changes in the ground thermal regime may cause degradation of the permafrost, resulting in total differential settlement and reduction in the bearing capacity of the pavement structure.

### 7.3.1 Roadway Design in Seasonal-Permafrost Areas

Design of roadways in areas of seasonal permafrost is based on two alternative concepts:

1. Control of roadway surface deformation resulting from frost action or seasonal thaw–freeze cycles
2. Maintenance of roadway bearing capacity during the adverse climate period

The first concept requires that the combined thickness of the pavement and the non-frost-susceptible base must contain the frost or thaw penetration. This approach can be labeled the *full-protection method*. The second concept anticipates subgrade frost penetration and the subsequent reduced strength of the subgrade during the melting period. Using these concepts, the U.S. Army Corps of Engineers has developed the following design methods:

1. Complete or full protection
2. Limited subgrade frost protection
3. Reduced subgrade strength

#### *7.3.1.1 Complete Protection*

The complete protection method will limit seasonal thaw or frost protection to the NFS base and subbase courses. This protection prevents the underlying frost-susceptible soils from freezing in seasonal frost areas or the frozen soil from thawing in permafrost areas. The required thickness of the base course can be computed by either of the methods outlined in Section 7.3.2. The lower 100 mm of the base must be designed as a filter. Use of high-moisture-retaining non-frost-susceptible soils, such as uniform sands, in permafrost areas decreases the required base-course thickness. The higher latent heat of these soils gives them higher resistance to thaw penetration. Use of insulation beneath the base course to prevent subgrade freezing or thawing is described in Section 7.6.1.

#### *7.3.1.2 Limited Subgrade Frost Protection*

The limited-subgrade-frost-penetration method attempts to confine deformations to small acceptable values by use of a combined thickness of pavement and non-frost-susceptible base and subbase courses, which limits the subgrade frost penetration. Added thickness may be needed in some cases to keep pavement heave and cracking within tolerable limits based on local experience and field data. Linell et al. (1963) recommended this method in seasonal frost areas for the following soils: (1) all silts, (2) very fine silty sands containing more than 15% fines less than 0.02 mm by weight, (3) clays with plasticity indexes of less than 12, and (4) varied clays and other fine-grained banded sediments. Exceptions include extremely variable subgrade conditions for which the complete-protection method must be used or when some nonuniform heave and cracking are not considered detrimental for flexible paved areas.

This design method uses an average air-freezing index for the three coldest years in a 30-year period (or for the coldest winter in 10 years of record) to determine the combined thickness of pavement and base required to limit subgrade frost penetration. To begin, determine the base thickness required for zero frost penetration by subtraction. The ratio $r$ is equal to the water content of the subgrade divided by the water content of the base. Enter Fig. 7–7 with the base thickness for zero frost penetration into the subgrade and, using the $r$ value, read the design base thickness on the left-hand scale and the allowable frost penetration on the right-hand scale. The $r$ value is limited to a maximum of 2, because part of the moisture in fine-grained soils remains unfrozen. The bottom 100 mm of the base should consist of non-frost-susceptible sand, gravelly sand, or similar material designed as a filter. When the combined design thickness of pavement and base exceeds 1.8 m, Linell et al. (1963) recommended consideration of the following alternatives: (1) limit the total combined thickness to 1.8 m, or (2) reduce the required combined thickness by use of a base of non-frost-susceptible uniform fine sand with high moisture retention in the drained condition in lieu of more free-draining material.

**Example 7–1**

Given:

- Frost penetration $c = 2$ m
- Water content of subgrade soils = 70%
- Water content of base material = 5%

Required: base thickness of the pavement.

### Solution

We have, $r = 0.70/0.50 = 1.4$. From Fig. 7–7, for $c = 2$ m and $r = 1.4$ we get: base thickness = 1.5 m.

#### *7.3.1.3 Reduced Subgrade Frost Penetration*

The reduced-subgrade-strength method is based on the anticipated reduced subgrade strength during the frost-melting period. The amount of heave is neglected with this method. Linell et al. (1963) and Hennion and Lobacz (1973)

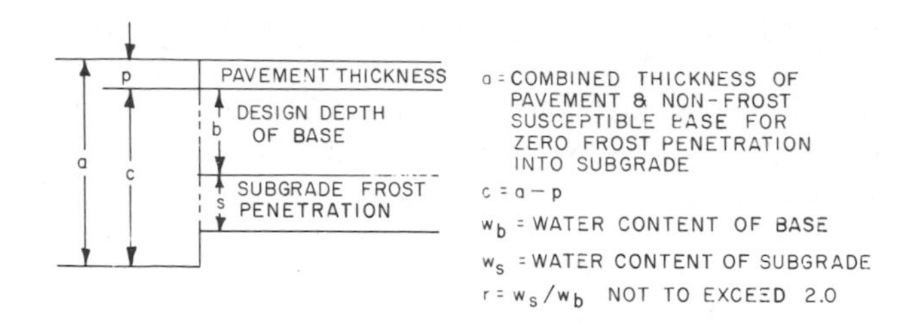


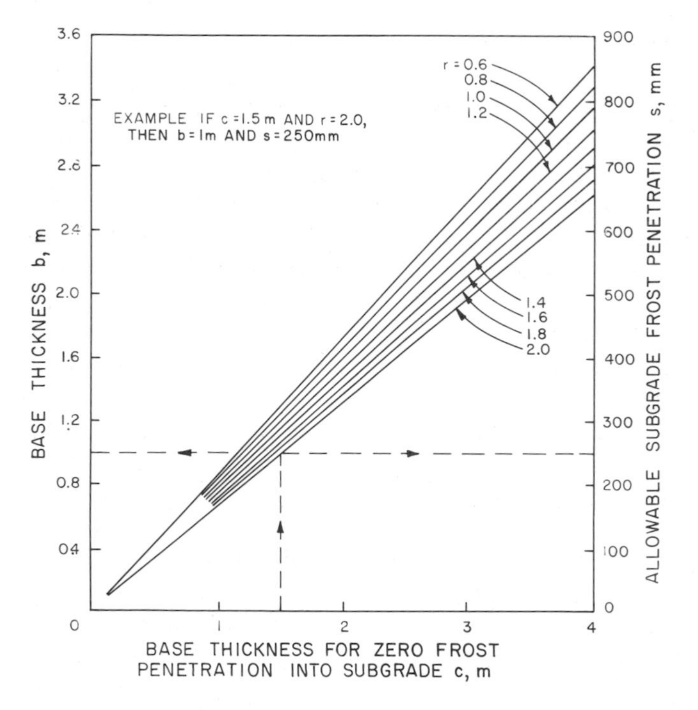


**Figure 7–7** Base thickness versus frost penetration. (After Lobacz et al., 1973.)

recommended this method for both seasonal frost and permafrost areas. This method may be used for flexible pavements on soil groups F1, F2, and F3 when subgrade conditions are sufficiently uniform to assure that objectionable differential heaving or subsidence will not occur. In certain cases nonuniform subgrade conditions are correctable by removal and replacement of pockets of more highly frost-susceptible or high-ice-content soils. For permafrost areas, an estimate should be made of the magnitude and probable unevenness that will result from future subsidence as thaw occurs in the existing frozen soil. This estimate should be based on a study of the area and on local experience. In some areas, living with long-term maintenance requirements may be more economical than providing adequate initial fill in areas where suitable subbase fill is limited or not available except under costly transportation conditions. A design that accepts a limited degree of frost heave or subsidence may be considered. This approach is referred to as *controlled subsidence*. For paved areas, base-course thickness may be increased as needed over the reduced-strength-design requirements to provide a surcharge load sufficient to reduce differential heave resulting from seasonal freezing of a frost-susceptible subgrade.

The design curves shown in Fig. 7–7 are used to determine the combined thickness of the flexible pavement and non-frost-susceptible base. The design index represents all traffic expected to use the pavement during its service life. The index is based on typical magnitudes and composition of traffic reduced to equivalents in terms of repetition of an 80-kN single-axle dual-tire load (U.S. Army, 1962). For the design index and soil group representative of the site, Fig. 7–7 gives the combined thickness of the pavement and base required by the reduced-subgrade-strength method. The combined thickness of the pavement and base should in no case be less than 230 mm where frost action is a consideration. Again, the lower 100 mm of the base should be graded to provide filter action against the subgrade.

Table 7–3 presents the properties of base and subgrade materials significant

**TABLE 7–3 Characterization of Material Properties for Flexible Pavements in Seasonal Frost Areas**

| Properties | Important for (Area of Design): | Dependent on Temperature, Frost, or Moisture? |
|---|---|---|
| Strength | | |
| Tensile (bound layers only) | Fracture under large load; slippage under braking load; thermal cracking | Temperature |
| Fatigue (bound layers only) | Fracture under repeated loading | Temperature |
| Shear | Plastic flow or shear under single or few excessive loads | Frost, moisture, temperature |
| Resistance to deformation—modulus (stress–strain) | Analysis of stresses and strains for application to fracture and distortion mode of distress | Asphalt-bound: temperature<br>Clean granular unbound base; moisture<br>Clean granular cohesionless soil: none<br>Silty and clayey soils: frost, moisture, temperature |
| Constancy of volume susceptible to frost heave | Distortion by differential heave | Frost, moisture |
| Expansive | Distortion by differential heave | Moisture, frost |
| Underconsolidated | Distortion by differential settlement | Moisture, frost (consolidation during thaw) |

to the performance of flexible pavements in seasonal frost areas. The key properties include shear strength, strain modulus, Poisson's ratio, and volume-change parameters.

## 7.3.2 Roadway Design under Permafrost Areas

In permafrost areas, roadway design must consider not only the effects of seasonal thawing and freezing cycles, but also the effects of construction on the existing thermal balance. In continuous-permafrost regions with high-ice-content frozen soils at shallow depths, the appropriate approach will be to restrict seasonal thawing to the pavement and non-frost-susceptible base course, thereby keeping the permafrost "frozen" as in the original undisturbed state. This method of preventing degradation of the permafrost is comparable to the complete-protection method used in seasonal frost areas; however, the critical factor here is depth of thaw penetration rather than depth of frost penetration.

The limited-subgrade-frost-penetration method discussed in the preceding section is impractical for most permafrost areas because of high freezing-index values, requiring thicknesses in excess of practical and economical limitations. Therefore, design, except in areas of continuous permafrost at shallow depth, is usually based on the reduced-subgrade-strength method and consideration of the effect of construction on the existing thermal regime. The limited-subgrade-frost-penetration method may be useful in areas of discontinuous permafrost with horizontal variables, highly frost-susceptible subgrade soils, and variable moisture conditions.

In areas where the soils are NFS or where precipitation and groundwater conditions preclude significant ice segregation, design principles are the same as in temperate zones. Pockets of frost-susceptible soils should be excavated to a depth of 1 to 2 m below the final grade and backfilled with NFS materials.

### *7.3.2.1 Complete-Protection Method in Permafrost Areas*

In continuous-permafrost regions, a design that limits seasonal thaw to a non-frost-susceptible base course keeps the subgrade frozen and prevents frost heaving or damaging subsidence. The required base-course thickness may best be computed using the modified Berggren equation (4–21). Other equations, presented in Chapter 4, may also be used. To use the minimum base-course thickness required to restrict seasonal thaw to the roadway section, the use of relatively high-moisture-retaining NFS soils, such as uniform sands, in the lower base should be considered. After initial freezing, such soils provide a "heat sink" that resists the thaw penetration because of high latent heat.

In areas where some heaving can be tolerated, the use of frost-susceptible soils of groups F1 and F2 in a subbase course may be permitted. The subbase course is treated as the subgrade in determining the upper base-course thickness by using the reduced-subgrade-strength method. An insulating layer may also be used to further limit seasonal thaw penetration.

#### *7.3.2.2 Reduced-Subgraded-Strength Method in Permafrost Areas*

In discontinuous-permafrost regions the roadway sections required to prevent seasonal thawing and freezing effects in the subgrade are generally greater than 1.8 m (except for extreme cold areas). The design must usually be based on the assumption that thawing and freezing will occur in the subgrade.

This method may be used for the design of flexible pavements on F1, F2, F3, and F4 subgrade soils and when subgrade conditions are significantly uniform to preclude objectionable differential heaving or subsidence (Fig. 7–8). This method may also be used for design of flexible pavements used for minor or slow-speed traffic over all subgrades when appreciable nonuniform heave on subsistence can be tolerated.

Thickness requirements for the reduced-subgrade-strength method provide adequate carrying capacity during the period of thaw weakening but may result in objectionable surface roughness and cracking due to heaving or subsidence. In such cases, design studies should include the frost heaving and settlement experience records from existing roadway pavements in the vicinity.

Design curves presented for seasonal frost areas may also be used for discontinuous permafrost areas to determine the thickness of roadways. A typical AASHTO recommended cross section for a two-lane highway is shown in Fig. 7–9.

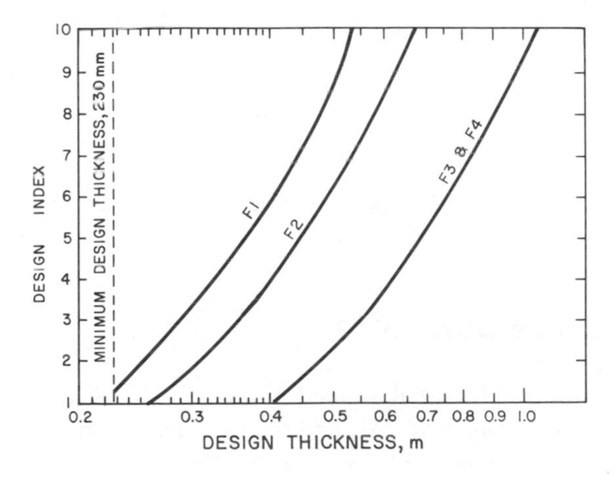


**Figure 7–8** Design index versus design thickness. (After Lobacz et al., 1973.)

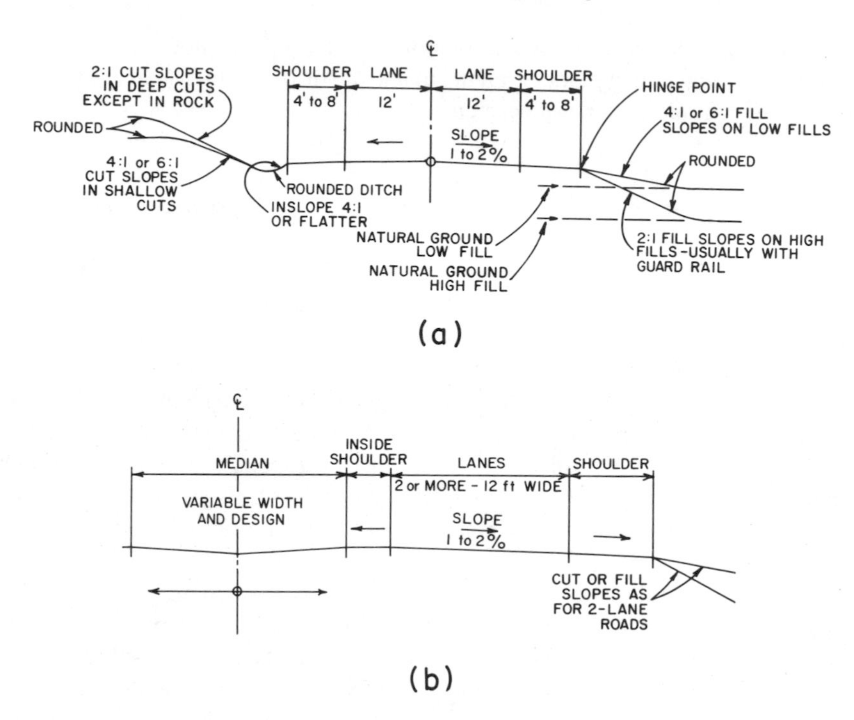


**Figure 7–9** (a) AASHTO-recommended cross section for two-lane highway; (b) AASHTO-recommended cross section for multilane highways and freeways.

**Example 7–2**

Given:

- Traffic index (T.I.) = 9
- Frost susceptibility of subgrade
- Soil classification = F3

Required: roadway thickness based on reduced subgrade strength.

### Solution

Use Fig. 7–8. Enter T.I. = 9 and F3. We get a total design roadway thickness of 0.95 m.

## 7.4 NEW DESIGN CONCEPTS

The latest design principles are described in this section. Based on field and laboratory studies of roadway materials, and the measurement of various data on the performance of newly built roadway sections, these new concepts may be improved. Construction design details may also incorporate these design principles, as appropriate.

### 7.4.1 Excess Fines Concept

Recently, the Alaska Department of Transportation and Public Facilities (1982) developed a method called the *excess fines* concept for the design of roadway sections on permafrost regions. The design procedure is based on:

1. The critical maximum fines contents throughout the pavement structure, as determined from an analysis of 120 Alaskan pavement structures. This analysis has demonstrated that under Alaskan conditions, pavement structures that have −No. 200 (fines) content less than maximums shown by Fig. 7–10 at depths from the bottom of the pavement to 48 in. will have maximum Benkleman beam deflection levels of less than 0.034 in. Calculations show that for this condition $D_{av}$ = 0.025 in. with a standard deviation of 0.0056 in. The upper 95th percentile is therefore 0.025 + 1.65(0.0056) = 0.034.
2. The reduction in stress with depth of a typical vehicle axle loading, as determined from Boussinesq's analysis of the stress reduction factor (SRF) beneath an equivalent circular loaded area on a homogeneous elastic solid, as shown by Fig. 7–11.

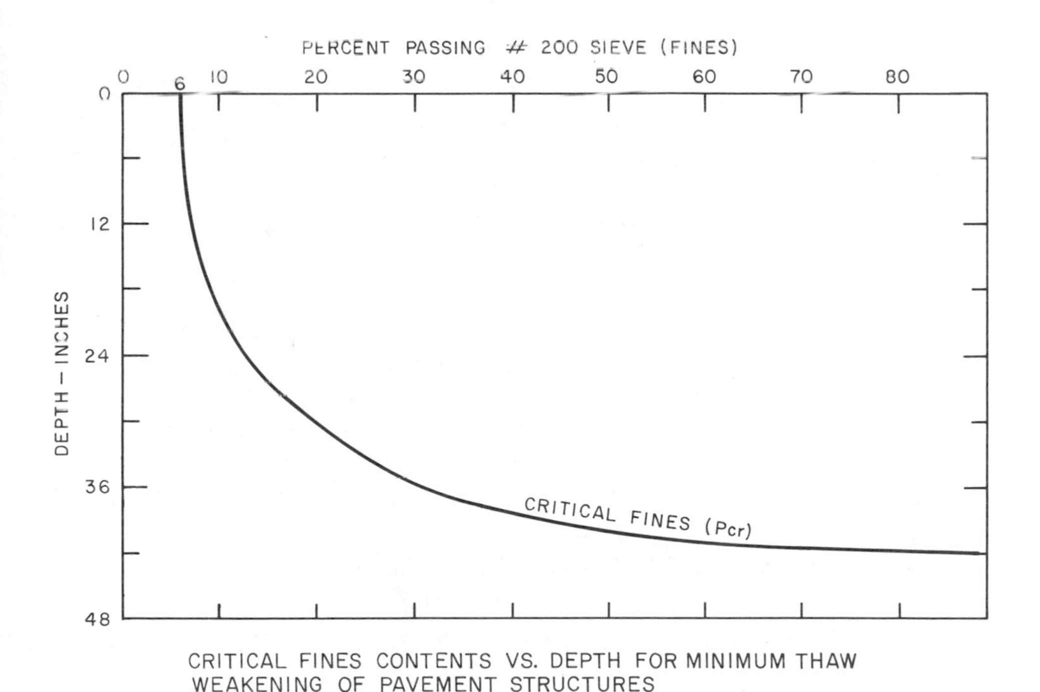


**Figure 7–10** Critical fines relationship. (After Alaska, 1981.)

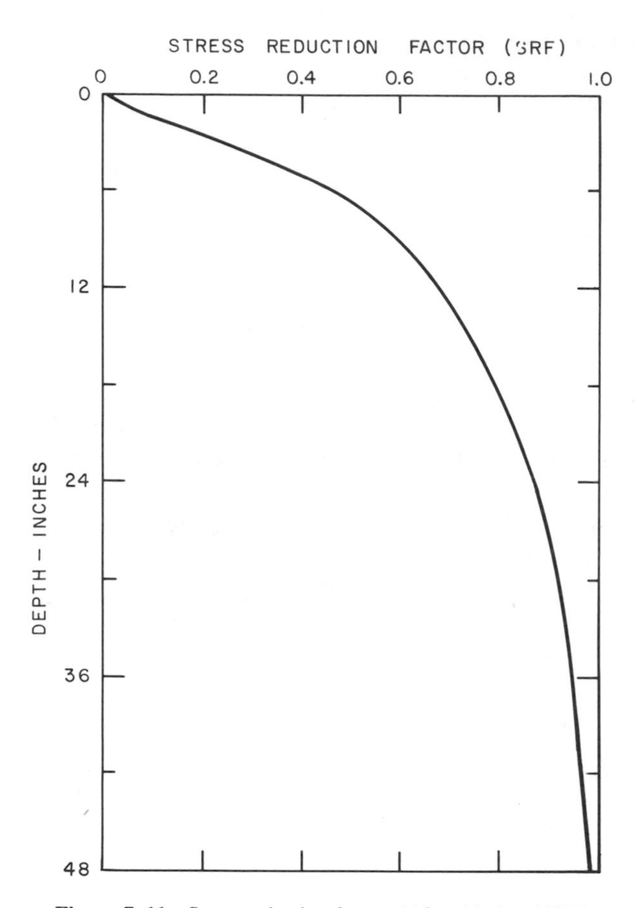


**Figure 7–11** Stress reduction factor. (After Alaska, 1981.)

3. Fines content in excess of those shown by Fig. 7–10 in any layer of the pavement structure are termed *excess fines* and have been found to contribute to increased frost susceptibility and much higher maximum Benkleman beam deflections than for structures without excess fines.
4. The effect of excess fines on deflections for all layers of a given pavement structure can be represented by a single number called the *excess fines factor* (EFF). The average deflection for pavement structures with excess fines has been found to be $D_{av} = 0.031 + 0.0035$ (EFF). The standard error of estimate was 0.015 in. The upper 95th percentile level of deflection is therefore 1.65 (0.015) + 0.031 = 0.056 + 0.0035 (EFF).
5. The contribution of excess fines in a given layer is determined by the following steps:
   (a) Separate the pavement structure to be analyzed into a series of layers of a thickness no greater than 15 cm (6 in. between the bottom of the

pavement) and a depth of 0.3 m (1 ft), and thicknesses no greater than 0.3 m beneath that depth.

(b) For each layer having an average fines ($P_{200}$) content in excess of the maximum value ($P_{cr}$) for the center of that layer from Fig. 7–10, calculate the excess fines: $P_{200} - P_{cr}$.

(c) Calculate the excess fines factor (EFF) for all layers by the following equation:

$$\text{EFF} = \sum_{j=1}^{n} \Delta\text{SRF}_j (P_{200} - P_{cr})_j^{0.8} \tag{7–2}$$

where $\Delta\text{SRF}_j$ = change in stress reduction factor for each layer, from Fig. 7–11
$j$ = layer number
$n$ = total number of layers to 48 in.
$P_{200}$ = average percent of −No. 200 size for each layer
$P_{cr}$ = critical percent of −No. 200 size for the middle of each layer, from Fig. 7–10

6. Calculate the probable maximum deflection level ($D_{max}$) from the following equation, which has been found to represent the upper 95th percentile:

$$D_{max} = 0.056 + 0.0035\,(\text{EFF}) \tag{7–3}$$

7. Calculate the design traffic number over the required pavement life (usually 20 years) from the total thawing season equivalent $18^k$ axle loadings (EAL) by the following equation:

$$\text{T.I.} = 8.87(\text{EAL}_{18k} \times 10^{-6})^{0.119}$$

or

$$\text{T.I.} = (4.93\text{DTN})^{0.119} \text{ for 12-month season (design traffic number = DTN)}$$

or

$$\text{DTN} = (\text{T.I.}/4.93)^{8.4} \tag{7–4}$$

8. Determine the required pavement thickness from Fig. 7–12 by entering the $D_{max}$ calculated in step 6 on the horizontal axis, moving vertically to intersect the appropriate DTN to determine the required asphalt concrete pavement thickness $T_p$. The example is presented in Table 7–4.

**Example 7–3**

Based on the excess fines concept, determine the minimum overlay thickness of a new pavement structure for the following route conditions:

Given:

| Depth (in.) | 0–6 | 6–12 | 12–24 | 24–36 | 36–42 |
|---|---|---|---|---|---|
| Fine content ($P_{200}$%) | 6 | 8 | 12 | 20 | 92 |

Traffic data: T.I. = 7.3.

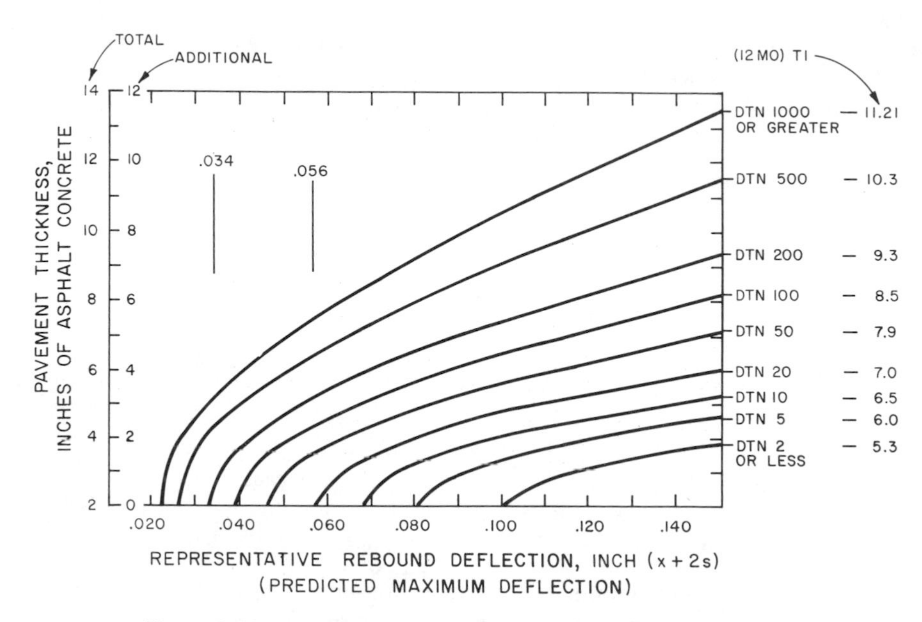


**Figure 7–12** Pavement thickness determination. (After Alaska, 1981.)

## Solution

1. A table (such as Table 7–4) is prepared and the given fines content at different depths are included in column 2, Table 7–4.
2. The critical fines ($P_{cr}$) using Fig. 7–10 are calculated and are presented in column 3, Table 7–4.
3. The excess fines for each layer are calculated using ($P_{200}$ - $P_{cr}$) and they are included in column 4, Table 7–4.
4. Using Fig. 7–11, the stress reduction factors (SRF) are calculated at the top and bottom of each layer and are included in columns 5 and 6 of Table 7–4, respectively.
5. The change in the stress reduction factor ($\Delta$SRF) for each layer is calculated from columns 5 and 6, Table 7–4. Column 7 represents these values.
6. The excess fines factor (EFF) as given by Eq. (7–2) is calculated for each layer. Column 8 includes the calculated values of EFF.
7. Using Eq. (7–3), the predicted maximum deflection $D_{max}$ is given by

$$\begin{aligned} D_{max} &= 0.056 + 0.0035(\text{EFF}) \\ &= 0.056 + 0.0035(1.32) \qquad \text{(EFF from column 8, Table 7–4)} \\ &= 0.0606 \text{ in.} \end{aligned}$$

8. The design traffic number DTN is calculated as

$$\text{DTN} = \left(\frac{\text{T.I.}}{4.93}\right)^{8.4} = \left(\frac{7.3}{4.93}\right)^{8.4} = 27$$

9. The required pavement thickness is determined from the Fig. 7–12 as 3½ in.

**TABLE 7–4**

| Column: Obtained from: | (1) Trial Dimensions | (2) Specifications or Field Data | (3) Fig. 7–10 | (4) (2)–(3) | (5) Fig. 7–11 | (6) Fig. 7–11 | (7) (6)–(5) | (8) (7) × 40.8 |
|---|---|---|---|---|---|---|---|---|
| | | | | | SRF | | | |
| Layer Number | Depth Interval (in.) | Fines Content, *P*200 | Maximum Fines (%) | Excess Fines (%) | At Top of Layer | At Bottom of Layer | ΔSRF | EFF |
| 1 | 0–6 | 6 | 6 | 0 | 0 | 0.45 | 0.45 | 0 |
| 2 | 6–12 | 8 | 6 | 2 | 0.45 | 0.68 | 0.23 | 0.40 |
| 3 | 12–24 | 12 | 9 | 3 | 0.68 | 0.87 | 0.19 | 0.46 |
| 4 | 24–36 | 20 | 20 | 0 | 0.87 | 0.95 | 0.07 | 0 |
| 5 | 36–42 | 92 | 42 | 50 | 0.95 | 0.97 | 0.02 | 0.46 |
| | | | | | | | | 1.32 |
| | | | | | | EFF = column 8 | | Total = 1.32 |

### 7.4.2 Design Based on Resilient Modulus

Numerous investigations of the resilient modulus (total stress divided by the recoverable strain) of soils subjected to freezing and thawing have been carried out. It has been reported that the resilient modulus can be determined in the laboratory by repeated load triaxial tests on samples obtained undisturbed while frozen. Field plate-bearing tests may also be carried out to obtain a range of values for different soils constituting the pavement profile (Johnson et al., 1978).

Once the resilient modulus properties of the materials in the roadway profile are characterized, the pavement can be analyzed as an elastic layered system to calculate the anticipated deflection. From the deflection, the potential damage or performance may be ascertained. The resilient modulus is a function of deviator stress, confining pressure, moisture content, dry density, and temperature. Because of these numerous parameters, it is very difficult to characterize a true representative value for a specific site. A statistical analysis approach may be used to obtain a range of modulus values for the theoretical computation of deflections under different loading configurations.

### 7.4.3 Frost Protection Layers

Phukan (1981d) reported that the fill material below the base course may be treated with portland cement admixtured in combination with various agents such as calcium lignosulfonate and hydroxylated carboxylic acid. Treatment of the subbase with these materials can reduce frost susceptibility and improve physical properties to resist thaw weakening. Such treated soils may be tested to check the potential frost heave and thaw weakening as well as reduction of shear strength due to freeze–thaw cycles followed by the CBR test. This CBR value may be used to design roadway sections under the design principle of limited subgrade frost protection.

This approach is useful in areas where the availability of NFS materials is limited and the hauling distance may be too far away from the construction area to be cost-effective. The marginal aggregates treated with cement and asphalt emulsions may be used as base or subbase materials. Typical design $k$ values for cement-treated soils are presented in Table 7–5.

**TABLE 7–5 Design *k* Values for Cement-Treated Subbases**

| Thickness (in.) | $k$ Value[a] (lb/in$^3$.) |
|---|---|
| 4 | 300 |
| 5 | 450 |
| 6 | 550 |
| 7 | 600 |

[a]Subgrade $k$ value is approximately 100 lb/in.$^3$

*Source:* After U.S. Army, CRREL, *Report 259* (1975).

## 7.5 ROADWAY CONSTRUCTION IN PERMAFROST REGIONS

There are two principal concepts to be considered with regard to the construction of roadways on permafrost; (1) the "passive" approach, where the roadfill is constructed with the intention of preserving the permafrost; and (2) the "active" approach, where preservation of permafrost is not possible or practical and the consequence of thaw is allowed for in the design. Another factor to be considered during the construction of roadways on permafrost is that the degree of thermal degradation during the construction periods should be minimal unless thaw acceleration is desired. Other factors influencing construction include topography, soil and rock conditions, permafrost conditions, drainage, economics, scheduling, and last but not least, environmental impact. The design and construction of any road is based, among other considerations, to some degree on the optimum utilization of available suitable construction materials and the cost of alternative sources. The availability of suitable construction materials may be a major factor in the design and construction of roadways on permafrost.

For any section of roadway the design should compare the cost of maintenance (due to settlement and frost heave) with the cost of reducing potential maintenance by (1) using thicker, wider fills and controlled construction techniques in continuous permafrost, or by (2) allowing the permafrost to melt to the maximum possible extent during the construction period in discontinuous permafrost.

In continuous-permafrost regions where it is intended that permafrost should be preserved to the maximum extent possible, winter construction operations may be advisable, particularly when no stripping or ground travel can be permitted within the right-of-way. All culverts may be placed during the winter, taking advantage of low water levels. The initial "pioneer" fill may be placed directly on undisturbed ground during the winter, when the active layer is completely frozen, minimizing both consolidation of the active layer and degradation of the permafrost. An initial fill layer is placed to the minimum height required to prevent permafrost from thawing during the following summer. Justification of winter fill construction is based on the thermal calculations described in Chapter 4. Winter construction decreases the volume of materials required and the subsequent settlement of these fills. Hence short-term maintenance costs may be expected. Winter construction also minimizes the potential of environmental damage. Placement of the remaining fill above the minimum required to preserve permafrost can be completed during the following summer. Where it is intended that permafrost is to thaw and settlement is anticipated, all operations should be scheduled with thermal factors in mind. The effects of clearing and stripping of existing natural vegetation conditions under a road section are shown in Fig. 7–13.

If the fill or embankment is properly designed and constructed, it should be done so as to satisfy the minimum standards of service and performance requirements for roadways. In preparing a roadway for construction, the right-of-way may be cleared by hand or Hydro-Axe. The embankment is then constructed by

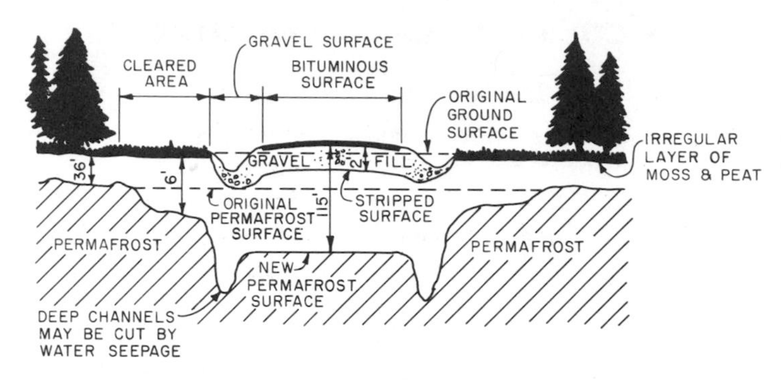


**Figure 7–13** Thermal degradation under road section.

the *overlay* or *end-dumping method* to minimize the disturbance of vegetative cover by hauling equipment. Typical cut or fill sections for roadways constructed on thaw-stable material is shown on Fig. 7–14. Materials brought from the borrow areas are placed on the end of the fill by the hauling equipment and pushed forward onto the undisturbed terrain by bulldozers. This action compacts the material as equipment moves back and forth. Whether placed in winter or summer, the construction must maintain the required thermal stability of the permafrost terrain. Figures 7–15 and 7–16 present a typical example of embankment construction with toe berm and insulation, respectively.

Some cracking and sloughing of the shoulders should be anticipated because the rate and depth of thaw under the side and at the toe of the slopes will be greater than under the main body of the embankment. In areas where thaw-unstable conditions are anticipated, the side slopes must be flattened, bermed, or insulated to ensure that the integrity of the main embankment is maintained, as shown in Figs. 7–15 and 7–16.

Esch (1978) presented a case study of the use of berm and insulation layers on an embankment section that was built on thaw-unstable soils. Conventional slopes can be constructed if the fill materials are dry and relatively well drained

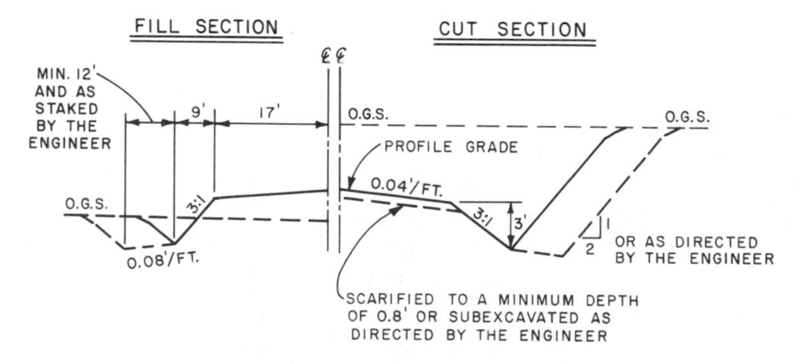


**Figure 7–14** Typical sections.

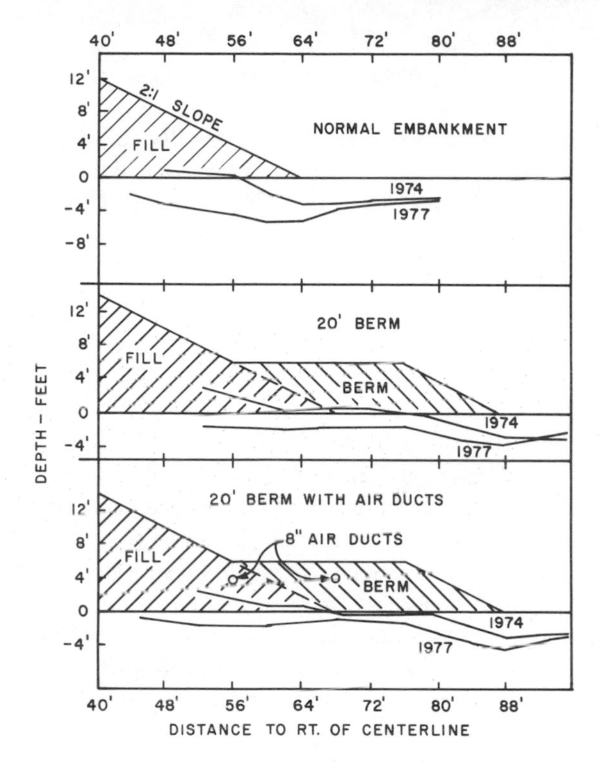


**Figure 7–15** Berm without insulation. (After Esch, 1978.)

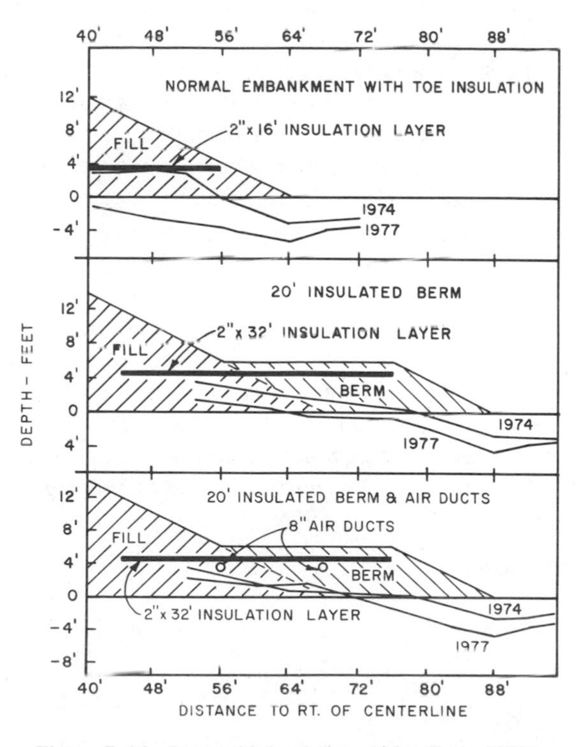


**Figure 7–16** Berm with insulation. (After Esch, 1978.)

and the foundation is stable when thawed. Generally, unstable conditions arise from the use of frozen soils that are difficult to compact. Problems in embankment construction are most often associated with inadequacy of either the quality of fill material or foundation conditions.

### 7.5.1 Subgrade Preparation

During construction operations, local adverse subsurface conditions may be exposed that may not be revealed by the design subsoil explorations. Construction inspectors must be aware of the design and take correct action when such situations occur, to avoid later problems. Visual identification of frost-susceptible soils encountered during construction should be checked and tested for gradation.

Pockets of frost-susceptible soils or sharp variation in the frost-heave potential of subgrade, within the active layer, can frequently be detected only during grading operations. Such conditions may be corrected by removal of undesired material and replaced with NFS materials or by providing transition zones between the areas of different heave potential. Installing transition zones will provide a surface that will remain acceptably smooth under the particular traffic use.

When unexpected wet areas are encountered in the subgrade, which may provide a source of moisture migration for potential frost action, additional drainage measures should be considered.

When isolated ice wedges are encountered in stripping and excavation in permafrost areas and the ice lies within the ultimate thaw zone under the completed pavement, the ice should be removed and replaced with NFS material. If the ice is extensive, the passive method discussed in Section 7.3 may be used to construct the roadway entirely on fill, or consideration may be given to relocating or changing the design grade. In discontinuous-permafrost areas where summer thaw penetration is deep or progressive degradation of permafrost cannot be prevented, such remedial measures may not be effective or economically feasible. It will be necessary to accept substantial summer maintenance of the surface, as it is unavoidable in such a case.

Bedrock in some areas may be a supply of large quantities of water to growing ice lenses; therefore, frost heaving may often be actually more severe in rock cuts than in adjacent soil areas. This is particularly the case where a concentration of fines exists at the surface of the bedrock, in mud seams, or in the base course itself. Rock cuts may therefore frequently require as much or more depth of base course than soil subgrades, to ensure pavement free of heaving and cracking.

### 7.5.2 Fill Materials

Typical gradations of base-course and subbase materials used for roadway fill are presented in Table 7–6 and 7–7, respectively.

**TABLE 7–6 Aggregates for Untreated Base**

| Sieve Designation | Percent Passing by Weight in Grading: | |
|---|---|---|
| | C–1 | D–1 |
| 1½ in. | 100 | — |
| 1 in. | 70–100 | 100 |
| ¾ in. | 60–90 | 70–100 |
| ⅜ in. | 45–75 | 50–80 |
| No. 4 | 30–60 | 35–65 |
| No. 8 | 22–52 | 20–50 |
| No. 40 | 8–30 | 8–30 |
| No. 200 | 0–6 | 0–6 |

**TABLE 7–7 Requirements for Grading for Subbase**

| Sieve Designation | Percent Passing by Weight in Grading: | | | | |
|---|---|---|---|---|---|
| | A | B | C | D | E |
| 4 in. | 100 | — | — | — | — |
| 2 in. | 85–100 | 100 | — | — | — |
| 1 in. | — | | 100 | — | — |
| ¾ in. | — | — | — | 100 | — |
| No. 4 | 30–70 | 30–70 | 40–75 | 45–80 | — |
| No. 10 | — | — | 25–55 | 30–65 | — |
| No. 200 | 10 max. | 3–10 | 4–10 | 4–12 | 0–6 |

There is a potential problem in the construction of fills and compaction of soil at freezing temperatures. It may not be possible to compact frozen fill materials properly, and if the fill contains ice layers or lumps of snow or ice, stability will be reduced and the course will settle upon thawing.

Construction of fills using quarry-run rock, crushed rock, and well-drained clean gravel at low moisture content may be compacted to about 95% of Modified Proctor Density. However, care must be taken to exclude any large chunks of ice or snow from the fills. Winter construction using preheated soils is also technically feasible but can be economically justified only when it is absolutely necessary to complete the work at an early date. Surface of fills may be protected overnight or over a weekend by placing calcium chloride (temperature limitation ≥ 25°F or −4°C) or insulating material. Salt may be used to depress the freezing point of soil moisture and keep it workable at subfreezing temperatures. It is generally found uneconomical to construct fill sections when the temperature drops to −7°C.

### 7.5.3 Cut Sections

It is a good practice to avoid cuts in permafrost wherever possible. Often it is impossible to locate linear structures such as roadways without making cuts and avoiding approaches to stream crossings and still maintain design gradients (Fig. 7–17). The behavior of cuts in frozen soils is directly related to the nature and composition of soil, the distribution of ground ice within it, and the season of construction. In ice-poor clean sand and gravels, slopes may be cut at angles comparable to those used in unfrozen soils. Slope and fill movements and erosion may develop in terrain containing ice wedges and other massive deposits if proper construction measures are not utilized.

Possible methods for the stabilization of cut slopes in ice-rich frozen ground are presented in Chapter 9 (Figs. 9–10 and 9–11). The cut-slope stabilization measures depend on the amount and type of ground ice, the soil type, and the depth of the cut. Other factors, such as the effect of the construction and cost of maintenance, potential environmental impact, and general aesthetics, will dictate the degree of slope protection required. Many cut slopes may self-stabilize within the first three thaw seasons, provided that reasonable care has been taken in the design and construction to ensure that unacceptable thawing and continuing retreat of the slope does not occur. The timing and procedures to follow for an excavation operation will depend on several factors, such as the size of the cut, soil conditions, and equipment available. The use of excavated frozen cohesive soils in adjacent fills may not be possible, and disposal of the material in a selected area must be considered in the design and scheduling of construction. In some areas (close to streams), exceptional measures such as ditch checks and settling ponds may be necessary to avoid erosion and stream siltation caused by meltwater and normal runoff from the cut.

Cuts in ice-rich frozen soils generally present a serious problem and a detailed geotechnical investigation is very useful to select appropriate design and construction measures for deep cuts. In cases where deep cuts are necessary, the

**Figure 7–17** Road cut close to Hess Creek, Alaska. (Courtesy of D. Esch.)

cut face may be made nearly vertical to reduce the area exposed to thawing and to obtain an equilibrium condition quickly. For shallow cuts (≤ 2 m depth), vertical cut slopes may be made at approximately the ditch line. In deeper cuts in ice-rich soils, ditches should be at least 3.5 m wide so that they can be cleared periodically by earth-moving equipment. Preferable back slopes are in a 1:4 ratio of horizontal to height. A wide ditch will also permit placement of a revetment of rock backfill at or against the bottom of the cut slope. Surface runoff from the slope above the top of a cut should be intercepted and diverted to the side by constructing small dikes (rather than ditches) on the ground surface to prevent water from running down and eroding the face of the cut. Trees should be hand-cleared from the top of the slope to a distance from one to two times the height of the slope.

The type of the slope protection method to be used on cuts in ice-rich frozen soils will vary depending on the slope angle, availability of granular materials, amount of excavation, and severity of erosion. Several possible techniques of slope stability are discussed in Chapter 9, and some design measures are illustrated in Figs. 9–10 and 9–11.

## 7.6 OTHER CONSTRUCTION TECHNIQUES

Thermal balance in the roadway subgrade may be maintained by installing a thermal barrier such as insulation or materials having very low conductive values. Another construction approach may be to prevent moisture from migrating to the freezing surface by means of moisture barriers. Thaw-weakening processes of subgrade soils may be counteracted by soil reinforcement to prevent any consolidation or subsidence of roadway sections. These techniques are discussed in the following sections.

### 7.6.1 Insulated Embankments

This method consists of conventional base and subbase layers above an insulating material of suitable properties and thickness to prevent the frost or thaw penetration into a frost-susceptible subgrade or ice-rich frozen ground. Layers of granular materials should be placed below the insulation to contain that part of frost or thaw penetration that occurs below the insulation. The thermal insulation is more effective when layers of granular materials are placed below the insulation than when the insulation is placed directly on the subsurface soils. The thickness of base materials placed above the insulation should be adequate to meet the structural requirements of the roadway section and may be determined by elastic-layered system analysis or other methods discussed in Section 7.3. The vertical stress on the insulation material caused by dead loads and wheel loads must be less than the compressive strength of the insulation. Generally, a safety factor of 3 is used to limit the maximum vertical stress on the insulation to not more than one-third of the compressive strength of the insulating material.

In seasonal permafrost areas, the total frost given by the calculation may be taken as the *a* value (Fig. 7–7) and a new combined thickness of pavement, base, insulation, and subbase is determined.

The thickness of granular material needed beneath the insulation is obtained by subtracting the previously established thicknesses of upper base and insulation. If the adopted design permits frost penetration below the insulation, the thickness of granular material below the insulation and subgrade should not be less than 10 mm (4 in.).

Insulation may also be used to resist thaw penetration into the permafrost. The case study Esch (1978) presented in Figs. 7–15 and 7–16 shows the effective use of insulation to prevent thaw penetration below the embankment. Figure 7–16 illustrates the significant difference in thaw penetration beneath an insulated embankment. The procedures that can be used to determine the thaw penetration below the insulating materials are discussed in Chapter 4. The thermal properties of different insulating materials are also included in the analysis.

### 7.6.2 Peat Overlay Construction Method

Esch and Livingston (1978) reported on the study of an experimental roadway section that was built with 4- and 5-ft thicknesses of peat placed beneath a roadway cut section in permafrost. Their study concluded that peat overlays can be effectively utilized beneath paved roadways to resist or prevent thawing of the permafrost. The large changes in thermal properties of peat that occur upon freezing results in wintertime conductivities roughly twice as high as during the thawing season. Those changes in thermal properties cause an altered thermal balance, and in the study it was demonstrated that the altered thermal balance results in more rapid annual refreezing and lowered permafrost temperatures. For climatic conditions similar to those of Fairbanks, Alaska, a minimum thickness of 76 mm of consolidated peat is required beneath a 1.2-m thickness of granular fill material to prevent thawing into the underlying permafrost. The peat must be consolidated to minimize long-term secondary consolidation.

The main drawbacks of this construction method are obtaining an adequate source of thawed peat and establishing the necessary construction control measures for the preconsolidation of peat prior to pavement construction.

### 7.6.3 Membrane-Encapsulated Soil Layer

The membrane-encapsulated soil layer (MESL) technique involves the containment of a frost-suspectible soil in a membrane wrapping system to prevent moisture gains and additional frost heaving. Various investigators (Burns et al., 1972; Smith, 1979) reported on the encapsulation of soils with membranes to restrict heave moisture.

Smith (1979) reported the field test results of two MESL test sections which were constructed into existing gravel surface roads at Elmendorf Air Force Base and Fort Wainwright in Anchorage and Fairbanks, Alaska, respectively. The Elmendorf AFB MESL contained a silty clay soil, whereas the Fort Wainwright MESL contained a nonplastic silt. Both sections were constructed with soil moisture contents approximately 2 to 3% below optimum for the Modified Proctor compacting efforts.

The Elmendorf silty clay, at an average moisture content of about 16% and an average dry density of about 1778 kg/m$^3$ (111 lb/ft$^3$) maintained a uniform moisture profile during closed-system freezing in the MESL section at a freezing rate estimated at approximately 2.54 cm (1.0 in.) per day. The prefreeze and after-thaw CBR values, in the range 10 to 24%, withstood low density, light vehicular traffic, and occasional heavy truck traffic with minimal surface maintenance.

The Fairbanks silt, at an average moisture content of about 14% and an average dry density of 1408 kg/m$^3$ (88 lb/ft$^3$) has maintained a uniform moisture profile during closed-system freezing in the MESL section at a freezing rate of approximately 3.8 cm (1.5 in.) per day. Traffic use of the MESL increased density about 3% from an average of 1408 to 1440 kg/m$^3$ (88 to 90 lb/ft$^3$). With a 20- to 25.4-cm depth of the sand and gravel surface layer, the MESL with a CBR value in the range 7 to 10% withstood medium-density light vehicular traffic with considerable heavy truck and construction equipment use with minimal surface maintenance. The degree of saturation of MESL soils placed was found to be an important factor for the performance of the road section against the potential heave. A critical question not yet resolved for MESL designs in permafrost regions involves the potential effects of embankment thermal cracking on the membrane, as these cracks may totally destroy membrane integrity.

### 7.6.4 Prethawing Methods

Various prethawing techniques such as clearing and stripping, gravel pad surface darkening with asphalt, and the use of clear polyethylene film may be used to prethaw the upper layers of permafrost so that thinner roadway embankments may be constructed in permafrost regions at a later date. Esch (1981) reported a recent study of various methods of surface modifications that may be used to prethaw upper layers of permafrost. These methods include stripping of vegetation, installing a polyethylene film covering, placing a thin (0.3-m) gravel pad, darkening the gravel surface, and hand-cleared vegetation. Based on the field results, it was concluded that thermally stable embankments may be constructed more economically by maximizing the thaw depth during a single summer prior to construction. As expected, significant increases in thaw depths resulted from machine stripping, asphalt surface darkening, and polyethylene surface coverings. The benefits of gravel pad installation were primarily for drainage, surcharge, and accessibility, to dissipate excess pore pressure buildup during the thawing process, thereby accelerating the consolidation process. Prevention of erosion and trafficability are also achieved by the use of gravel pads.

An economic analysis was also performed (Esch, 1981) and this analysis shows excellent benefit/cost ratios for the various prethaw modes used. Results of this type of economic analysis must be considered relative, as they will vary with various construction techniques, material costs, climatic factors, and soil types. The economic aspects of distress and increased maintenance resulting from these thermally unstable designs have not been evaluated.

### 7.6.5 Soil Reinforcement

Goughnour and DeMagio (1978) reported on soil reinforcement methods of highway projects. Methods such as reinforced earth, stone columns, permanent soil anchors, and arrays of small-diameter cast-in-place piles have provided positive and cost-effective solutions to many geotechnical problems in unfrozen soils. Experience using these methods in permafrost areas related to roadway construction is limited. Therefore, the appropriateness of their applications under permafrost conditions remains to be proven.

As used in moderate climate, the earth reinforcement methods appear to have potential for road embankment construction in permafrost regions. Especially at bridge abutments and high cut-and-fill sections, the earth reinforcement method should reduce the volume of construction materials to be placed. For a thaw-unstable subgrade, a layer of soil reinforcement at the bottom of the "overlay" fill section may improve the CBR value for the thawing condition and the performance of the roadway during the spring thaw. Field tests are needed to finalize the actual benefits of the soil reinforcement technique and to assess the appropriateness of these techniques in the construction of roadways in permafrost regions.

### 7.6.6 Construction Cost

The cost of a roadway section will include surveys, engineering design, right-of-way acquisition and permit applications, supervision of construction, testing of materials, and exploration of sites. Construction costs are described in this section. On highway projects, cost per mile is generally used.

The construction elements of a typical roadway section are shown in Table 7–8. The items shown may be broken down further and individual costs may be calculated in terms of labor, materials, and equipment. Three types of cost estimates that are commonly used in highway projects are conceptual or preliminary, final engineers' estimates, and contractors' bids. The more detailed and accurate the estimate, the more costly and time consuming the estimation process. Several factors should be considered before a particular cost estimate process is selected. Some of these factors are: the degree of accuracy required at each stage of estimation, the lowest cost estimate, the amount of load involved in the project, and the availability of time.

The in-place cost of excavation, hauling, and placing borrow material for embankment construction have been found to be generally within the range $10

**TABLE 7–8**
**Roadway Construction Elements**

| Site Preparation | Grading | Drainage | Fill or Overlay | Subbase | Base Course | Asphalt Pavement | Misc. Items |
|---|---|---|---|---|---|---|---|
| Clear | Excavation | Trench excavation | Borrow material | Borrow material | Aggregate preparation | Aggregate preparation | Borrow material |
| | Borrow material | Underdrain | Spreading | Spreading | Spreading | Mix and spread aggregate | Placement |
| | | Culvert | Compaction | Compaction | Compaction | Compaction | |

to \$30 (U.S.—based on the 1982 cost) per 0.76 $m^3$ (cubic yard) depending on the location, hauling distance, and availability of materials. The material transportation and placement cost for the synthetic insulation material used in an insulated embankment can be expected to be in the range of \$0.50 to \$1.50 (U.S.) per 0.093 $m^2$ (square foot), depending on the type of insulation, thickness, and method of transportation. The cost of insulation material varies from \$0.35 to \$0.50 (U.S.) per 2.54 cm, per 0.09 $m^2$ (inches per square foot).

Considerably less volume of embankment material will be required to prevent thaw penetration into the subgrade permafrost if synthetic insulation material is used compared to utilizing gravel as the insulating material. Where NFS material costs are relatively high or where the in-place cost of insulation per square foot is low, the synthetically insulated embankments could cost significantly less than their all-NFS-material counterparts.

## 7.7 AIRFIELD DESIGN AND CONSTRUCTION

Recently, the construction activities related to resource development, including petroleum, mining, and hydroelectric works in permafrost regions of the world, have demanded the development of northern airports to meet new aviation technology using larger and heavier aircraft for the transportation of cargos. A photographic view of an airport terminal built on continuous permafrost is shown in Fig. 7–18.

Airport pavement built on embankment sections must provide adequate support for the loads imposed by aircraft using an airport; and they should be constructed to produce a firm, stable, all-year, all-weather surface free of particles that may be blown or picked up by propeller wash or jet blast. In addition, pavement and embankment sections must possess sufficient inherent stability to withstand, without damage, the abrasive action of traffic and other deteriorating influences from frost action and adverse weather.

**Figure 7–18** Longyearbyen, Norway, Airport on continuous-permafrost zone.

Typical plan and cross-sectional drawings for runway pavements to serve jet-powered aircraft are shown in Fig. 7–19.

Some of the design concepts discussed for roadways in seasonal permafrost and permafrost areas may be used for the airfield embankment. The following design methods are generally used for seasonal permafrost conditions:

1. The complete-protection method involves the removal of frost-susceptible material to the depth of frost-penetration and replacing the material with non-frost-susceptible material.
2. The limited-subgrade-frost-penetration method allows the frost to penetrate a limited depth into the frost-susceptible subgrade. This method holds deformations to small acceptable values.
3. The reduced-subgrade-strength method usually permits less pavement thickness than the two methods discussed above and should be applied to pavements where aircraft speeds are low and the effects of frost heave are less objectionable. The primary aim of this method is to provide adequate structural capacity for the pavement during the frost-melt period. Frost heave is not the primary consideration in this method.
4. The reduced-subgrade-frost-protection method provides the designer with a method of statistically handling frost design. This method should be used only where aircraft speeds are low and some frost heave can be tolerated. The statistical approach of this method allows the designer more latitude than the other three methods discussed above.

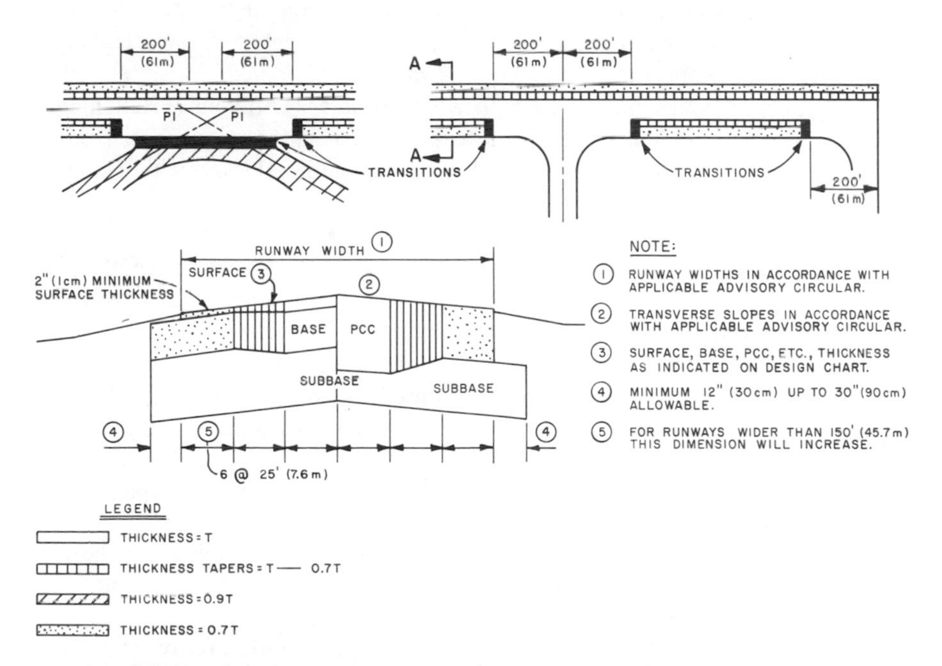


**Figure 7–19** Typical plan and cross section for runway pavement. (After FAA, 1978.)

The design of pavements in permafrost areas requires efforts to restrict the depth of thaw. Thawing of the permafrost can result in the loss of bearing strength. If thawed permafrost is refrozen, heaving can result and cause pavement roughness and cracking. Two methods of design are available for construction in permafrost areas; complete-protection methods are somewhat similar to the methods discussed for seasonal frost design.

The depth of frost penetration can be computed using the modified Berggren equation (4–21). The Berggren equation requires several inputs concerning local soil conditions and local temperature data (Examples 4–2 and 4–3).

When the complete-protection method is not economically feasible, design is usually based on the reduced-subgrade-strength method and consideration of the effect of construction on the existing thermal regime. The use of insulation to prevent freezing and thawing of the subgrade is an alternative design measure for many problem areas (Esch and Rode, 1977). Specifically, it offers a satisfactory design alternative where NFS granular base material is unavailable or is so expensive that insulation is more economical.

The depth of thaw penetration under a bituminous pavement may be reduced by painting the surface white (Fulwider and Aitken, 1962; Tobiasson, 1978). A reduction in thaw penetration on the order of 35% during one thawing season results from such surface treatment in an area having an air thawing index of 780 degree F-days.

Good drainage is necessary at airfield sites in permafrost. The same drainage provisions used for roadways may also be used for airfields (U.S. Army/Air Force, 1965). Special attention must be given to drainage at the toe of embankment slopes, where ponding of water may increase the depth of thaw leading to slope failures and embankment problems. Also, particular attention is required for the use of culverts under the airfield in permafrost areas. Differential settlement is generally observed around the culverts in such areas. Moreover, special design measures are necessary to prevent culverts from becoming blocked with ice during the winter (Fig. 7–2).

Crory et al. (1978) reported case histories of design and construction problems of two airfields located in the northern regions of Alaska. Several design options were considered. These included (1) gravel over sand, (2) gravel over insulation or sand, (3) landing mat with insulation, and (4) landing mat without insulation. The assumed soil characteristics and design criteria for the sites are presented in Table 7–9. Crory et al. (1978) used the modified Berggren equation to calculate the depth of thaw penetration under various air indexes and the results are presented in Fig. 7–20. Figure 7–20 indicates that the design under 61 cm (24 in.) of gravel, 2.54 cm (1 in.) of sand insulation, gave the least anticipated depth of thaw penetration.

For design purpose, Berg (1974) prepared several charts showing the relationship of the thickness of pavement and base, the gross weight of aircrafts, and the characteristics of subgrade soils in terms of frost-susceptibility soils. These charts (Fig. 7–21) are useful guidelines for the determination of pavement thickness for airfields.

**TABLE 7–9 Upper Boundary Conditions, Initial Condition, and Soil Characteristics for Thaw-Depth Calculations**

| | | | | | | |
|---|---|---|---|---|---|---|
| Surface thawing index (degree F-days): | | 500 | 1000 | 1500 | 2000 | 3000 |
| Length of thawing season (days): | | 100 | 104 | 108 | 112 | 120 |
| Mean annual temperature (°F): | | 14 | 14 | 14 | 14 | 14 |
| *n* factors for thawing conditions | | | | | | |
| Gravel and sand surfaces | 1.4 | | | | | |
| White-painted landing mat on sand | 1.0 | | | | | |
| White-painted landing mat on insulation | 2.0 | | | | | |

| | Soil Characteristics | | | | |
|---|---|---|---|---|---|
| Material | Dry Unit Weight (lb/ft$^3$) | Moisture Content (%) | Thermal Conductivity (Btu/ft$^3$ hr °F) | Latent Heat (Btu/ft$^3$) | Volumetric Heat Capacity (Btu/ft$^3$ °F) |
| Gravel | 130 | 8 | 1.86 | 1498 | 29.9 |
| Sand | 100 | 23 | 1.51 | 3312 | 34.2 |
| Insulation | 4 | 0 | 0.02 | 0 | 1.8 |
| Landing mat | 34 | 0 | 15 | 0 | 20 |

*Source:* After Crory et al. (1978).

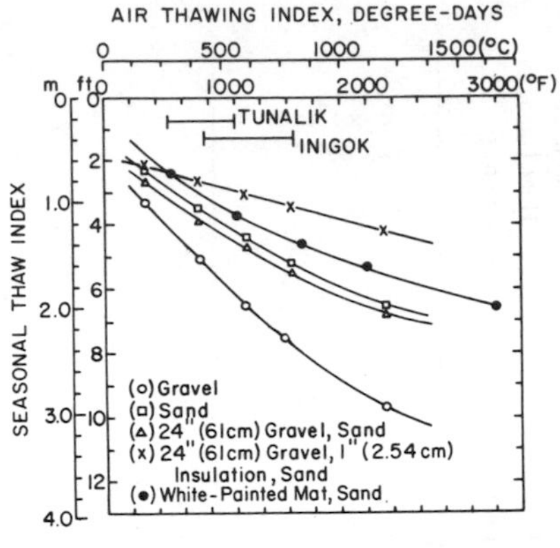


**Figure 7–20** Calculated depths of thawing. (After Crory et al., 1978.)

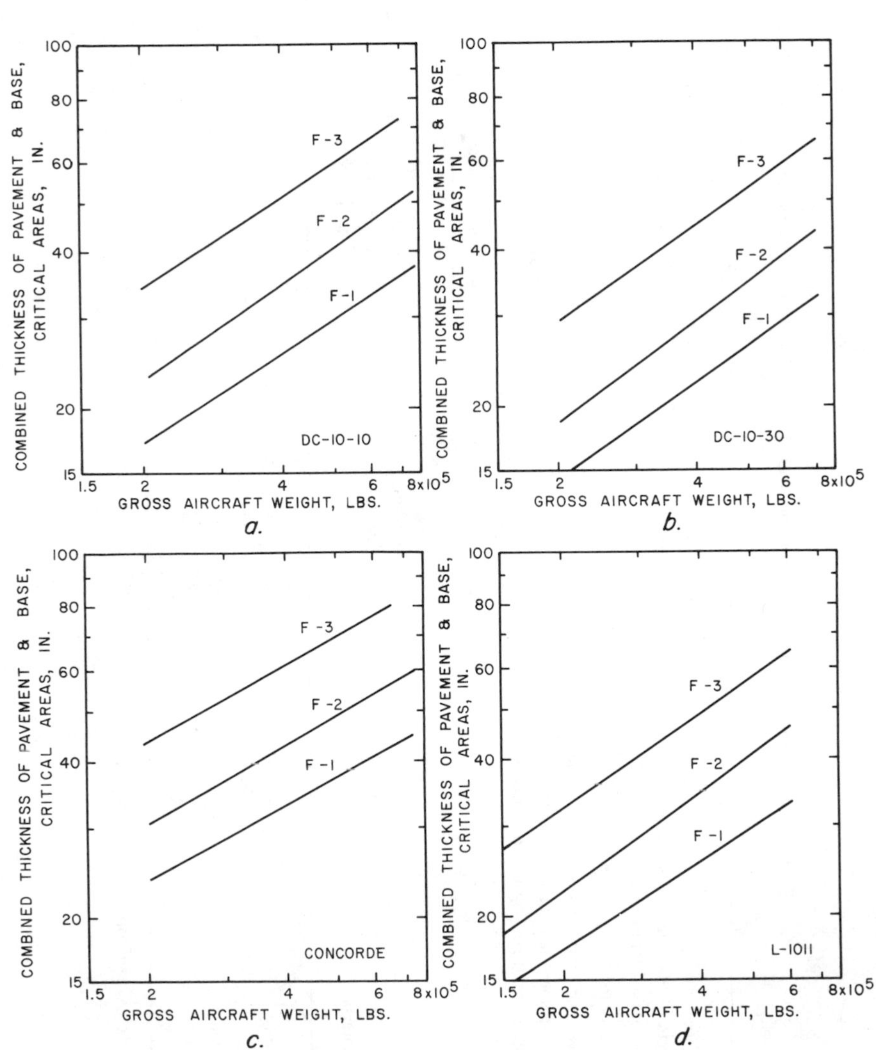


**Figure 7–21** Reduced subgrade strength design charts for flexible pavements. (After Berg, 1974.)

## PROBLEMS

**7–1.** A road is to be designed on permafrost areas for heavy truck loads. Discuss the different design methods that can be used to construct the road.

**7–2.** Estimate the thickness of a clean, well-graded sand and gravel pad required to prevent thaw penetration into the original ice-rich ground. The pad is to be used as **(a)** a roadway and **(b)** an airstrip. The following is the site information:

- Air thawing index = 1000 degree-days
- Length of thawing season = 80 days
- Mean annual temperature = −9°
- Soil conditions: organic mat of thickness 0.5 m underlain by silt; dry density of silt = 1040 kg/m$^3$; water content = 55%; mean wind velocity = 6 mile/hr

**7–3.** Estimate the thickness of the design roadways in Problem 7–2 based on **(a)** limited frost penetration and **(b)** reduced subgrade strength.

**7–4.** If the gravel fill in Problem 7–2 were paved with a bituminous concrete, what additional fill thickness will be required to prevent thaw penetration into the permafrost soils?

**7–5.** For the data given below, determine the design thickness of a residential road on a discontinuous permafrost region.

| Water Content of Subgrade Material (%) | Water Content of Base Material (%) |
|---|---|
| 25 | 3 |
| 35 | 5 |
| 45 | 8 |
| 10 | 6 |
| 15 | 10 |

**7–6.** If subgrade materials in Problem 7–5 are classified as F2, F3, and F4, what should be the design thickness of the road?

**7–7.** Given:

| Depth (in.) | Fines Content |
|---|---|
| 0–6 | 6 |
| 6–9 | 8 |
| 9–15 | 10 |
| 15–18 | 10 |
| 18–30 | 20 |
| 30–40 | 20 |

Determine the thickness of pavement required for the DTN of 200. Use the excess fines concept.

**7–8.** Estimate the thaw penetration after 1, 5, and 10 years beneath an airstrip consisting of 6 in. of concrete and 3 in. of Styrofoam insulation on a sand pad of 3.5 ft thick. The original ground is frozen silt. The mean annual air temperature at the site is $-2°C$ and the length of the thawing season is 100 days. The following additional data are available:

- Sand pad: $\rho_d = 125\ lb/ft^3$, $\omega = 10\%$
- Frozen silt: $\rho_d = 70\ lb/ft^3$, $\omega = 45\%$

# CHAPTER 8

# UTILITY SYSTEMS

## 8.1 INTRODUCTION

Several factors, such as frozen ground, climate, remoteness, lack of planning for services, and limited economic base cause special problems to the development of northern communities. These factors contribute to the problem of providing utility systems in the arctic and subarctic regions. The design of utility systems in permafrost regions is discussed in this chapter. The scope of the chapter is to highlight some of the design considerations regarding the design of systems for water supply and sewage disposal in permafrost regions. In addition, foundation problems that may be encountered in the design and construction of utility systems in frozen ground are addressed. No attempt has been made to cover all design aspects and construction details of the actual utility systems.

Considerations must be given to the physical, economic, and environmental impact factors in the design of utility systems. Also, the design and construction plans must be socially acceptable to the communities to be served. Generally, the basic engineering principles of utility systems used in temperate zones may be used in the arctic and subarctic regions with some additional features. These include thermal analysis and balance between energy supply and system losses. The planned utility systems must be simple and operable by persons with minimum qualifications. Compact water and sewer systems are necessary if they are to be maintained and supplied with energy economically. Special attention must be given to disposal of all forms of wastes to avoid polluting nearby land, air, and bodies of water. A comprehensive design manual for utilities in cold regions has been published (Canada Environment, 1979).

## 8.2 WATER SUPPLY

Various investigators (Alter, 1969; Slaughter et al., 1975; Reed, 1970; Smith and Justice, 1975; Foster, 1975; James and Suk, 1977) have reported on the various design aspects of water supplies and systems for northern communities. A safe, adequate supply of water that is economically feasible to build and operate is essential. Sources of water supply in northern communities are often limited due to the presence of frozen ground or permafrost, thick ice cover on lakes and rivers, contamination of surface water, floating ice, and debris carried by the flood water.

The northern cold climate conditions require that special attention be given when selecting and developing water sources. Particular attention must be given to selecting potential sources of water supply. Hydrological data on northern lakes, streams, and groundwater are scarce and typically cover periods of short duration. The lack of adequate historical data makes it difficult to calculate reliable yields for water supply purposes. Detailed preliminary engineering studies and direct observations of the potential water sources must be developed prior to final selection.

Several options are available to facilitate development of water supply systems. The sources of water supply in cold regions may be either surface water or groundwater. They are discussed in the following sections.

### 8.2.1 Surface Water Sources

Surface water sources include snowmelt, lakes, and rivers. Most lakes and rivers in the northern regions are shallow and they may be expected to freeze completely in the winter. They are usually not a satisfactory source of continuous water supply. Only the larger streams and deeper lakes remain liquid beneath a winter ice cover. The deeper lakes may be suitable for year-round supply.

A streambed forces water flows to the surface through cracks and along the shores, where it then freezes, causing an "icing" situation. Frozen water of this type is called *aufeis* and is essentially not available until breakup and thaw occur. Ice or snow may be stored and melted as required. However, a large storage area is required to collect snows [approximately 2 $ft^3$ of snow = 1 gal (U.S.) of water = 3.8 liters of water]. Near Barrow, Alaska, the U.S. Army Corps of Engineers augmented the freshwater supply in that community by using snow fences to accumulate large quantities of snow.

Water intakes on lakes and rivers must be designed against ice damage, bank erosion, and high water levels during the spring. The distribution of frozen ground adjacent to the water body is also an important factor for the location of pipelines and pumping stations.

Seawater is a potential source along the arctic mainland coast and in the Arctic Islands in the continuous-permafrost zone, but it must be desalinized for domestic use. The desalinization process is expensive, requiring sophisticated equipment and skilled operators. As such, the use of seawater as a source will be limited until further advancements improve the cost-effectiveness of the process.

### 8.2.2 Groundwater

Groundwater may be considered one of the most desirable sources of water in the cold regions. Normally, groundwater maintains a nearly year-round constant temperature and is warmer than surface water in the winter.

Groundwater in continuous-permafrost regions may be found in three locations: (1) above permafrost, within the active layer (suprapermafrost); (2) within permafrost in thawed areas (intrapermafrost); and (3) beneath permafrost (subpermafrost). Water found within the active layer generally needs extensive treatment before use. Such water also has a high mineral content. The quantity of this water is often low and unreliable. A system for the use of water found in the suprapermafrost layer was developed for Port Hope, Alaska (McFadden and Collins, 1978).

Intrapermafrost water is unreliable and not a suitable source of water supply, due to impurities caused by its high mineral content.

Subpermafrost is the most satisfactory groundwater source in permafrost regions, although the water tends to be hard. Water may be found just below or at a considerable depth beneath the bottom of permafrost. The water must be protected as it is brought up through the permafrost. Protective measures include providing insulation to reduce the potential of heat transfer from the water being pumped through the permafrost layer. This often requires special well casings, grounding methods, and heat-tracing water lines. A typical occurrence of groundwater in a permafrost region is shown in Fig. 1–6.

### 8.2.3 Water Distribution Systems

Various methods of distributing water from source to user that are in use in northern communities include:

1. Self-haul system
2. Watering points
3. Small heated building containing a water tank, water piping, and valves
4. Truck or vehicle delivery system
5. Piped systems

The self-haul system has a limited application and the mode is used only for a small community where no mechanization exists. Water points are usually operated in conjunction with other forms of water distribution. They are often located in older parts of a community where the houses are not equipped for a trucked or piped system. The trucked water system is preferred where possible over a central pickup water point. The trucked water system is a more positive means of supplying housing units with clean, safe water on a regular basis. The system allows increased water usage, with better hygiene and the possibility of connecting to a piped system with minimum problems.

Piping systems are the most efficient, safe, and economical means of distributing water. They must be designed to prevent water freezing in pipes exposed to cold air or ground temperatures. Generally, insulation of piping and the addition of heat prevents such freezing from occurring. Also, the structural integrity of piping systems must be maintained against permafrost-induced damage. Design measures often used to minimize frost heave for the foundations of structures in permafrost areas can be applied to maintain the structural integrity of water piping systems. The use of non-frost-susceptible granular soils as backfill and bedding materials, insulation, pile foundations, stronger pipe materials to withstand greater differential settlement, and the placing of the pipe system above ground are the most common design measures used in dealing with water pipe and other piping systems.

The piping system may be placed above or below ground; the option selected will depend on specific site conditions (Fig. 8–1). Generally, buried systems are preferred over aboveground systems where feasible. Aboveground systems have been used where ground conditions and potential thaw from the piping system do not permit the use of a below-ground system.

In areas where there is no permafrost, pipes can be buried below seasonal frost penetration. However, this procedure may be impractical or very expensive due to the presence of deep frost or the need for excavation through bedrock. Grainge (1969), Ryan, (1977), and Dawson and Cronin (1977) discuss the various factors to be considered in the design and construction of pipeline systems in the northern regions.

Some factors, such as minimum ground temperature; the distribution of permafrost and its composition, including ground ice; depth of frost penetration; and ease of excavation dictate the use of a bare-pipe buried system or an above-

**Figure 8–1** Aboveground piping system at roadway crossing, Thule, Greenland. (Courtesy of B. Miller.)

ground system. In some cases, an insulated buried system may be placed just above the permafrost table to minimize the cost of deep burial. However, consideration must be given to the potential of frost-heave forces and the thaw-settlement characteristics of the subsoils. The ability of the pipe to withstand traffic loads if lines are placed under roadways must also be evaluated. A typical cross section of a buried system under a road is shown in Fig. 8–2.

**Figure 8–2** Below-ground road crossing of utilidors. (Reproduced with permission of the Minister, Supply and Services, Canada, *Design Manual*, 1979.)

Because of problems related to foundations and other elements, a buried system may not be feasible in many cases. Where a buried system is not feasible, the pipes may be insulated and placed on a berm. Berms must be designed and constructed to eliminate erosion and unacceptable pipe movements due to frost heave and thaw settlement. In addition, the surface drainage should not be obstructed by the placement of the berms.

In most continuous-permafrost regions where the thickness of the active layer is shallow and the depth of the permafrost is thick as well as widespread, services are usually placed in utilidors (Figs. 8–3 and 8–4). Utilidors are containers enclosing a number of pipes and placed on or above the ground. They may carry steam or hot water heating pipes, fuel distribution lines, and electrical cables in addition to sewer and water mains. Utilidors are discussed further in Section 8.4.

## 8.3 SEWAGE COLLECTION AND DISPOSAL

In the North, sewerage systems generally present fewer freezing problems than water systems do because the sewage is warmer than the water in the water mains. Sewage will generally freeze in branch lines having very low flow or in staggered lines where the sewage collects or blocks the line at any low points. The most commonly used sewer system is the gravity system. This system is used to carry sewage either in buried or aboveground lines in utilidors. Considerations regarding freezing and maintaining structural integrity of the sewer lines in permafrost regions are similar to those factors discussed in Section 8.2.3.

Buried sewers are either insulated or placed in the same trench as the water main; this is a major departure from the design standards in nonpermafrost areas. Intermittent flow systems can be used where the flow is limited to keep the sewer line open. The system is designed to carry periodic high flows.

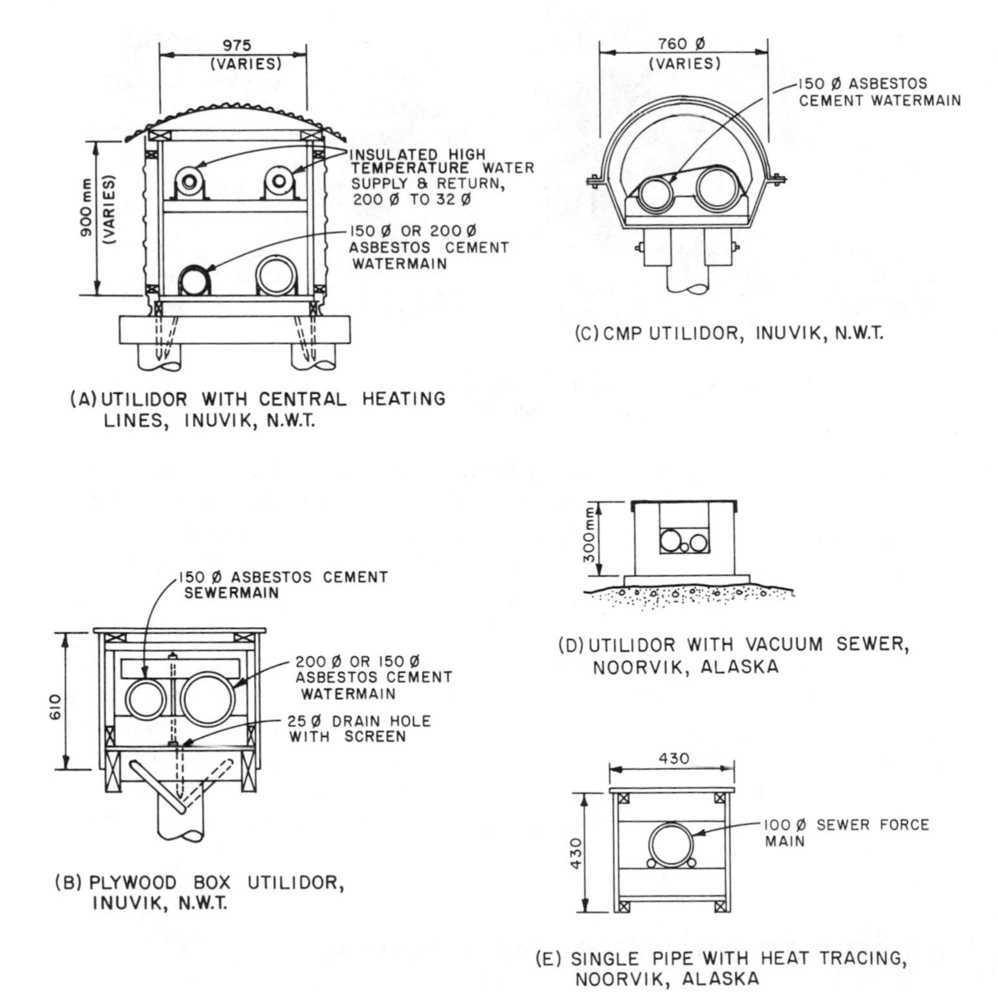


**Figure 8–3** Various utilidors installed in cold regions. (Reproduced with the permission of the Minister, Supply and Services, Canada, *Design Manual,* 1979.)

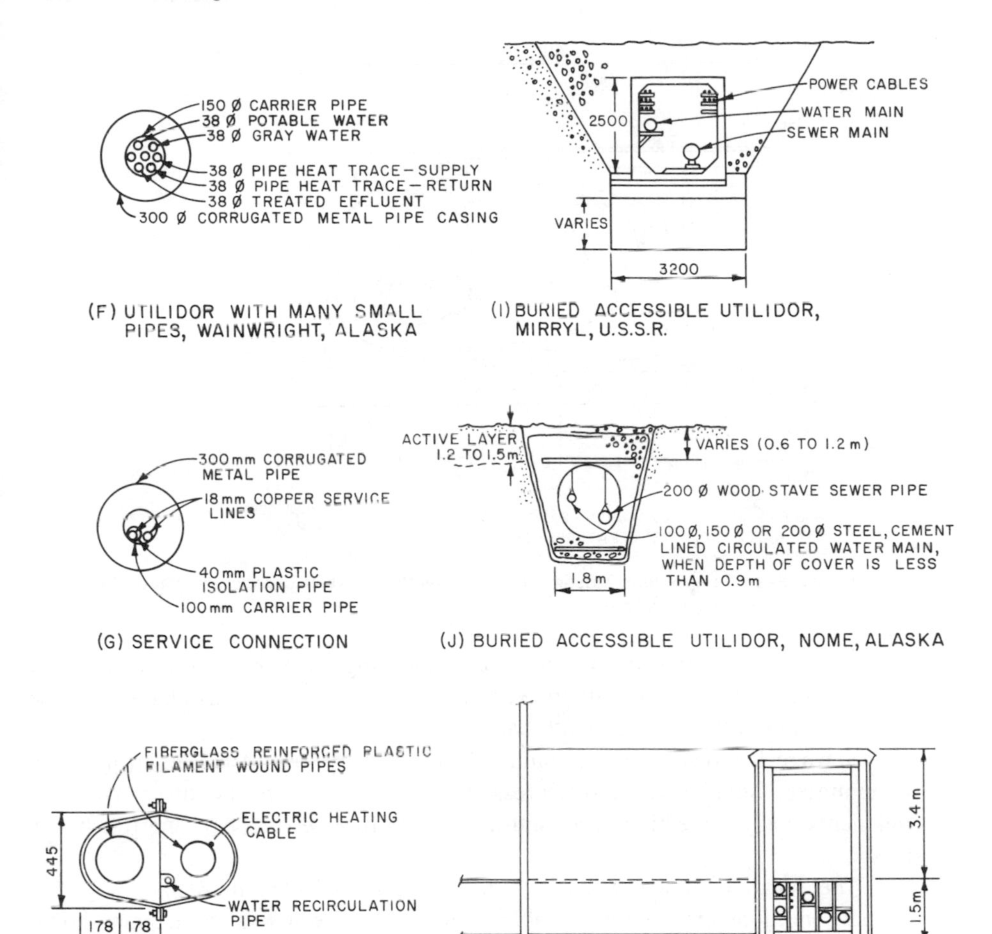


**Figure 8–3** *continued*

Vacuum and pressure systems may be considered where the water supply is limited and the installation of gravity lines presents problems (Averill and Heinke, 1974; Rogness and Ryan, 1977). Both of these systems minimize the problems encountered in deep cuts in permafrost and layout restrictions in communities.

Given the impervious nature of frozen ground, individual septic tanks with leaching fields cannot be used to dispose of sewage in continuous-permafrost regions. In discontinuous-permafrost areas, septic systems may be feasible in some areas where unfrozen soil conditions are encountered and the septic tanks can be placed below the depth of frost penetration.

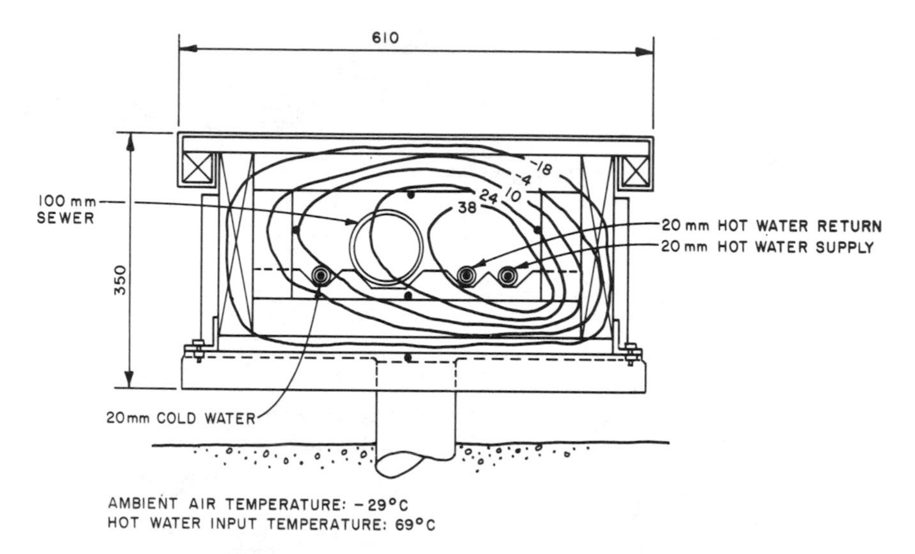


**Figure 8–4** Temperature variation in utilidor with hot water distribution, Fairbanks, Alaska. (After Reed, 1977.)

If sewage is dumped onto the river ice during the winter, it will generally not disappear during the spring break. Often, the sewage will wash back onto the shore, causing pollution and health hazards.

Sewage lagoons are most commonly used to dispose of wastes in northern communities. Small lakes and depressed landforms may be modified with embankments and the erection of a control tower to receive, process, and retain raw sewage.

The location of a sewage lagoon should be selected in such a way that ground conditions are relatively stable. The depth of water in the lagoon should be at least 2 m, to minimize the effect of a thick ice formation. Inlets and outlets to the lagoons require special attention, as their elevation relates directly to the formation and thickness of ice in the lagoon.

There are numerous mechanical treatment plants in Alaska that serve pipeline construction, mining, and drilling camps. The requirement to enclose and heat biological plants causes the problem of high humidity in the building. Smith and Given (1977) have published a critical evaluation of the extended aeration systems in arctic and subarctic areas.

## 8.4 UTILIDORS

As discussed in the preceding section, utilidors are conduits that enclose utility piping which, in addition to water and sewer mains, may include central heating, fuel oil, natural gas, and electrical and telephone conduits. Various types of utilidors used in cold regions are illustrated in Figs. 8–3 and 8–4.

Utilidors are insulated to reduce heat loss and may include other freeze-protection measures, such as a circulation system to maintain a constant flow in the pipes. Most utilidors have some mutually beneficial heat-transfer component between the enclosed pipes. Utilidors are most commonly constructed above ground in permafrost areas (Fig. 8–6). Design principles discussed in Chapters 4, 5, and 6 are applicable to utilidor structures built in frozen-ground conditions.

Foundation considerations will be different for utilidors that are below ground, at ground level, or above ground. Each will require site investigations and design to consider the effects of frost heaving, settlement, and surface and subsurface drainage.

Where anticipated ground movements are within acceptable limits, utilidors can be installed directly on the ground surface or on a berm, earth mounds, sleepers, or posts. In unstable areas, utilidors are commonly supported on piles that are adequately embedded into the permafrost to resist the upward and downward loads discussed in Chapter 6. Utilidors are typically lightweight and thus the vertical loads on the piles are relatively smaller than the uplift forces due to frost heave that will govern the effective embedment length of piles Lateral forces may be significant on some permafrost slopes, and lateral thermal expansion and hydraulic stresses must be considered at bends. Various types of piles may be used to support aboveground piping and their selection depends on the availability of local material, the length of pile required, and the cost. Frost heave must also be considered in shallow-buried utilidors. Thawing must be prevented in ice-rich frozen ground.

### 8.4.1 Below-Ground Utilities in Permafrost Areas

A typical below-ground utility system is shown in Fig. 8–5. Heat loss from buried, warm pipelines and thermal disturbances resulting from their installation will cause thawing that is greater than the natural thawing process in undisturbed areas. Special attention must be given to understanding the thermal regime and the stability of the soils with regard to the piping contained in such buried systems in frozen ground. The degree of concern and design measures to be taken will depend on the thermal sensitivity and ice inclusion in frozen ground. For example, at locations where the mean annual ground temperature is just below freezing, it is impractical to maintain the thermal equilibrium condition once the surface vegetative cover has been disturbed. As such, utilities in the discontinuous-permafrost zones are generally designed for thawing and possible settlement (Stanley, 1965; Klassen, 1965). In continuous-permafrost regions, ground temperatures are colder and thawing can be prevented more easily. A distinction must be made between the relatively small cool water and sewer pipes and larger utilidors or high-temperature heating pipes. With the latter, the foundation of nearby structures may also be adversely affected. Measures that can be used to reduce thaw settlement are reduction of thermal influence, replacement of ice-rich soils, anchoring pipes, and freezing of foundation soils.

**Figure 8–5** Below-ground utilidor, Noorvik, Alaska. (Courtesy of Alaska Area Native Health Services.)

The thermal influences of water and sewer pipes can be minimized by placing insulation around or below them. Heat loss and thermal influence can be reduced by lowering the operating temperature of water pipes and restricting the temperature of wastewater discharges. It may be necessary to install utilities only during periods when the air temperature is below freezing in order to reduce the thermal disturbance, including heat input from the backfill and open excavations.

Natural or prethawing may be used in ice-rich foundation soils in discontinuous-permafrost areas. Generally, soils are mechanically excavated and replaced with compacted granular unfrozen soils in such conditions.

Buried pipes and their appurtenance placed within the active layer must be designed against frost heave. Two primary design measures to reduce the effects of frost heaving on pipes are deeper burial, and overexcavation and backfill with non-frost-susceptible granular soils within the trench. Flexible pipes and joints may be used where differential movement is expected, such as at building connections. To reduce the heaving forces, the appurtenances may be wrapped with polyethylene or can be encased in oil and wax to break the bond between the soil and the appurtenance surface. Manholes are also fabricated in an inverted cone shape to minimize heaving forces.

### 8.4.2 Aboveground Utilities in Permafrost Areas

Below-ground utilities are usually preferred, but local conditions, operating requirements, and economics may dictate the attractive use of aboveground utilities and utilidors. A typical aboveground utilidor is shown in Fig. 8–6.

**Figure 8–6** Aboveground utilidors, Longyearbyen, Norway.

Aboveground utilidors should be compact and as close to the existing ground surface as practical to reduce obstruction to traffic. Low-level utilidors also reduce the elevation of buildings necessary for gravity sewer drainage. To minimize permafrost disturbances, utilidors are commonly installed on piles or gravel pads. It is a common practice to use the smallest utilidors possible, to reduce heat losses and costs.

In unstable areas utilidors are usually supported on piles which are adequately embedded into the frozen ground (Figs. 8–4 and 8–6). Because of the light weight of utilidors, the vertical loads on the piles are relatively small and frost heaving will be the most significant design consideration for calculation of the embedment length of piles. Lateral forces may be significant on some permafrost slopes, and lateral thermal expansion and hydraulic stresses must be considered at bends. As discussed in Chapter 6, various methods may be used for the installation of piles. The selection of the pile type will depend on the availability of local material, length of the pile required, and economics.

Small utilidors are commonly placed on a single pile, but large utilidors may require double piling for stability. Moreover, pile caps are often used to allow for poor alignment and lateral movement.

CHAPTER 9

# Slope Stability in Frozen Ground

## 9.1 INTRODUCTION

Due to increasing construction and resource development activities in the arctic and subarctic regions, slope-stability analyses of frozen ground as well as thawing frozen ground is becoming very important to the engineering profession. The subject of slope stability is concerned with the downward movement of slope-forming materials, which may consist of soil, rock, artificial fill, or a combination of these materials. Frozen ground may creep at a very slow rate under its own weight or externally applied loads. Due to surface disturbances or heat input, frozen ground may undergo thermal degradation and a mass movement may cause slope failure. Movements may also occur in fine-grained soils in cold regions due to the action of freezing and thawing. Slope movement in frozen soils and in thawing soils may introduce failure processes, resulting in flowing, sliding, and falling, or by their combination. These failure characteristics are to be considered in slope-stability analyses and subsequently, when evaluating alternative slope stabilization design measures.

Many early investigators (Capps, 1919; Taber, 1943; Washburn, 1947), and later workers (Issacs and Code, 1972; Hughes, 1972; Hardy and Morrison, 1972; McRoberts and Morgenstern, 1973, 1974 a,b; Phukan, 1976; Pufahl et al., 1974; and others) provided a substantial quantity of observations and qualitative information regarding slope instability in periglacial regions. It is apparent from the findings of earlier investigations that major mass movements or landslide activities could occur in both frozen and thawing soils. Furthermore, it was found that development is invariably associated with disturbance of the active layer and the existing ground thermal regime, resulting in increased depths and rates of melting. These increased melting rates cause increased incidents of instability. With increased knowledge of the factors responsible for slope failures, greater attention has been given to movements in thawing soils as well as with the instability of frozen ground.

The development of roads, highways, and airfields of increased geometric standards in the northern regions requires that cuts be made through frozen soils and still maintain suitable vertical and horizontal alignment. Lotspeich (1971) described some of the initial concerns regarding the behavior of cut slopes in Alaska in relation to their impact on the environment. Smith and Berg (1973) reported the experience with cut-slope construction and behavior on the Trans-Alaska Oil Pipeline system haul road from Livengood to the Yukon River. Pufahl et al. (1974) did an extensive reconnaissance of northern Canada and Alaska and reviewed many existing cut slopes. During this reconnaissance, it was attempted to relate the performances of cut slopes to the geological history of the terrain, the landform, soil type, ice content of the sediments, the nature or type of ground ice, and the initial geometric cross section of the back slope. McPhail et al. (1976) reiterated the importance of maintaining the stability of cuts and fills in permafrost terrain.

The notation used in this chapter is given in Table 9–1.

## 9.2 CLASSIFICATION

The soil mass movements in frozen ground may be classified into the following groups:

1. Flow
   (a) Skin
   (b) Bimodal
   (c) Progressive or multiple-retrogressive flows
2. Slide
   (a) Rotational
   (b) Block
   (c) Progressive or multiple-retrogressive flows
3. Fall
4. Solifluction

This slightly modified classification is based on the work reported by McRoberts (1973) and McRoberts and Morgenstern (1973). The following summarizes the findings taken from their work.

The classification system above is based on the morphology of the failed mass. The primary level of classification of flow, slide, or fall is used as a de-

**TABLE 9–1 Notations**

| Symbol | SI Unit | Definition |
|---|---|---|
| $C$ | $kN/m^2$ | Cohesion (Eq. 9–2) |
| $C'$ | $kN/m^2$ | Effective cohesive strength |
| CU or $R$ | — | Consolidated undrained triaxial test |
| CD or $S$ | — | Consolidated drained triaxial test |
| $C_v$ | $m^2/yr$ | Coefficient of consolidation (Eq. 9–19) |
| $d$ | m | Thaw depth (Eq. 9–13) |
| $F$ | — | Safety factor |
| $H$ | m | Height of slope (Eq. 9–4) |
| $H_c$ | m | Critical height of slope (Eq. 9–8) |
| $K$ | — | Seismic coefficient (Eq. 9–7) |
| $k$ | cm/s | Coefficient of permeability (Eq. 9–19) |
| $K_u$ | W/m K | Thermal conductivity of thawed soil (Eq. 9–18) |
| $L$ | $J/m^3$ | Volumetric latent heat of frozen soil (Eq. 9–18) |
| $m_v$ | $(kN/m^2)^{-1}$ | Coefficient of volume compressibility (Eq. 9–9) |
| $P$ | m | Perimeter of thaw bulb (Eq. 9–9) |
| $R$ | — | Thaw-consolidation ratio (Eq. 9–14) |
| $S$ | — | Degree of saturation (Eq. 9–8a) |
| $S$ | $kN/m^2$ | Available shear strength (Eq. 9–2) |
| $S_{uu}$ | $kN/m^2$ | Undrained shear strength (Eq. 9–8) |
| $T_s$ | °C | Step temperature (Eq. 9–18) |
| $U$ | $kN/m^2$ | Pore pressure |
| UU or $Q$ | — | Unconsolidated undrained triaxial test |
| $W$ | kN | Weight of slice |
| $\phi$ | — | Angle of internal friction |
| $\phi'$ | — | Effective angle of internal friction |
| $\theta$, $\beta$ | — | Slope angle |
| $\sigma$ | $kN/m^2$ | Normal stress (Eq. 9–2) |
| $\rho$ | $kg/m^3$ | Unit weight |
| $\rho_b$ | $kg/m^3$ | Buoyant weight |
| $\rho_s$ | $kg/m^3$ | Saturated bulk density |
| $\rho_\omega$ | $kg/m^3$ | Unit weight of water |
| $\tau$ | $kN/m^2$ | Applied shear stress (Eq. 9–1) |
| $\omega$ | — | Water content |

scriptor and implies none of the mechanistic meanings normally attributed to these terms. The descriptive approach has been taken since many landslides are complex and it is desirable to avoid a mechanistic classification that may stress one process at the expense of another. A typical slope failure in frozen ground is illustrated in Fig. 9–1.

### 9.2.1 Flow

The term *flow* describes a movement having the characteristics of a viscous fluid When soil flows occur in a slide, there is substantial mobility and the preslide

**Figure 9–1** Road-cut slope failure in frozen ground.

topography is destroyed. Flow slides have been subdivided into skin, bimodal, and multiple retrogressive flows.

Skin flows are the detachment of a thin veneer of vegetation and mineral soil followed by rapid downhill movement (Fig. 9–2). They are commonly very shallow, long, and ribbonlike but may coalesce into broad unstable areas. Al-

**Figure 9–2** Skin flow near Root River, N.W.T., Canada. (After McRoberts and Morgenstern, 1974b.)

**Figure 9–3** Bimodal flow failure. (After McRoberts and Morgenstern, 1974.)

though skin flows develop on steep slopes, they are common on slopes of 6 to 9°. In the MacKenzie Valley, Canada, skin flows frequently occur in burnt-over areas where forest fires have reduced the insulation provided by vegetation and have allowed a deeper penetration of summer thaw. Skin flows show up clearly on air photos by changes in vegetation cover on slopes.

Bimodal flows describe a mass movement that has a biangular profile. The term implies two distinctly different modes of mass movement (Fig. 9–3). Well-developed forms of this type of slide have a low-angle tongue and a steep head scarp. The flow begins at a roughly semicircular head scarp which is the source area of the colluvium, which, in turn, forms the tongue. The tongue is an elongated, shallow, lobate flow mass that ends in the terminal area. The planar scarps may have inclinations up to 40° and are usually rich in ground ice. They are bare of vegetation but moss overhangs are common. Thin veneers of soil from the rapidly melting head scarps flow down the scarp face and then down the tongue. The tongues can develop slopes as low as 3°, and slopes as high as 14° have been noted. The moss or vegetated overhangs were not observed to be any longer than about 3 m (9 ft). Hence head scarp thaw would be expected to progress upslope until a scarp height of 3 m (9 ft) or so occurs.

Multiple-retrogressive flow is a subdivision that is needed as a transition between flows and slides. The movement presents an overall flow form, but within the slide mass a portion of the prefailure relief is preserved. There is a series of arcuate or bow-shaped ridges derived by failures of the receding head scarps. The classification as a subdivision of "flow" or "slide" is one of judgment and depends largely on the predominance of either the flow feature or the rotational feature.

### 9.2.2 Slide

Landslides, considered under the heading "slides," may be subdivided into multiple retrogressive, block, or rotational modes of failure.

A block slide involves a predominantly outward movement with a minor downward movement of a large soil mass. The failed mass remains essentially intact and the surface is capable of supporting upright living vegetation. Large V-shaped gullies often form on either flank of the slide mass due to runoff erosion.

The rotational failures are often small and generally occur in unfrozen deposits but may also be found in the head scarps of multiple retrogressive slides. They are similar to rotational slope failures in cold-temperate areas.

### 9.2.3 Fall

Falls occur in areas where lateral river erosion undercuts the frozen banks. These undercuts are often referred to as a thermal-erosion niche. The result of this erosion is a cantilevered soil block that fails in tension and topples into the river. This mode of failure may be the dominant mechanism in the lateral migration of rivers in permafrost areas.

As in many natural phenomena, the classification of mass movement that has been briefly presented does not have clear distinctions between categories. In general, the definitive categorization is of less consequence than an appreciation of the method of failure. If corrective or stabilizing measures are being considered, an understanding of the failure mechanism is the key issue.

### 9.2.4 Solifluction

Solifluction features have been reported by Washburn (1947, 1973), Benedict (1970), Carson and Kirkby (1972), and Embleton and King (1975). For geotechnical engineering purposes, solifluction is defined as a type of mass movement that occurs in the active layer of permafrost regions due to reduction of the shear strength of the thawing soils. Generally, fine-grained soils are required for solifluction movements to take place. It is often noted that solifluction is particularly active when ice-rich frozen ground thaws. During the summer-thaw season, most of these solifluction downward movements are activated and other movements might occur during the freeze-back time in early winter. Surface vegetation plays an important role in the appearance of solifluction movements. Generally, bulbous or lobate features are seen when the solifluction movements are restricted by vegetation. On the other hand, patterned ground features are observed if the slope has minimal surface vegetation or is barren.

## 9.3 GENERAL SLOPE ANALYSIS

Some of the slope-stability methods of analyses used in unfrozen soils may be used in frozen ground and in thawing frozen ground. As a starting point, the most simplified approach is termed *infinite slope analysis,* in which the height of the slope is insignificant in comparison to the length of the slope. As shown in Fig. 9–4, infinite slope analysis is applicable to those problems in which the failure plane is approximately parallel to the slope angle. This method may also be applied to shallow instability problems, such as solifluction slopes and tongues of bimodal flows.

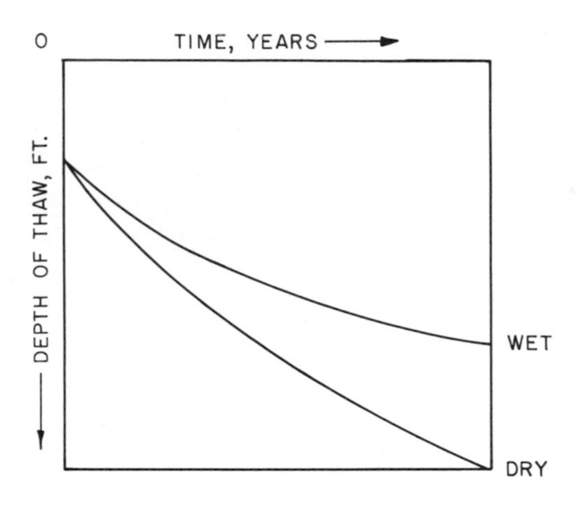


a.) TIME vs. DEPTH OF THAW

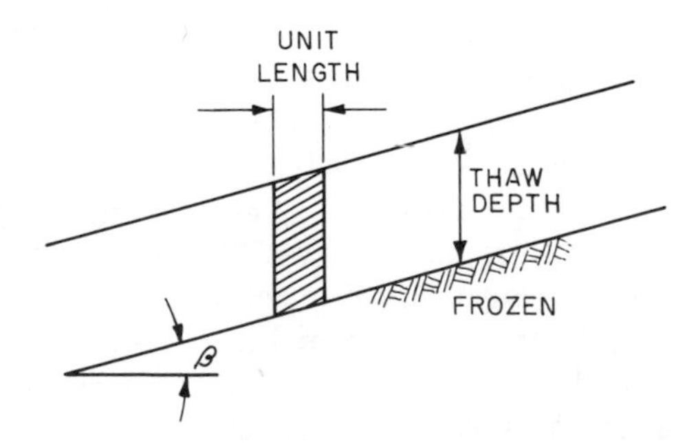


b.) INFINITE SLOPE CONDITION

**Figure 9–4** Typical thaw geometry.

The general safety factor ($F$) for an infinite slope may be expressed as

$$F = \frac{\text{available sheer strength}}{\text{applied shear stress}} = \frac{S}{\tau} \qquad (9\text{–}1)$$

The available shear strength is given by Terzaghi's effective stress principle as

$$S = C + (\sigma - U) \tan \phi \qquad (9\text{–}2)$$

where $C$ = cohesion
$\sigma$ = normal stress
$U$ = pore water pressure
$\phi$ = angle of internal friction

Referring to Fig. 9–4, the applied shear stress on a unit length of an infinite slope is given by

$$\tau = \rho H \cos\theta \sin \theta \qquad (9\text{–}3a)$$

and the available shear strength

$$S = C + (\rho_s H - \rho_w H) \cos^2 \theta \tan \phi \qquad (9\text{–}3b)$$

From (9–1) we have

$$F = \frac{C + (\rho_s H - \rho_w H) \cos^2 \theta \tan \phi}{\rho H \cos \theta \sin \theta}$$

where $H$ = height of slope
$\rho$ = unit weight of soil
$\rho_s$ = saturated unit weight of soil
$\rho_w$ = unit weight of water

or

$$F = \frac{C + \rho_b H \cos^2 \theta \tan \phi}{\rho H \cos \theta \sin \theta} \qquad (9\text{–}4)$$

where $\rho_b$ = buoyant unit weight = $\rho_s - \rho_w$. For cohesionless soils such as sands and gravels, $C = 0$, and if there is no groundwater level, Eq. (9–4) becomes

$$F = \frac{\tan \phi}{\tan \theta} \qquad (9\text{–}5)$$

If the groundwater level is considered, Eq. (9–4) becomes

$$F = \frac{\rho_b \tan \phi}{\rho \tan \theta} \qquad (9\text{–}6)$$

Since the $\rho_b/\rho$ ratio is approximately 0.5, the seepage condition reduces the safety factor to one-half of that slope without seepage (dry).

The most commonly used method of slope analysis in slopes of finite height is the *method of slices,* in which the failure surface is assumed to be an arc of a circle. Detailed analyses of the slope involve the division of the trail failure mass into a series of vertical slices. The safety factor of the slope is given by the ratio between the sum of the resisting forces and the activating forces of all slices. Details of this method are given by Terzaghi and Peck (1967). The circular arc failure surface is generally applied in homogeneous soil deposit conditions.

In stratified soil deposits, the failure surface may be observed along the contact between the dissimilar strata or along any weaker plane dipping at angles other than that of the bedding plane. In such conditions, the method of noncircular analysis (Morgenstern and Price, 1965) is generally used.

## 9.4 SLOPE STABILITY IN FROZEN GROUND

As described in Chapter 3, the deformation characteristics of frozen ground are a function of soil type and soil-type composition, applied stress, temperature, and time. In most cases, maintaining the existing thermal condition of the frozen ground averts slope instability. However, external loading conditions or change in environmental factors may increase the probability of slope failure to occur in frozen ground. For example, road construction on the side of a slope consisting of frozen ground may lead to slope movements causing significant damage to the roadway. High roadway embankment built on ice-rich frozen ground may also lead to slope movements due to creep of frozen ground. Morgenstern (1981) reviewed the creep in a natural frozen ground. Referring to the glaciated terrain of the MacKenzie Valley, Canada, he reported that slope failures occurred both through frozen ground and through thawing ground. The failures through the thawing ground were caused by high rates of thaw generating pore pressures, high rates of ablation at ice-rich faces, or a variety of more conventional mechanisms in previously thawed material. Failure through frozen ground generally was restricted to large-scale features.

As discussed in Chapters 3 and 6, the creep of ice will provide a sensible upper bound to the creep of ice-rich frozen soil. McRoberts (1975) adopted infinite slope analysis to calculate downslope velocities as a function of depth of ice-rich soil and slope inclination. For relatively warm ice (say, warmer than $-4°C$) the analysis indicated that surface velocities of about 10 cm/yr might be expected on a slope with 10 m of ice-rich soil inclined at 15° to the horizontal. This is a very aggressive geomorphological process. Based on a case study, Morgenstern (1981) reported the creep movement of a natural slope consisting of ice-rich permafrost (Fig. 9–5).

Figure 9–5 shows the stratigraphic cross section of the subsoil condition of the slope studied; the clay layers overlying the glacial till are fissured throughout. The clay layers contain reticulate ice veins. The case history showed that the slope movements are typically smaller and much less abrupt where zones contain ice

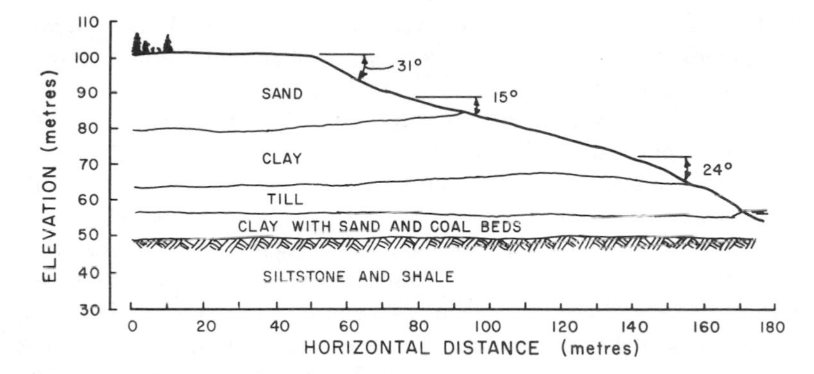


**Figure 9–5** Stratigraphic cross section of subsoils conditions. (After Morgenstern, 1981.)

lenses separated by less than 1 m and the natural moisture content of soil is at least 25 to 30%. The velocity at the top of the clay layer (Fig. 9–5) is between 0.25 and 0.30 cm/yr. Above the 29-m depth where ice lenses are large and closely spread, the velocity gradient is almost uniform. The shear strain rate through this zone is approximately $2 \times 10^{-4}$ per year. At depths of approximately 20 to 34 m, where large ice lenses are more widely separated, the velocity is erratic, with proportionally more movement associated with the large ice lenses. Below the 34-m depth, where only small ice lenses are present, the velocity gradient becomes more uniform, with a shear strain rate of about $0.4 \times 10^{-4}$ per year.

Slope failures by falls occur in riverbank areas (or along the seacoast) where lateral erosion undercuts the frozen banks. These undercuts are often referred to as a *thermal erosion niche*. The result of this erosion is a cantilevered soil block that fails in tension and topples into the river or sea. This mode of failure may be the dominant mechanism in the lateral migration of rivers in permafrost areas.

Of prime importance when assessing slope-stability questions in frozen ground is the soil type, ice content, and temperature of the frozen ground. If the frozen state of soil can be maintained, the slope movement may not be significant and the slope stability may not be of major concern (creep may cause some movement, as mentioned earlier). Only external disturbances may cause the frozen ground to thaw, depending on the climate, surface vegetation, hydrology, nature and composition of soils, and other factors. Thawing of frozen ground may lead to major slope-instability problems. The specific slope stability of thawing ground is discussed in the following section.

McRoberts (1973) describes slope failures in discontinuous-permafrost regions. On rivers, ice jams and high flows may flood parts of the valley walls. Subsidence of high water may bring about rapid draw-down failures.

Heavy rains, even of short duration, can be a major cause of skin flows and speed up the action of bimodal failure in frozen ground. The warm rain increases the weight of the active layer and reduces the effective stress in the soil. There may also be a thickening of the active layer from the warmth of the rain, and this thawing may produce pore pressure. These factors contribute further to slope instability in frozen ground.

The presence of vegetation is a major consideration in slope behavior in frozen ground. It has been noted that there is an increase in the incidence of skin flows following a forest fire. After fires occur, there is a decrease in the insulation value of the vegetative mat and an increase in the thickness of the active layer. A fire also embrittles the surface mat of moss and lichens, inhibiting its ability to drape over a thawing slope and to re-root itself. Clearing for construction purposes has the same effect as fire, even if care is taken to remove trees by hand. Equipment can compress the vegetation and decrease its insulation value. The mechanistic method of slope analyses discussed in the preceding section cannot be applied in frozen ground. These analyses may be used only in thawing ground or in discontinuous frozen-ground conditions.

Evidence of deep-seated landslides in permafrost in the MacKenzie River Valley, Canada, is documented by McRoberts and Morgenstern (1973, 1974b). Their work can be summarized as follows. Based on field inspections of a wide range of landslides, toe erosion is one of the main elements of landslide activity. Block and multiple retrogressive slides can be found throughout the MacKenzie River Valley and its tributaries, where intense toe erosion is occurring. These slides in the MacKenzie Valley are essentially restricted to glaciolacustrine soils, although they can be found in certain fill materials.

Another necessary condition for block and multiple-retrogressive slide development is related to the morphology of the slides. The minimum bank height that failed in this mode ranges from 30 to 60 m (98 to 197 ft), and overall shape angle at failure ranged from 9.5 to 20°. From morphologic evidence, it appears that block or multiple-retrogressive slide movements in the MacKenzie Valley are unlikely in frozen or partly frozen slopes less than 30 m (98 ft) high and less than 9° in slope. The final condition is that permafrost bottoms out near river level. These landslides are also readily apparent on aerial photographs. The air photographic interpretation is one of the most useful techniques for identifying landslides occurring on natural slopes. Aerial photographs give very useful information for the initial slope analysis, especially in remote areas where access is limited.

The long-term strength of frozen soils must be considered when studying the deep-seated slides found in frozen ground. The methods discussed in Chapters 3 and 10 may be used to determine the long-term strength of frozen soils, which essentially assumes the cohesive response. McRoberts and Morgenstern (1973) proposed that the long-term strength of ice-poor or structural soil is frictional. Both cohesive and frictional responses were used to analyze the stability of a landslide. It was concluded that stability analyses that used a frictional response for the frozen soil gave reasonable agreement. Knowledge of effective stress of unfrozen soils encountered at river level will be required for a complete analysis.

## 9.5 SLOPE STABILITY IN THAWING FROZEN GROUND

Two primary factors that cause thawing of frozen ground are (1) disturbance of soil surface conditions by destruction, removal, altering, or covering the surficial vegetative mat; and (2) heat input to the frozen ground from a heated structure, resulting in a rise in the ground operational temperature to above 0°C or 32°F. These disturbances are often the result of road construction, heated oil pipeline construction, removal of natural vegetation cover or snow cover, and increased temperature variations in the underlying soil. These disturbances can result in higher mean annual ground temperatures and the amplitude of seasonal variations in the soil-surface temperature. The dominant variables involving the thaw of a frozen ground are the ground-surface temperature, the thermal properties of soils and its composition, and the total quality of unfrozen water in the frozen soils.

Accordingly, thawing soils can be classified into three categories: (1) thawing soils of no concern, such as thaw-stable sand and gravel; (2) thawing soils of concern such as silty soils; and (3) thawing soils of great concern, such as ice-rich silty soils or fine-grained soils. The geological background of soil deposit formation is very helpful in classifying thawing soils and in determining the nature of ice inclusions in a particular frozen-ground location. Determination of thaw depth under different conditions is discussed in Chapter 4.

When a frozen soil begins to thaw, producing a *thaw bulb,* the generation of melting water may cause excess pore-water pressure, which reduces the shear strength of the thawed soil. Thus an instability within the thaw bulb may be created where the rate of melting is in excess to the flow of water. In addition, loss of shear strength during thawing may cause overstressing of weak soils. Especially in fine-grained soils, where the hydraulic conductivity and dissipation of excess pore water are slow, reduction or complete loss of shear strength which can occur during the thawing period can result. Therefore, the loss of strength in thawing soil depends mainly on the amount of ice in the soil, the type of soil, and its relationship between the rate of thawing and the potential rate of dissipation of excess water.

Potential instability of thawing frozen ground may be analyzed by a simplified procedure (Phukan, 1976). A sophisticated method (McRoberts and Nixon, 1977) may also be used. The sophisticated method requires more detailed field investigation to collect required data.

### 9.5.1 Simplified Procedure

In the simplified method developed by the author, the stability of the thaw bulb produced by surface disturbances as well as heat input into the ground is analyzed. A typical thaw geometry is presented in Fig. 9–4.

It is apparent from Fig. 9–4 that the stability of a thaw bulb will be critical where the ground slope is significant. If thawing occurs at a slow rate, any excess water generated will flow from the thawed soil at about the same rate as it is produced. No excess pore pressure will be sustained. However, if thawing occurs at a faster rate in relation to the ability of the thawed soil to drain away excess water, excess pore-water pressure will be generated near the thaw front. The phenomena of rate of thawing, water drainage of flow, and generation of excess pore-water pressure are related to several factors, such as amount of ice in the frozen soil, compressibility, and hydraulic conductivity. Instability of a thaw bulb may lead to failure of the slope and associated structures. Considerations must be given to assure that the instability of the thaw bulb will not have any significant adverse effect on the structure and the surrounding environment.

The physical model selected for the stability analysis is an infinite slope (Fig. 9–4) where the equilibrium of a unit length of section is considered. The thaw depth to be considered will depend on various factors, such as the soil type and its thermal properties and the surface temperature. A typical thaw-depth value of 6 m (20 ft) is considered in the design chart for silt (Fig. 9–8).

The thaw-bulb stability analysis consists of four steps:

1. Establishment of procedures and of a physical model for the analysis.
2. Determination of the shear stress required to resist the driving forces tending to cause instability. These driving forces are the component of the gravity force of a unit section considered and earthquake forces acting in a horizontal direction.
3. Determination of the ultimate shear strength available at the thaw face; this strength is dependent on the flow conditions in the thaw bulb.
4. Determination of the safety factor, which is defined as the ratio of the resisting forces to the driving forces, or the ratio of the available shear strength to the stress required to maintain equilibrium.

The shear stress required to maintain equilibrium of the thaw bulb for a given ground slope in the potential failure direction depends on the size, shape, and unit weight of the thaw bulb and on the maximum earthquake effects to be sustained.

The gravity-force component tending to cause thaw-bulb instability equals the weight of the soil in a 1-ft section length of thaw plug multiplied by the sine of the slope angle (Fig. 9–4). The effect of the horizontal seismic force is calculated as the slice mass $W$ times the design acceleration $K$, times the cosine of the slope angle $\beta$, as shown in the following equation:

$$\text{horizontal seismic force} = WK\cos^2\beta \tag{9–7}$$

Possible laboratory soil tests to represent the shear strength at the thaw face are the unconsolidated-undrained (UU or $Q$), the consolidated-undrained (CU or $R$), and the consolidated-drained (CD or $S$) triaxial compression tests.

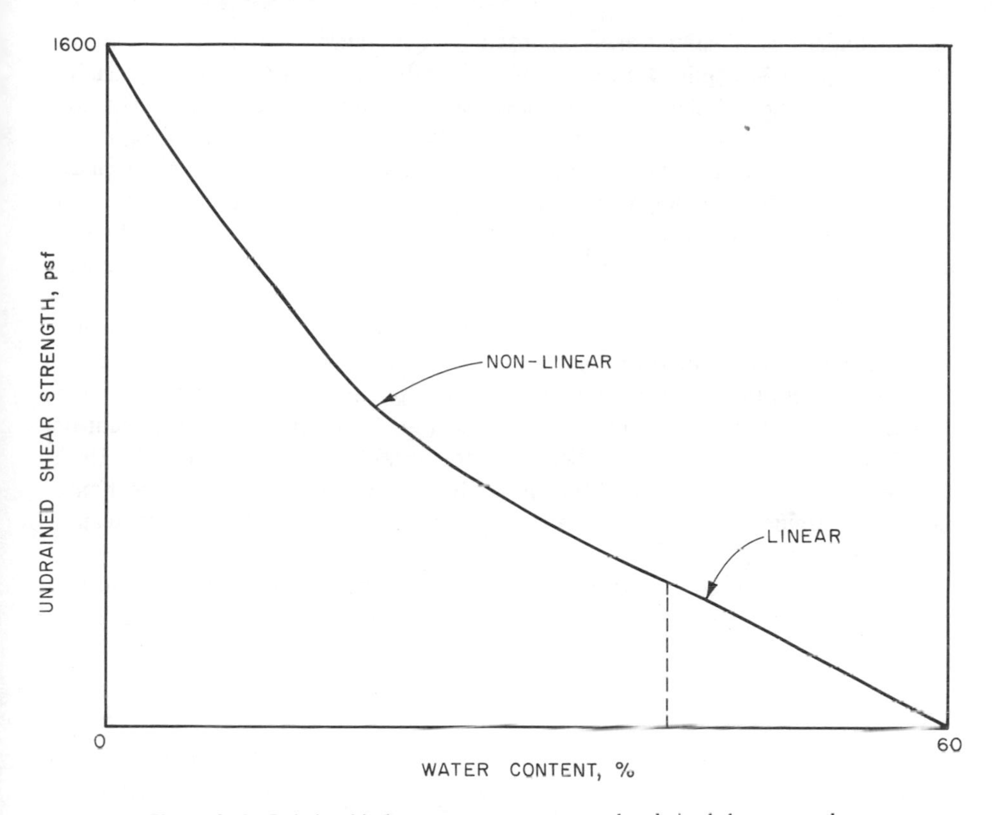


**Figure 9–6** Relationship between water content and undrained shear strength.

The UU test condition corresponds to undrained thawing followed by undrained shear; water is not allowed to drain from the sample during any phase of the test. The UU test condition would represent quite closely the limiting case of a completely undrained thaw soil, followed by undrained shearing in soil at its face.

The CU test is made by allowing full consolidation after thawing followed by shearing with no drainage permitted. This condition represents the limiting case of slow thaw with complete dissipation of excess water followed by rapid shearing: as, for example, during an earthquake.

The CD test is made by allowing full drainage after thaw, followed by slow shearing with full drainage allowed at all times. This test represents the limiting case of slow thawing without buildup of excess pore pressures followed by slow, essentially static shearing. The test approximates conditions of gravity forces without earthquake effects. It can be expected that of the three strength values, the UU strength would be lowest, the CU strength intermediate, and the CD strength highest.

Because a considerable portion of the total shear stress applied on the thaw-bulb boundary is dynamic stress imposed by earthquake action, the use of an undrained shear strength to assess thaw-bulb stability is appropriate and usually conservative. Further, the in situ undrained strength must lie between the UU and CU values, depending on the degree of consolidation achieved at the thaw front. If full consolidation is achieved during thaw, such as when thawing takes place slowly enough that no appreciable excess pore pressures are created, the CU strength applies. On the other hand, if no significant consolidation is achieved during thaw, such as for the case of very rapid thawing in fine-grained soil, the UU strength applies.

The safety factor is determined from the ratio of the available shear strength to the shear stress required to maintain equilibrium. From various field observations, it is found that there is a definite relationship between the water content and undrained shear strength of soil, as shown in Fig. 9–6. As the water content of soil is decreased, the undrained shear strength of soil will increase linearly up to a point, after which the relationship is nonlinear. It is apparent from the figure that the undrained shear strength of soil is practically negligible at high water-content levels.

To expedite design, curves may be prepared relating slope inclination (β, critical thaw depth ($H_c$), and the water content of soil (Fig. 9–7). These relation-

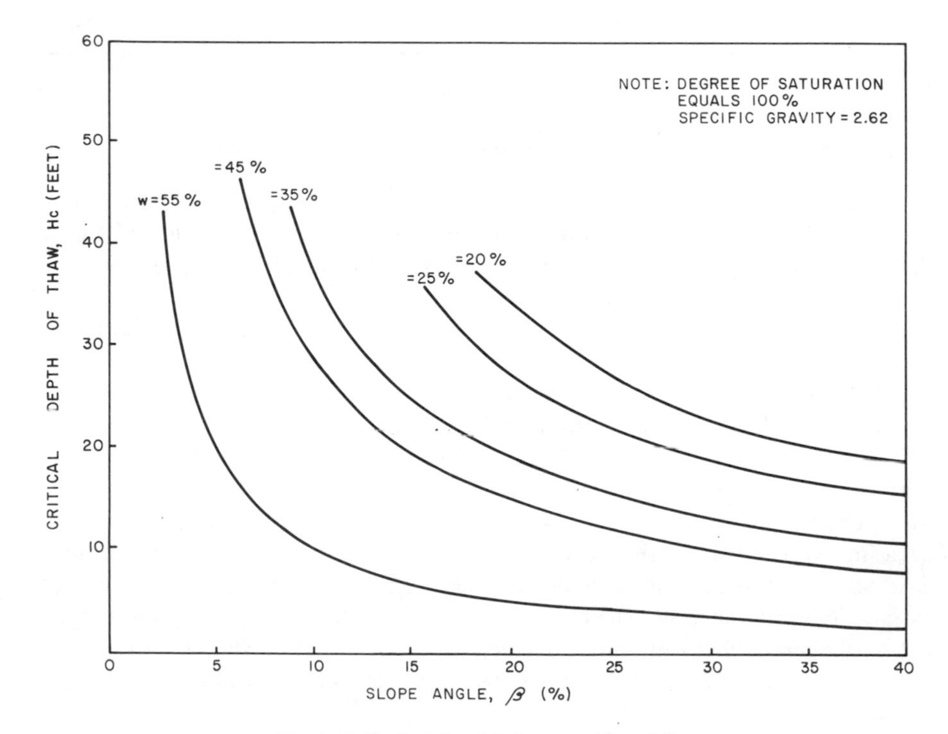


**Figure 9–7** Relationship between $H_c$ and β.

ships may be obtained from the infinite method of analysis as discussed in Section 9.3.

$$\text{safety factor} = \frac{\text{available shear strength}}{\text{required shear stress}}$$

or

$$F = \frac{S_{uu}}{\rho H_c \cos\beta \sin\beta} = \frac{2S_{uu}}{\rho H_c \sin 2\beta}$$

or

$$H_c = \frac{2S_{uu}}{F\rho \sin 2\beta} \tag{9–8a}$$

where ρ is given by

$$\frac{(1 + \omega)\rho_w}{\rho_w/S + 1/G_s} \tag{9–8b}$$

where $S_{uu}$ = undrained shear strength
$\beta$ = slope angle
$\omega$ = water content
$\rho_w$ = unit weight of water
$S$ = degree of saturation
$G_s$ = specific gravity of soil particles

The relationship between the undrained shear strength of soil and the water content of soil as presented in Fig. 9–6 is used to determine different values of $H_c$ for various values of slope angle and water content (Fig. 9–7).

Generally, the thaw-bulb instability is anticipated in silty soils. The shear strength of silt may be represented by

$$S = \begin{cases} 250\ \text{lb/ft}^2 + \sigma \tan 11° & \text{(loose silt)} \\ 500\ \text{lb/ft}^2 + \sigma \tan 12° & \text{(dense silt)} \end{cases}$$

where σ is the normal stress in $\text{lb/ft}^2$. The shear-strength values above are typical for Alaska's silty soils.

Considering a pseudostatic dynamic force and a water level at thaw line, the safety factor can be expressed as

$$F = \frac{CP + W\cos^2\beta \tan\phi}{(\sin\beta\cos\beta + K\cos^2\beta)W} \tag{9–9}$$

where $C$ = cohesion
$P$ = perimeter of the thaw bulb

$W$ = weight of slice considered
$\phi$ = angle of internal friction
$K$ = coefficient of seismic acceleration

Considering seepage parallel to the slope and the water level at the surface, the safety factor can be expressed as

$$F = \frac{CP + \rho_b \cos^2 \beta \tan \phi}{(\sin \beta \cos \beta + K \cos^2 \beta)W} \tag{9–10}$$

where $\rho_b$ is the buoyant weight of the soil.

Equations (9–9) and (9–10) are solved for different values of silt and the results are shown in Fig. 9–8.

**Example 9–1**

Given: an infinite slope angle of 40° in homogeneous frozen sandy silt with the undrained shear strength ($S_{uu}$) of 1000 lb/ft$^2$, $\phi = 25°$, $\rho = 100$ lb/ft$^3$, and $\rho_S = 120$ lb/ft$^3$. Required: the critical depth of thaw and the safety factor.

### Solution

Critical depth of thaw from Eq. (9–8a):

$$H_c = \frac{2S_{uu}}{F\rho \sin 2\beta}$$

$$= \frac{2\ (1000)}{(100) \sin 2(40)} \quad \text{(generally, } F = 1 \text{ is used to determine the most critical factor)}$$

$$= 20.3 \text{ ft}$$

Safety factor:

$$F = \frac{\rho_b \tan \phi}{\rho \tan \beta} = \frac{(120 - 62.4) \tan (25°)}{100 \tan (40°)} = 0.32$$

**Example 9–2**

Determine the safety factor of a thawing slope for the following conditions:

- Slope angle $\beta = 25°$
- Angle of internal friction of soil $\phi = 35°$
- Water content $\omega = 35\%$
- Degree of saturation = 90%
- Specific gravity of soils $G_s = 2.76$

### Solution

Unit weight of soil:

$$\frac{(1 + \omega)\ \rho_w}{\omega/S + 1/G_s} = \frac{(1 + 0.35)(1 \text{ g/cm}^3)}{0.35/0.9 + 1/2.76}$$

$$= 1.80 \text{ g/cm}^3 = 1800 \text{ kg/m}^3$$

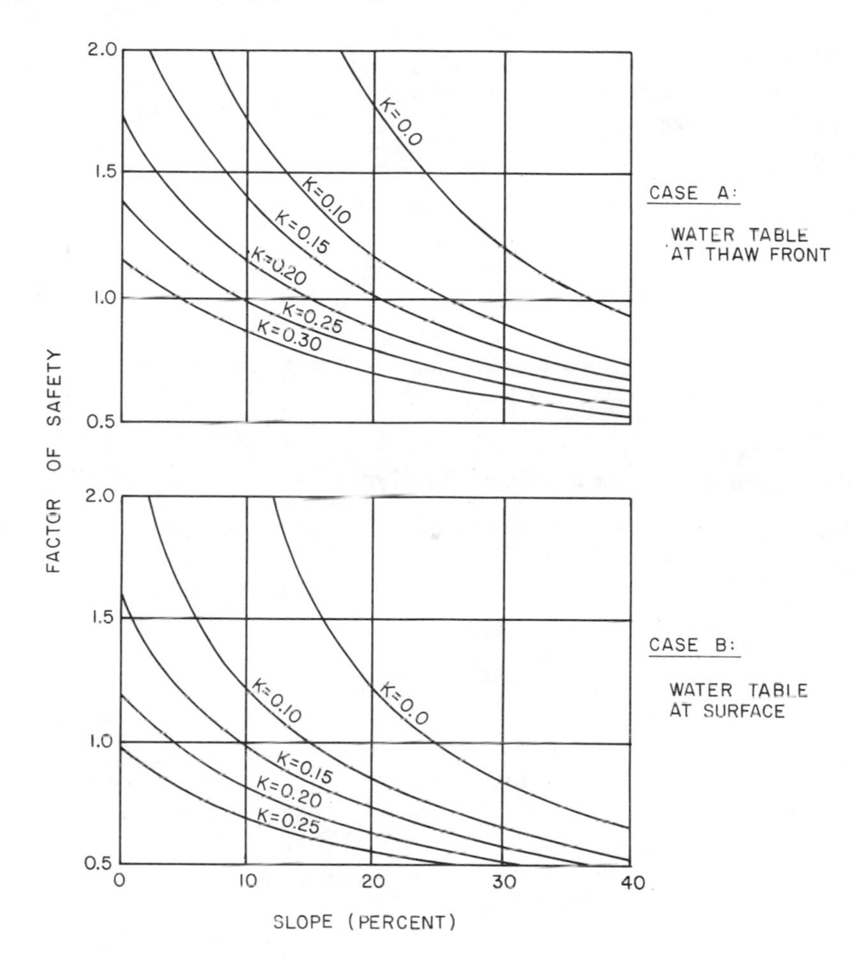


**Figure 9–8** Relationship between safety factor and slope for loose silt.

Assuming water content at saturation = 40%,

$$\rho_s = 1800(1 + 0.4) = 2520 \text{ kg/m}^3$$

Safety factor:

$$\begin{aligned} F &= \frac{(\rho_s - \rho_w) \tan \phi}{\rho \tan \beta} \\ &= \frac{(2520 - 1000)}{1800} \frac{\tan (35)}{\tan (25)} \\ &= 1.3 \end{aligned}$$

**Example 9–3**

In Example 9–2, if seismic forces are considered with $K = 0.2$, estimate the change in the safety factor of the slope.

Solution

From Eq. (9–9) we have (water level at thaw line)

$$F = \frac{\cos^2 \beta \tan \phi}{\sin \beta \cos \beta + K \cos^2\beta}$$

$$= \frac{\cos^2(25) \tan 35°}{\sin (25) \cos (25) + (0.25) \cos^2 (25)}$$

$$= 0.97$$

## 9.5.2 Other Methods of Slope Analysis

McRoberts and Nixon (1977) presented the following solution for the stability of thawing slopes (Fig. 9–9):

CASE 1: SEEPAGE SLOPE ON PLANE A–A

$$\text{pore pressure} = \rho_w d \cos \beta$$

$$\text{total stress} = \rho_s d \cos \beta \qquad (9\text{–}11)$$

$$\text{effective stress} = (\rho_s - \rho_w) d \cos \beta$$

$$= \rho_b d \text{ Cos } \beta$$

$$\text{safety factor } F = \frac{\rho_b \cos \beta \tan \phi'}{\rho \sin \beta}$$

$$= \frac{\rho_b \tan \phi}{\rho \tan \beta} \qquad (9\text{–}12)$$

where $\phi$ = effective angle of internal friction
$\beta$ = slope angle
$\rho$ = unit weight of soil
$\rho_b$ = buoyant unit weight of soil

Equation (9–12) is the same as Eq. (9–6).

CASE 2: THAW SLOPE ON PLANE A–A′. The moving thaw boundary is defined by the Neumann solution:

$$d = \alpha\sqrt{t} \qquad (9\text{–}13)$$

where $d$ = thaw depth
$\alpha$ = constant (discussed in Chapter 4)
$t$ = time

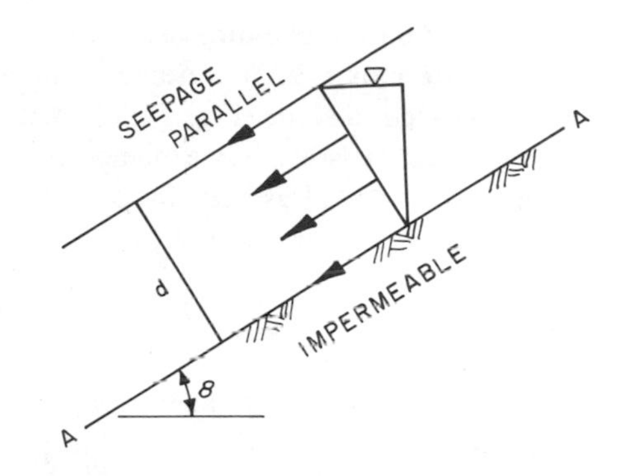


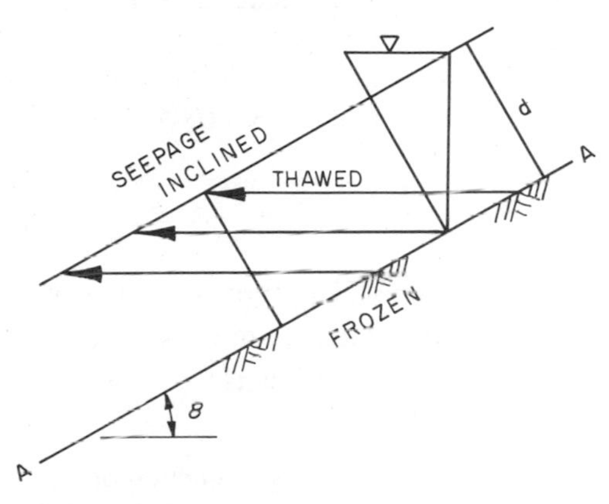


**Figure 9–9** Thawing slope analysis. (After McRoberts and Nixon, 1977.)

A solution may be obtained in term of the thaw consolidation ratio $R$ given by

$$R = \frac{\alpha}{2\sqrt{C_v}} \tag{9–14}$$

where $\alpha$ is defined by Eq. (9–13) and $C_v$ is the coefficient of consolidation (discussed in Chapter 4). The ratio expresses the relative influence of the rate at which water is produced by thaw and the rate at which it may be squeezed out of the thawed soil overlying the moving thaw interface. For an infinite soil mass thaw consolidation under self-weight conditions the excess pore pressure is given by

$$U = \frac{\rho_b d}{1 + 1/2R^2} \tag{9–15}$$

where $\rho_b d$ is the effective stress after the completion of excess pore pressures.

Based on the one-dimensional thaw-consolidation theory discussed in Chapter 4, McRoberts (1972) presented a method to predict the stability of a thawing slope. Consider a slope (Fig. 9–9b), where a thaw front has penetrated to a depth, $d$. The effective stress on a plane A–A, after the dissipation of excess pore is given in Eq. (9–11) as $\rho_b d \cos \beta$. It is analogous to Eq. (9–15), that a measure of the excess pore pressure, $U$, on A–A′ is

$$U = \rho_b d \cos \beta \frac{1}{1 + 1/2R^2} \tag{9–16}$$

Applying a statical balance of forces, the safety factor becomes

$$F = \frac{\rho_b}{\rho} \frac{1}{1 + 2R^2} \frac{\tan \phi'}{\tan \beta} \tag{9–17}$$

It can be seen that if no thaw occurs or if no excess pore pressures are generated, Eq. (9–17) reduces to Eq. (9–12). Equation (9–17) can be solved in terms of $R$ and water content.

The solution to Eq. (9–17) can be extended if it is assumed that $\alpha$ is governed by the Stefan solution (Nixon and McRoberts, 1973) as

$$d = \sqrt{\frac{2k_u T_s}{L}} \sqrt{t} \tag{9–18}$$

where $T_s$ = step temperature causing thaw
$K_u$ = conductivity of thawed soil
$L$ = volumetric latent heat of the frozen soil
$\alpha = \sqrt{2k_u T_s / L}$

In Eq. (9–18), $\alpha$ is a function of $K_u$ and $L$. If it is assumed that all the water in a soil of water content $\omega$ is frozen and if the unfrozen conductivity is defined as a function of water content, then both $k_u$ and $L$ are uniquely defined at a given $\omega$. As the solution for $\alpha$ is more-or-less independent of frozen-ground temperature (Nixon and McRoberts, 1973), it is possible to isolate the dominant effects of temperature and coefficient of consolidation $C_v$, in reducing equilibrium slope angles.

The rate at which excess pore fluid is liberated in a thawing soil is governed by the rate of advance of the thaw front. If thaw proceeds in accordance with Eq. (9–13), the thaw-consolidation theory derived by Morgenstern and Nixon (1971) could be applied to thawing problems in natural active layers. The constant $\alpha$ is discussed in Chapter 4 and McRoberts (1972) gave a range of values that lie between 0.1 and 1.0 $\text{mm/s}^{1/2}$.

The value of $C_v$ may range over a wide spectrum and may be obtained from the direct results of consolidation tests or can be estimated as

$$C_v = \frac{k}{m_v \rho_w} \tag{9–19}$$

where $k$ = coefficient of permeability
$m_v$ = coefficient of volume compressibility
$\rho_w$ = unit weight of water

$m_v$ is given by Eq. (4–34). So the value of $C_v$ depends on both permeability and compressibility of soil.

The rate at which excess pore fluid can be squeezed out of the thawed soil above the thaw interface is governed by the coefficient of consolidation $C_v$. The value $C_v$ may easily range from $10^{-1}$ cm²/s for sandy soils to $10^{-5}$ cm²/s or less for fine-grained clayey soils. While $C_v$ enters the $R$ value under the square root, its potential variation and the factors affecting it make the degree of uncertainty associated with obtaining the correct $C_v$ value much higher. Therefore, it may be concluded that the analysis of thaw slope stability resolves itself into a requirement for a detailed knowledge of geotechnical properties of thawing soils such as $C_v$ and the effective stress parameters ($C'$ and $\phi'$). It is less dependent on a detailed knowledge of the thermal solution.

**Example 9–4**

Given: the slope angle 30° in a homogeneous frozen sandy silt layer with $C = 500$ lb/ft³, $\phi = 20°$, $\rho = 90$ lb/ft³, and $\rho_s = 110$ lb/ft³.
Required: the safety factor of the slope if the slope thaws at a rate $\alpha = 4$ ft/yr$^{1/2}$ and $C_v = 200$ ft²/yr.

Solution

Thaw consolidation ratio:

$$R = \frac{\alpha}{2\sqrt{C_v}} = \frac{4}{2\sqrt{200}} = 0.14$$

Safety factor:

$$F = \frac{\rho_s - \rho_w}{\rho} \quad \frac{1}{1 + 2R^2} \quad \frac{\tan\phi}{\tan\beta}$$

$$= \frac{110 - 62.4}{90} \quad \frac{1}{1 + 2(0.14)^2} \quad \frac{\tan 20}{\tan 30} = 0.32 < 1$$

The slope will fail.

## 9.6 SLOPE STABILIZATION IN FROZEN SOILS

The type of slope-stabilization methods selected will depend primarily on the amount and type of ground ice, the soil type, the depth of the cut and the slope angle, availability of granular material, and potential environmental impact

Some of the recommended practices for slope stabilization in cut slopes are discussed in the following sections.

### 9.6.1 Self-Healing Cut Slopes

Many cut slopes will be self-stabilizing over the period of one to two thaw seasons. If the maximum retreat is in the order of 2 to 4 m and if the results of the degradation do not seriously interfere with the functional aspects of the project or pose a potential hazard to the environment, a self-healing approach may be applied.

Pufahl and Morgenstern (1980) reported that self-healing will be common in land forms that are composed of till, colluvial or alluvial silt and sand, and other combinations of rock and soil that are relatively free-draining and stabilize at reasonably steep angles. A number of helpful design and construction procedures that can be used for the self-healing process are as follows.

1. Use steep and short slopes that will intersect fewer ice wedges than will long flat slopes.
2. Use wide ditches at the toe of the cut slope, which will provide space for accumulating debris and will allow the placement of rock revetments.
3. Place rock revetments at the toe of the slope to retain the soil and allow the water released from the melting of frozen soils to drain into the ditch. The material may be crushed rock or gravel. The rock revetments may be placed mechanically and suitable dimensions are approximately 2 m wide and 1 to 1.5 m high.
4. Trees should be hand-cleared back from the top of the slope to a distance equal to one to two times the height of the slope. The stumps, cut no higher than 0.3 m, can be tied together with a coarse wire mesh to prevent or reduce damage of the organic mat as it moves into the slope (Fig. 9–10). Seeding of the lower portions of the slope should commence as soon as the slope becomes relatively stable as illustrated in Fig. 9–10.
5. Divert the surface runoff above the back slope by the use of dikes rather than ditches. Special care must be taken not to disturb the organic cover above the slope during all construction activity.
6. Limit the depth of cut to 4 to 5 m whenever possible if the self-healing techniques are to be successful.
7. Do rapid excavation with immediate replacement of granular subgrade material if construction is occurring during the summer period.

### 9.6.2 Other Measures for Slope Stabilization

In terrain units that possess less favorable goetechnical properties related to soil and ground ice conditions, a more positive approach must be taken to prevent uncontrolled and long-term recession of slopes. Several possible techniques are illustrated in Fig. 9–11.

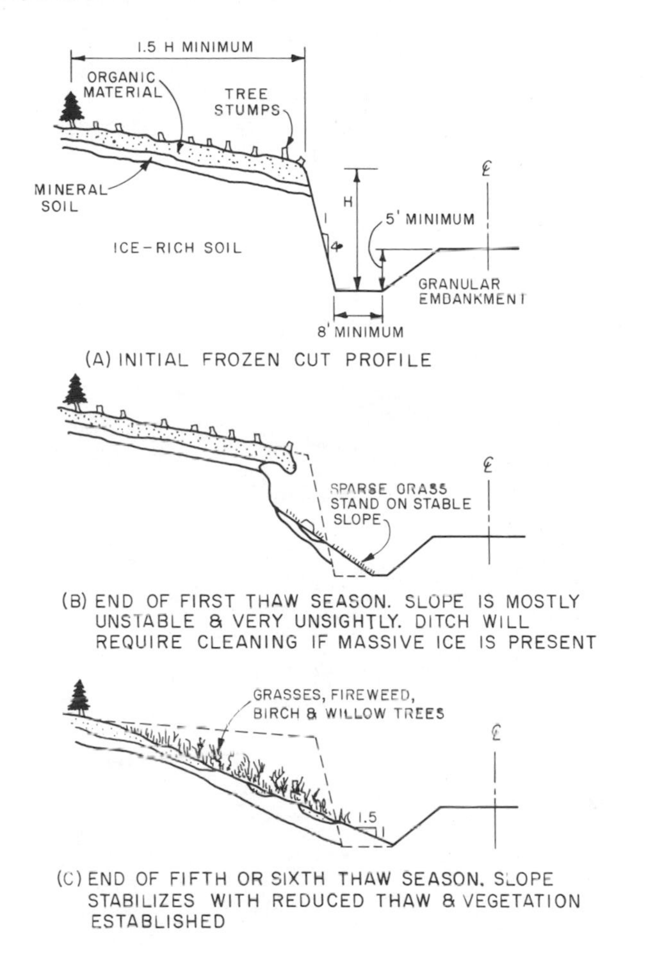


**Figure 9–10** Stability of ice-rich cut slopes. (After Berg and Smith, 1976.)

A suitable stabilization technique for steep slopes or exposed head scarps with established biangular profiles is illustrated in Fig. 9–11a. A large quantity of gravel is required for this design measure. Styrofoam insulation (or its equivalent) may be used near the top of the slope to reduce the thickness of gravel in that part of the exposure.

Figure 9–11b illustrates a plausible design for steep cut slopes of 15 to 50°. The choice of lower bound was based on friction angle of gravel on frozen ground containing significant quantities of massive ice. This aspect of the design may require further consideration by some model and field tests of sand/gravel on melting ice. The cut surface is steeped to provide mechanical interlocking of the frozen soils and the gravel cover. The inclusion of artificial insulation is an optional feature in this design and will depend on the abundance of granular materials. Figure 9–11c illustrates a stabilization technique for slopes of 15 to 20° or less.

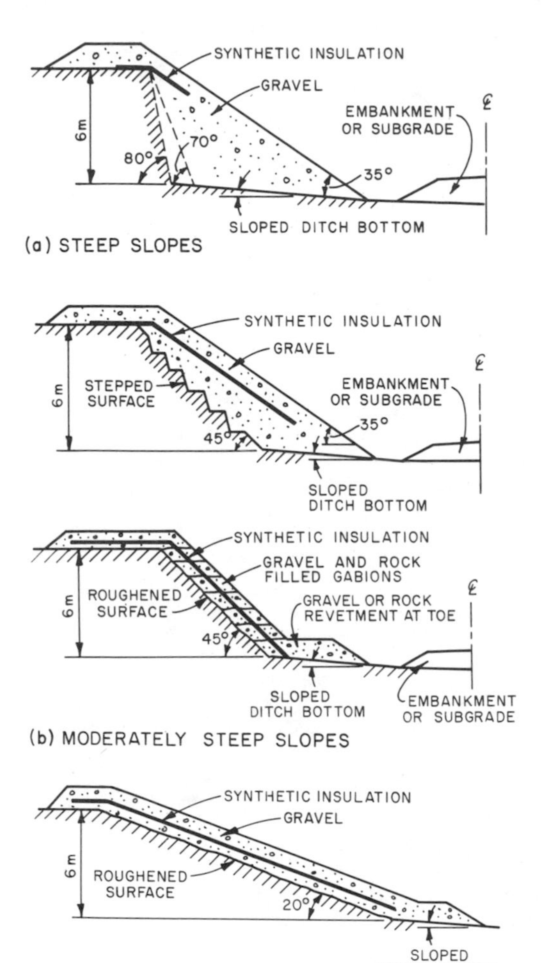


**Figure 9–11** Stabilization of cut slopes. (After Pufahl and Morgenstern, 1980.)

Again, the inclusion of insulation is optional in this design. The surface of the cut should be roughened to ensure some degree of mechanical interlocking between the gravel and frozen ground.

All of the foregoing approaches should provide drainage of meltwater, both on the slope and at the toe. Sufficient strength, flexibility, stability, and durability of the cover as well as insulation must be provided. The thickness of insulation provided must be sufficient to prevent thawing or reduce the rate of thawing to be an acceptable value. It is important to emphasize that all construction and stabilization or preservation measures should be undertaken during the months of the year when freezing air temperatures prevail.

## PROBLEMS

**9–1.** Discuss different modes of slope failures in frozen ground. What design measures are effective to stabilize thawing slopes in ice-rich frozen soils? Draw typical recommendations for stabilization of slopes in frozen ground.

**9–2.** An infinite slope 2 (horizontal) to 1 (vertical) in homogeneous frozen soil has the following data:

$$S_{uu} = 0.072 \text{ MPa}$$

$$\omega = 30\%$$

$$S = 80\%$$

$$G_s = 2.7$$

Estimate the critical depth of thaw.

**9–3.** A cut at an angle of 25° is to be made in frozen ground with $C = 0.08$ MPa, $\phi = 20°$, and $\rho = 1600$ kg/m$^3$. Estimate the safety factor of the slope if the height of the cut is equal to 5 m.

**9–4.** If seismic forces are considered (say, $K = 0.15$), find the change in safety factor of the slope given in Problem 9–3.

**9–5.** What is a thaw consolidation ratio $R$? How does it affect the stability of a thawing slope? Discuss the use of $R$ in the slope-stability analysis of thawing slope.

**9–6.** Estimate the safety factor of a thawing slope whose soil parameters are as follows: $C = 35$ kPa, $\phi = 25°$, $\rho = 1280$ kg/m$^3$, $\alpha = 0.02$ cm/s$^{1/2}$, and $C_v = 60$ m$^2$/yr.

CHAPTER 10

# Drilling, Sampling, and Testing in Frozen Ground

## 10.1 INTRODUCTION

For many projects constructed in locations where frozen ground conditions occur, drilling, sampling, and testing of the soils are required for foundation design and construction, road and utility route or site selection, and feasibility studies. Many of the same drilling and sampling techniques used in unfrozen soils may be used in frozen ground. However, additional complications arise in drilling frozen soils because of greater penetration resistance. Also, greater efforts are required to obtain undisturbed frozen samples because of potential thermal degradation to the samples. Determination of frozen soil type, distribution, and its properties are essential, and lack of information on the frozen ground condition may lead to increased construction costs and unacceptable performance of structures. Special attention is required in frozen ground where ground ice is encountered. Depending on the nature of the project; its location in continuous or discontinuous permafrost; and potential frozen soil type, composition, and temperature, the drilling and sampling techniques in frozen ground will vary. Other factors, such as logistics, economics, and environmental factors at the site, must also be considered.

Many investigators (Phukan, 1979; Johnston, 1962, 1981; Hvorslev, 1949; Lange, 1973; Linell and Johnston, 1973) reported about drilling, sampling, and testing of frozen soils. Although air-photo interpretation and remote sensing methods may give reliable information about the subsoils conditions at a site, drilling and sampling are generally carried out to confirm these predictions. This chapter describes the methods that are commonly used to drill through frozen ground and the corresponding ways of sampling it. Field and laboratory tests, which are essential, are also discussed.

## 10.2 DRILLING METHODS

The drilling system consists of three main functions: (1) penetration of the ground material, (2) removal of surplus materials, and (3) stabilization of the drilled hole wall surface.

The penetration of frozen ground may be achieved by rotation, percussion, augering, and vibration. Accordingly, the drilling methods are categorized as rotary drilling, percussion drilling, augering, and vibrating. The choice of the particular type of drilling technique in frozen ground will depend on the scope and objective of the project; anticipated site conditions including ground ice; quality, size, and depth of sampling required; accessibility, and ground and air temperature during drilling operations.

Various types of rotary drilling equipment, such as mobile B-61, CME 750, and Acker AD II, may be used to obtain information to depths of more than 100 ft (30.5 m). Drills may be mounted on trucks or tracks. Recently, a hydraulically operated jackhammer has been used to investigate subsoil conditions at remote sites (Fig. 10–1). In some cases (Pihlainen and Johnston, 1954; Veillette, 1975),

**Figure 10–1** Marlow hydraulically operated jackhammer. (Courtesy of Cold Regions Consulting Engineers.)

helicopter-transportable drill rigs are also being used to carry out investigation in remote areas. The rotary drilling method is widely used in frozen gravels and bedrock. Various types of core bits, barrels, and auxiliary equipment are used with the rotary drilling techniques. Drilling fluids are used to lubricate the bit as well as to remove cuttings. Also, the fluid helps to remove excess heat generated in cutting the frozen material. Water, brine solutions, diesel fuel, special fuel, and muds and compressed air have been used as circulating media. Drilling fluids should meet the following specifications (Issacs and Code, 1972):

1. Fluids should be nontoxic and should not disturb the ecological balance
2. Fluids should possess low freezing points
3. Fluids should be immiscible with water
4. Fluids should not easily penetrate the voids in the soil
5. Fluids should possess good heat-transfer characteristics
6. Fluids should be fairly easily pumped at low temperatures even if suspensions form with water or crystals of ice
7. They should be inexpensive

Generally, freshwater and brine solutions are used as drilling fluids when the ambient air temperature is above 6.6°C and soil thawing is acceptable. Below −6.7°C or 20°F, these drilling fluids will freeze, and diesel fuel or water with antifreeze solution to depress the freezing point can be used as the drilling fluid. Drilling fluids should be refrigerated naturally or artificially to obtain thermally undisturbed samples. Dewaxed diesel fluid must be used when the drilling operation is in extremely cold conditions. Recently, the use of diesel fuel as a drilling fluid has been restricted due to various environmental concerns in permafrost areas.

At ambient air temperatures below −12°C (10°F), compressed air can be used as a drilling fluid when drilling with rotary rock bits such as a tricone bit. Above −12°C (10°F), the compressed air must be artificially refrigerated to recover material in frozen condition. Wyder et al. (1972) reported the use of compressed air at the rate of 75 $ft^3$/min at 40 psi in a drilling and sampling program in Tuktoyaktuk arca, N.W.T., Canada.

The augering drilling technique is commonly used in frozen fine-grained and sandy soils. Drill rigs that have augers on extendable/retractable Kelly bars are convenient for shallow-frozen-ground investigations. Frozen ground generally does not need casing to stabilize the drilled hole surface unless the moisture content is so low that the material is not well bonded. If the hole is to be open a long time or is to be fairly deep, casing should be provided. Hollow-stem augering is suitable in such cases because the augers provide for the casing and the soil samples to be recovered through the hollow stem.

Frozen silts and sandy silts require that the cutting teeth be sharp and set at a fairly shallow angle, whereas frozen gravels and sandy gravels should be augered with bit teeth that have vertical or nearly vertical faces. The shallow bit angle allows the frozen fine-grained material to be shaved off at the bottom of the hole in thin layers with the rotation of the bit. The nearly vertical bit angle loosens the matrix of frozen gravelly soils as the bit rotates. Frozen sand can be drilled with either type of teeth.

The percussion method of drilling (Fig. 10–1) may also be used in frozen ground, and it is very efficient to use when probing a bedrock profile. Recently, this method of drilling and sampling was found to be cost-effective in remote places. Cores of large diameter can be obtained very rapidly by advancing and retracting steel pipe using a vibratory (sonic) pile driver (Bendz, 1977).

## 10.3 SAMPLING FROZEN GROUND

Techniques for sampling frozen ground are similar to those used for testing unfrozen soils but with some modifications to facilitate recovery of the samples in their frozen state. Special attention must be given in handling frozen samples to avoid significant thermal disturbance of the samples. The methods used to obtain frozen samples depend on various factors such as soil type and composition presence of ground ice, ground temperature, the ultimate use of samples, design requirements, and the nature of the project. The following primary sampling method can be used in frozen ground:

1. Driven soil samples
2. Soil samples recovered from test pits
3. CRREL ice-coring auger
4. Core barrel samples from drilling

Drive sampling consists of pounding a split or solid tube sampler into the ground with a drop hammer (Fig. 10–2). As commonly used in unfrozen soils, a 140-lb hammer falling 30 in. may be used to drive a sampler of 1.4 in. I.D., 18 in. long. In unfrozen soils, the number of blows needed to drive the sampler the last 12 in. gives the standard penetration resistance (blow count or SPT), which indicates the relative density of granular soils. However, such blow counts should not be used to determine the relative density in frozen soils. Often, resistance of the sampler occurs due to the presence of large cobbles or gravels. To avoid problems in driving samples, large drive samplers (2.5 to 6 in. I.D.) with 350-lb hammers are commonly used.

Test pits or trenches to a depth of 3 to 5 m (10 to 15 ft) permit the visual examination of materials at the site. Useful disturbed or undisturbed frozen soil samples may also be recovered from the walls or bottom of the test holes. Test pits are generally excavated by powered hand tools, such as a jackhammer, and heavy equipment, such as a bulldozer, backhoe, ripper, and power shovel.

**Figure 10–2** Drive sampling arrangement. (Courtesy of Alaska Area Native Health Services.)

The Cold Regions Research and Engineering Laboratory (CRREL) has designed an electromechanical device to core ice (Ueda et al., 1975), and this device may be used to sample frozen fine-grained soils. The unit consists of a 4-in. O.D. 1-m-long core barrel with a small continuous flighting connected to the outside. The CRREL sampler is driven by an electrical motor that is located on the ground surface and is designed to sample to depths of 100 m in ice. In frozen fine-grained soils, the unit's range is approximately 3 to 4.6 m (10 to 15 ft).

A combination of rotary drill and drive sampling with heavy walled tubes gives reasonable results in frozen fine-grained soils with temperatures above −4°C or 25°F (Kitze, 1956; Davis and Kitze, 1967). Also, a dry coring device consisting of a 5-ft length of BX casing (2.5 in. I.D., 2.9 in. O.D.) with a carbide cutting shoe on the bottom edge may be rooted or forced into the ground at a constant rate to recover frozen soil samples (Reimers, 1980). The water generated by thawing frozen soils at the tip of the core barrel serves to lubricate and cool the cutting edge. The core is extracted from the casing with a hydraulic extractor.

The Shelby tube is the common name for thin-walled samplers made of hard-drawn seamless steel tubes and are pushed into the ground to recover silty or clayey soil samples. The modified Shelby tube is a Shelby tube that has had a small rectangular piece of carbide welded to the cutting edge, or the tip of the tube is shredded. The modified Shelby tube may be rotated or forced down with a drilling rod connected to the rotating drive of a drill rig. This unit has proven to be an excellent way to recover suitable samples of frozen sands and silts.

To obtain undisturbed samples, coring of frozen soils is best done with double core barrels that are designed in a way that the circulating fluid flows down the annulus between the outer shell and the tube that receives the core. The fluid then emerges and flows upward between the outer shell and the wall of the hole, carrying the cuttings with it. Out of the available bottom discharge and internal discharge bits, the former is preferred in frozen ground as it reduces the contamination of samples.

Special core-drilling equipment and techniques may be used to obtain frozen soil and rock samples from any depth (Hvorslev and Goode, 1966; Lange and Smith, 1972; Sellman and Brown, 1965; Roggensack, 1979). Equipment has been designed for taking samples at particular locations to solve certain sampling requirements.

### 10.3.1 Postsampling Treatment of Frozen Cores

Precautions must be taken to transport recovered frozen soil samples and maintain their frozen state until they arrive at the testing laboratory. During the transportation, the frozen-soil samples may be affected by thermal and mechanical disturbances, and sublimation of specimens may occur. Any changes in the condition of the samples may alter the properties and deformation characteristics of the recovered materials.

Samples should be placed in deaired and tightly sealed plastic bags and transported in well-insulated boxes. Crushed ice or dry ice and additional insulation should be placed around the samples in the boxes to reduce or prevent sublimation. Considerable care should be taken to maintain the original sample temperature to whatever extent possible. Thermal shock of the samples must be prevented by proper placement of ice around samples (Chamberlain, 1981). The core should be wrapped in cellophane and placed in a double polyethylene bag or polyvinyl chloride (PVC) core container in order to reduce the loss of the moisture of the frozen samples. Samples may also be shipped in portable freezers, but the cost is generally high.

## 10.4 GROUND-TEMPERATURE PROFILE MEASUREMENT

As discussed in previous chapters, the temperature of frozen ground is one of the most important parameters required for design and construction of building roadways or utility projects in permafrost regions. Generally, the ground temperature is measured during a site investigation. The thickness of the active layer or the portion of 0°C isotherm can be obtained by relatively simple methods. For most projects, ground-temperature measurements at prescribed depths are obtained to determine the thermal regime or the whiplash curve that gives the depth of zero amplitude not only for thermal design, but also to assist in testing frozen soils at appropriate temperatures. Various methods ranging from simple manual readings to fully automated and sophisticated remote transmission devices may be used to obtain ground temperatures at various prescribed depths. The selection of a ground-temperature measurement device depends on the location of the site, nature of the project (short term or long term), depth at which temperatures are required, the frequency of observations, what is to be measured, and accuracy and cost of equipment.

The active layer and the seasonal frost penetration depth can be measured by hand probing with a mild steel rod about 0.5 in. (12.5 mm) in diameter and 5 to 9 ft (1.5 to 3.0 m) in length. Care must be taken when probing ice-rich fine-grained soils in warm temperatures where the probe can be pushed as much as 1 ft or more below the frost layer without much penetration resistance. In gravelly soils, however, the active depth may be underestimated due to refusal at shallower depths. More reliable results may be obtained by using hand-coring tools that take a small-diameter sample core 15 to 20 mm in diameter and 150 to 300 mm long (Hughes and Terasmae, 1963).

The frost tube reported by Gandahl (1963), Rickard and Brown (1972), and Mackay (1973) is commonly used to determine the depth and rate of thaw and frost penetration in the active layer and seasonal frost zone. The simple device consists of an outer casing containing a removable, transparent tube filled with a methylene indicator solution that changes color upon freezing (Fig. 10–3). The instrument is easy to install, inexpensive, easy to read, and accurate to about 50 mm.

The probes and frost tubes indicate the position of the 32°F or 0°C isotherm and no information on ground-temperature profile is obtained. Various types of temperature sensors are available to measure ground temperatures. Thermocouples and thermistors are widely used depending on the accuracy of the readings required and the particular conditions. Many investigators (Hansen, 1963; Johnston, 1963, 1973; Mackay, 1974; Judge, 1974) have reported on the merits and demerits of their use. In most engineering projects, an accuracy of 0.2°C is desirable and can be obtained either with thermocouples or thermistors provided that precautions are taken during the installation and fabrication of probes and connectors, switches, and selection of readout instruments.

Multiconductor cables with sensors placed at selected positions (preferable spacing of 0.25 to 0.5 m to a depth of 3 to 6 m and 1 to 3 m thereafter) can be fabricated and installed in a drilled borehole that is backfilled carefully to ensure intimate contact with the surrounding material. Care should be taken to prevent the percolation of surface water down the installed hole. Adequate time must be allowed for the installed sensors to reach thermal equilibrium of ground which is disturbed due to drilling and installation of instruments. Depending on the project requirements, readout equipment ranging from simple to manually operated potentiometer or resistance bridges to automatic, power-operated data acquisition sys-

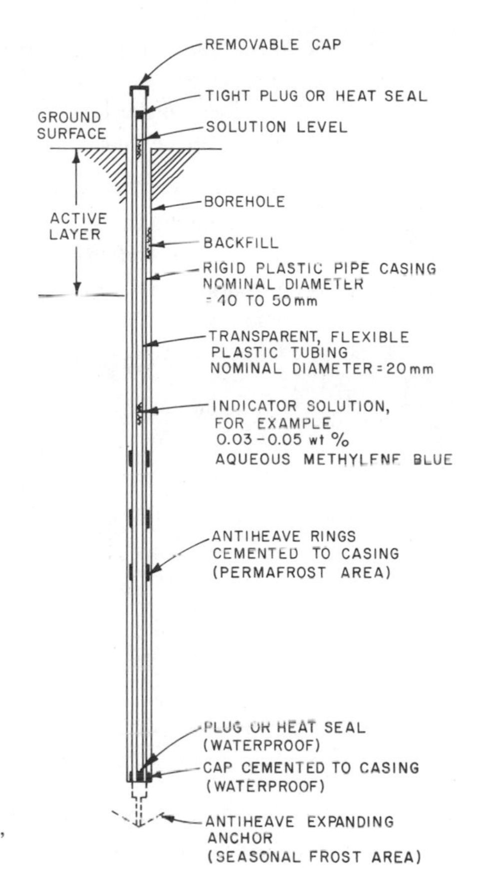


**Figure 10–3** Frost tube. (After Johnston, 1981.)

tems that record data on charts or magnetic tapes at preselected times may be installed for the measurement of ground temperatures.

## 10.5 LABORATORY TESTING OF FROZEN SOILS

In Chapter 3 the physical, mechanical, and thermal properties of frozen soils are discussed. In this section some details are given regarding the laboratory testing of frozen soils that are categorized into two main groups. No standard procedures are available regarding the laboratory testing of frozen soils.

GROUP 1: CLASSIFICATION AND INDEX TESTS. As described in Chapter 2, frozen soils are classified according to unfrozen soil classification using the Unified Soil Classification System (USCS), ice description, and surface characteristics. Grain size distribution and Atterberg limits tests are performed for the USCS. Water content ($\omega$), dry density ($\rho_d$), and unfrozen water content ($\omega_{uw}$) are important index properties of frozen soils. The relationship between $\omega$ and $\rho_d$ may be used in estimating thermal properties and then settlement characteristics. Thermal properties can also be determined in the laboratory by direct measurement (Penner et al., 1975; Williams, 1964; Slusarchuk and Foulger, 1973).

GROUP 2: STRENGTH AND DEFORMATION PROPERTIES. Important parameters for these tests are temperature, time, and ice inclusions in the frozen soil samples. Representative undisturbed frozen soil samples must be obtained to perform strength and deformation tests. Vyalov (1966a), Sayles (1968), Akili (1971), Haynes (1978), Roggensack and Morgenstern (1978), and Phukan (1980b) have reported on the unconfined compression test ($q_{uu}$) and triaxial compression and

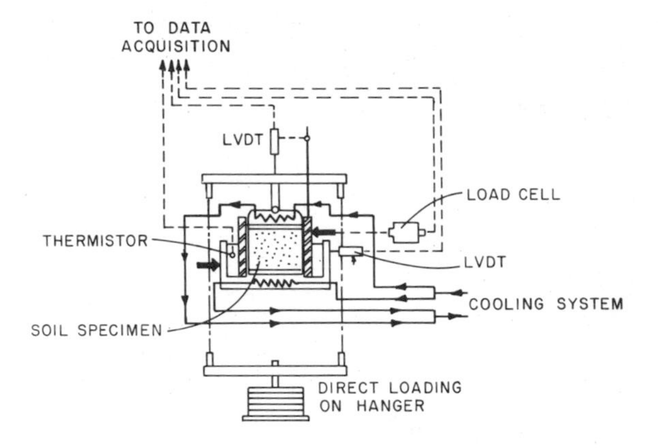


**Figure 10–4** Direct shear test device for frozen soils. (After McRoberts et al., 1978.)

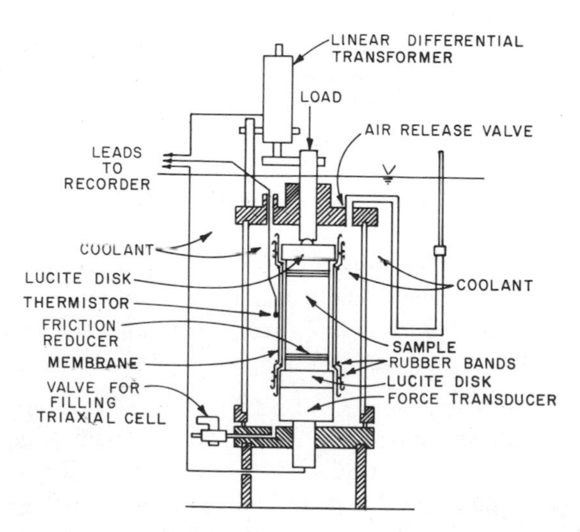


**Figure 10–5** Triaxial cell for testing frozen soil. (After Goughnour and Andersland, 1968.)

direct shear laboratory test techniques on frozen soil samples. Figures 10–4 and 10–5 illustrate the laboratory apparatus used in direct shear and triaxial testing of frozen soils, respectively. Most of these tests were carried out at different constant strain rates and temperatures. Tests can also be performed at different stress levels to investigate the creep characteristics of frozen soil. A photographic view of a special triaxial cell used to test frozen soil samples is shown in Fig. 10–6.

Thaw settlement and consolidation tests can be carried out in the laboratory (Nixon, 1973; Nixon and Morgenstern, 1974; Crory, 1973a), and many test results have been reported.

**Figure 10–6** Triaxial cell for testing of frozen soil sample.

## 10.6 GEOPHYSICAL TECHNIQUES USED IN FROZEN GROUND

Conventional geophysical methods such as seismic-refraction surveys, galvanic-resistivity, and electromagnetic methods may be used in frozen ground. They are used effectively to map the distribution of permafrost, for delineating frozen-unfrozen ground profiles, and for determining the thickness of the active layer. The application of geophysical methods requires an understanding of the dependence of the physical parameters on temperatures, and consequently on the depth of strata below the ground surface. Seismic compressional wave velocity and the electrical resistivity are the most common parameters measured by the geophysical methods.

Applications of geophysics to frozen-ground problems can be categorized into three main groups. The first is the detection and delineation of frozen ground in the discontinuous permafrost regions. The second is the detection of massive ice or ice-rich ground. The third is the identification of subsurface materials within frozen-ground regions. Depending on the scale and details required, either ground or airborne geophysical techniques may be utilized for delineation of permafrost terrain.

As a result of pore-water freezing, the physical, mechanical, and electrical properties of soils and rocks may change. The most common properties that may change are density, elastic modulus, and conductivity. As such, electrical and seismic methods are the most useful geophysical methods for frozen ground. Both these methods are also sensitive to the presence of ice in the soil or rock pores. But they do not give information on the temperature profile of frozen ground.

### 10.6.1 Detection and Delineation of Frozen Ground

The geophysical methods that can be used for the detection and delineation of frozen ground in discontinuous permafrost regions are electrical resistivity, electromagnetic, and seismic-refraction surveys.

Figure 10–7 illustrates typical resistivity-temperature data for several soils and one rock type derived from the work of Hoeskstra and McNeill (1973). It can be seen from the figure that the resistivity of unfrozen and frozen soils up to a temperature of $-10°C$ differs by a factor of 10. Differentiating between frozen and unfrozen ground solely on the basis of resistivity is not always possible, as the resistivity of frozen clay can be less than the resistivity of unfrozen silt, sand, or rock. Therefore, resistivity surveys alone will not form the basis for mapping permafrost. They must be accompanied by other geological information (Ferrians et al., 1969). Vanyan (1965), Ogilvy (1967), Morgan (1968), and Parkhomenko (1967) published resistivity data on several soils.

Two electromagnetic methods used to measure the electrical properties of soils and rocks are wave tilt of radio surface waves and inductive coupling between low-frequency dipole antennas. The main difference between these methods is that the wave tilt technique uses propagating plane waves with the receiver located in the far field of the transmitter, whereas the inductive coupling technique locates the transmitter and receiver dipoles in close proximity so that at the relatively low frequency used, the instrument functions as an electromagnetometer. Figure 10–8 illustrates a typical inductive coupling method that is commonly used for the detection of frozen ground. Both methods are well covered in many publications (Hoekstra and McNeill, 1973; Keller and Frischknecht, 1966).

The hammer seismic refraction method, which is based on the travel of

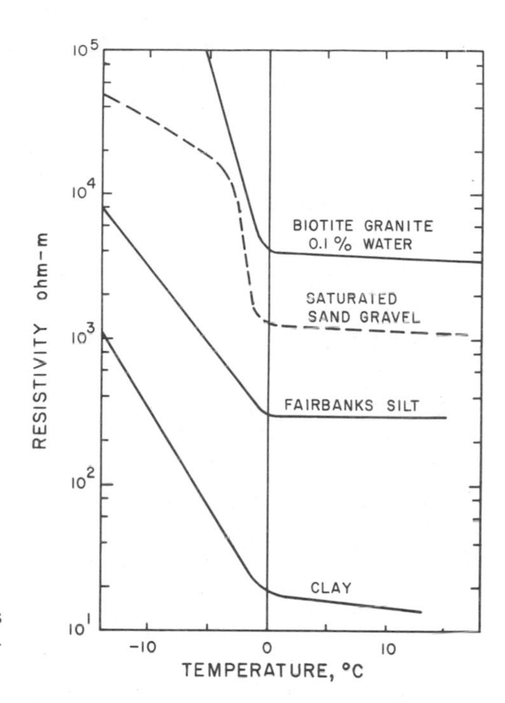


**Figure 10–7** Resistivity of several soils and one rock type as a function of temperature. (After Hoekstra and McNeill, 1973.)

vibrations through the ground, is illustrated. The parameters that influence the compressional wave ($V_p$) and shear wave velocities ($V_s$) are lithology and grain size, total moisture content and the nature of pore fluid, temperature of pore water, porosity and pore structure, matrix composition, and degree of cementation. Porosity is usually the dominant parameter affecting $V_p$ in dry and/or saturated rock at a given pressure (Garg, 1969). There is a general increase in $V_p$ and $V_s$ with the decreasing temperature over the range of 0 to $-2$°C in saturated material.

**Figure 10–8** Resistivity of soils measurement by EM31. (Courtesy of R. Jurick.)

In this method, vibrations or shocks are produced by hitting a steel plate on the ground with a suitable weight. The ground shock produces a variety of waves, such as the longitudinal or compressional waves (which are faster), followed by slower shear or transverse waves and the much slower transverse surface waves. The compressional wave velocity $V_p$ will penetrate to a higher velocity layer, travel along this layer, and send a longitudinal refracted wave continuously back to the surface. This wave is easily recognized as the first-arrival geoplanes placed on the ground surface. Various investigators (Barnes, 1963; Garg, 1969; Dobrin, 1960; Hans, 1960; Hobson and Hunter, 1969) have described this method, and wave velocity data were reported. Table 10–1 presents some of the seismic compressional wave velocities measured in permafrost.

**TABLE 10–1 Typical Values of Seismic Compressional Velocities in Permafrost**

| Soil Types | Locality | Compression Wave Velocity (km/s) | | Estimated Ground Temperature (°C) |
|---|---|---|---|---|
| | | Frozen | Unfrozen | |
| Coarse-grained | | | | |
| Floodplain alluvium | Fairbanks area, Alaska | 2.3–4.3 | 1.9–2.1 | −1 |
| Gravel | Fairbanks area, Alaska | 4.0–4.6 | 1.8–2.3 | −1 |
| Glacier moraine | Delta Junction, Alaska | 2.3–4.0 | | −2 |
| Aeolian sand | Tetlin Junction, Alaska | 2.4 | | −3 |
| Outwash gravel | Tanacross, Alaska | 2.3–3.0 | | −3 |
| Glacial outwash | Thule, Greenland | 4.5–4.7 | | −11 |
| Glacial till | Thule, Greenland | 4.7–4.8 | | −11 |
| Glacial till | McMurdo Sound, Antarctica | 3.0–4.3 | 0.5–1.5 | −20 |
| Till | Norman Wells, N.W.T. | 2.2–3.6 | | |
| Frozen ground | Lake Fryxell, Lake Vanda, Lake Bonney, Antarctica | 3.8–4.5 | | |
| Sand and clay | Norman Wells, N.W.T. | 3.1–3.4 | | |
| Saturated sand | | 3.2–4.0 | | |
| Water-saturated gravel | | 3.6–4.0 | | |
| Gravel | Klondike area, Yukon | 5.5 | | |
| Outwash with relatively thick ice lenses | Lake Vida, Antarctica | 5.7–5.9 | | |
| Fine-grained | | | | |
| Silt and gravel | Fairbanks area, Alaska | 2.3–3.0 | | −1 |
| Silt and organic matter | Fairbanks area, Alaska | 1.5–3.0 | 0.6–1.2 | −1 |
| Silt with ice lens | Eielson AFB, Alaska | 2.0–2.8 | | −1.5 |
| Alluvial clay | Northway, Alaska | 2.4 | | −2 |
| Silt | Glen Creek Valley, Alaska | 2.7–3.3 | | −1.5 |
| Tundra silts, sands, and peats | | | | |
| Gubik Formation, probably saline | Barrow area, Northern Petroleum Reserve, NPR-4, Alaska | 2.4–2.7 | | −9 |
| | Skull Cliff area, NPR-4, Alaska | 2.3–2.7 | | −9 |

**TABLE 10–1 Typical Values of Seismic Compressional Velocities in Permafrost**

| Soil Types | Locality | Compression Wave Velocity (km/s) | | Estimated Ground Temperature (°C) |
|---|---|---|---|---|
| | | Frozen | Unfrozen | |
| Fine-grained *(continued)* | | | | |
| Gubik Formation, less saline | Topagoruk area, NPR-4, Alaska | 2.4–3.7 | | −9 |
| Unclassified sediments | Isachsen, Canada | 2.7 | | −10 |
| Silt and clay | Norman Wells, N.W.T. | 3.1 | | |
| Clay | | 1.5–2.1 | | |
| Ice-saturated silts | | 1.8–3.1 | | |
| Clay | Norman Wells, N.W.T. | 2.5–2.8 | | |
| Rocks | | | | |
| Iron ore unaltered | Schefferville, Canada | 6.1 | 3.0 | −1 |
| Iron ore altered | Schefferville, Canada | 5.5 | 1.4 | −1 |
| Shales and sandstone | Alaska | 2.5–2.6 | 1.5–2.1 | −9 |
| Limestone | Bedford, U.K. | 6.1 | 4.8 | −8 |
| Sandstone | Boise, Idaho | 5.1 | 3.3 | −1 |

*Sources:* After Vinson (1977); Garg (1973); King et al. (1974).

### 10.6.2 Detection of Massive Ice or Ice-Rich Frozen Ground

Osterkamp et al. (1980) reported the use of galvanic electrical resistivity measurement for the detection of excessive ice (segregated ice and massive ice) in permafrost. The galvanic resistivity (GR) method is commonly used to measure soil resistivity (Keller and Frischknecht, 1966). In four-terminal methods, current is driven through two electrodes and the potential established in the earth by this current is measured with the other two electrodes. This allows determination of an apparent resistivity of the soil. The application of GR measurements in permafrost terrain has been described by Barnes (1963) and Arcone et al. (1979). It should be noted that the GR method suffers from problems in obtaining good electrode contact with frozen ground.

Mackay (1967) and Rampton and Walcott (1974) reported the use of gravity measurements over pingos in the MacKenzie Delta region and indicated that the large ice cores of pingos can be readily detected with a sensitive gravimeter. Kawaski et al. (1983) described the gravity measurements in permafrost terrain containing massive ground ice. Their investigation indicates that it is possible to detect isolated massive ground ice deposits by means of commercially available field instruments with sensitives well below 0.005 mgal and with quicker response time setup to reading, the gravity technique is not suitable for rapid permafrost terrain reconnaissance. However, the technique appears to be feasible for detailed work such as a preconstruction investigation of building sites where massive ground ice may occur in an overburden or for investigating selected sites along airfields, roadways, and pipelines. At present, drilling and coring, galvanic electrical resistivity and gravity methods are the only available methods for obtaining information on excessive ice content of permafrost.

### 10.6.3 Identification of Subsurface Materials within Frozen Ground

Many available borehole geophysical methods and tools may be used to identify materials within frozen ground. Methods that are commonly used for delineating permafrost formation are (1) acoustic or sonic log, (2) electrical-resistivity log, (3) induction log, and (4) gamma-gamma log.

In the acoustic or sonic log method, the compressional and shear-wave velocities of the formation are measured. They are correlated with changes in lithology and porosity of formation adjacent to the borehole, which is filled with either water or drilling mud before the sonic log is run. As discussed in Chapter 3, the dynamic elastic constants of the formation can be calculated from the compressional and shear wave velocities measurements. More detailed information on this method are given by King et al. (1974) and the Commission of Standardization of Laboratory and Field Test (1976).

In electrical-resistivity log methods, the apparent resistivity of the formation along the uncased and conductive mud-filled borehole is measured by sending a current into the formation through two electrodes and measuring the potential difference in the other two (Seguin and Garg, 1974; Commission on Standardization of Laboratory and Field Tests, 1976).

The induction log method may be used to identify bed boundaries, to make lithological determinations, and to measure the formation resistivity. These can be achieved by recording the signal received by a coil and induced by eddy currents in the conductive formations, which in turn are electromagnetically induced by a transmitter coil through which a high-frequency alternating current is sent (Commission on Standardization of Laboratory and Field Tests, 1976).

In gamma-gamma log, the back-scattered gamma rays emitted by a source are measured and the bulk density of the formation can be determined, as the scattering depends on the electron density. The method uses a sonde as a gamma-ray source and one or two detectors of scattered gamma rays (Commission for Standardization of Laboratory and Field Tests, 1976). This method may also be used to determine indirectly the ice content of permafrost, as the bulk density is proportional to the ice content discussed in Chapter 3.

APPENDIX A

# Glossary of Terms

**Active layer.** The top layer of ground above the permafrost table that thaws each summer and refreezes each fall.

**Axle load.** The total transmitted by all wheels whose centers may be included between two parallel transverse vertical planes 40 in. apart, extending across the full width of the vehicle.

**Base course.** The layer or layers of specified or selected material of designed thickness placed on a subbase or subgrade to support a surface course.

**Beaded stream.** A drainage pattern of individual streams in which pools or small lakes are connected by short stream reaches.

**Depth of thaw.** The distance from the ground surface downward to frozen ground at any time during the thawing season.

**Depth of zero annual amplitude.** The distance from the ground surface downward to the point beneath where there is virtually no annual fluctuation in ground temperature.

**Flexible pavement.** A pavement structure that maintains intimate contact with and distributes loads to the subgrade and depends on aggregate interlock, particle friction, and cohesion for stability.

**Freezing index.** The number of degree-days (the difference between the mean temperature each day and 0°C, either positive or negative) between the highest point in the autumn and the lowest point the next spring on the cumulative degree-day time curve and for one freezing season. The air freezing index is determined from temperatures measured about 1.4 m above the ground surface, while that determined from temperatures measured at or immediately below a surface is known as the surface freezing index.

**Frost-susceptible soil.** Soil in which significant detrimental ice segregation occurs when the requisite moisture and freezing conditions are present.

**Frozen ground.** Soil or rock having a temperature below 0°C.

**Ground.** Earth materials, including all types of soil and rock and their constituents.

**Ground heave.** Upward movement of the ground causing a raising of the ground surface as a result of the formation of ground ice in excess of pore fillings.

**Ground settlement.** Downward movement of the ground causing a lowering of the ground surface resulting from the melting of ground ice in excess of pore fillings.

**Ice, buried.** Ice formed at the surface and later covered with soil.

**Ice, excess.** The ice in the ground that exceeds the total pore volume that the ground would have under natural unfrozen conditions.

**Ice, ground.** Ice in pores, cavities, voids, or other openings in soil or rock, including massive ice.

**Ice, intrusive.** Ice formed from water intruded or injected under pressure into a porous earth material.

**Ice, massive.** A comprehensive term used to describe large (with dimensions measuring at least 10 to 100 cm) masses of underground ice, including ice wedges, pingo ice, and ice lenses.

**Ice, pore.** Ice occurring in the pores of soils and rocks.

**Ice, reticulate.** Network of horizontal and vertical ice veins forming a three-dimensional rectangular or square lattice commonly found in frozen glaciolacustrine sediments.

**Ice, segregated.** Ice formed by the migration of pore water to the freezing plane, where it forms into discrete lenses, layers, or seams ranging in thickness from hairline to greater than 10 m.

**Ice, vein.** A comprehensive term for ice in cracks where it occurs in bodies of various shapes, including tabular forms and wedges.

**Ice content.** The ratio, expressed as a percentage, of the weight of the ice phase to the weight of dry soil.

**Ice lens.** (1) A dominantly horizontal lens-shaped body of ice of any dimension.

(2) Commonly used for layers of segregated ice that are parallel to the ground surface. The lenses may range in thickness from hairline to as much as about 10 m.

**Ice segregation.** The process of formation of segregated ice by freezing of water in mineral or organic soil.

**Ice wedge.** A massive generally wedge-shaped body with its apex pointing downward, composed of foliated or layered, vertically oriented, commonly white ice; from less than 10 cm to 3 m or more wide at the top, tapering to a feather edge at the apex at a depth of 1 to 10 m or more. Some ice wedges may extend downward as far as 25 m and may have shapes dissimilar from wedges. They may be "active" or "inactive," depending on whether they are or are not growing by repeated, but not necessarily annual (winter), cracking.

**Icing.** A sheetlike mass of ice, either on the ground surface or on the surface of river ice.

**Modulus of subgrade reaction (*k*).** Westergaard's modulus of subgrade reaction for use in rigid pavement design (the load in pounds per square inch on a loaded area of the subgrade or subbase divided by the deflection in inches of the subgrade or subbase psi/in.).

**Non-frost-susceptible soil.** A soil that does not display significant detrimental ice segregation during freezing.

**Palsa.** A Fennoscandian term for a round or elongated hillock or mound, maximum height about 10 m, composed of a peat layer overlying mineral soil. It has a perennially frozen core that extends from within the covering peat layer downward into or toward the underlying mineral soil.

**Patterned ground.** A general term for any ground surface of surficial soil materials exhibiting a discernible, more or less ordered and symmetrical, microphysiographic pattern.

**Pavement performance.** The trend of serviceability with loads application.

**Pavement structure.** A combination of subbase, base course, and surface course placed on a subgrade to support the traffic load and distribute it to the roadbed.

**Peat.** An unconsolidated, compressible soil consisting of partially decomposed, semicarbonized remains of plants (such as mosses, sedges, and trees), some animal residues, and commonly some mineral soil.

**Perennially frozen ground.** *See* Permafrost.

**Periglacial.** (1) The area, geomorphological processes, and deposits characteristic of the frost-affected immediate margins of existing and former glaciers and ice sheets.

**Permafrost.** The thermal condition in soil or rock of temperatures below 0°C persisting over at least two consecutive winters and the intervening summer; moisture in the form of water and ground ice may or may not be present. Earth materials in this thermal condition may be described as perennially frozen, irrespective of their water and ice content.

**Permafrost, continuous.** Permafrost occurring everywhere beneath the exposed land surface throughout a geographic regional zone, with the exception of widely scattered sites (such as newly deposited unconsolidated sediments) where the climate has just begun to impose its influence on the ground thermal regime and will cause the formation of continuous permafrost.

**Permafrost, contemporary.** Permafrost in thermal equilibrium or newly formed permafrost in an area where surface temperatures have fallen below 0°C.

**Permafrost, discontinuous.** Permafrost occurring in some areas beneath the ground surface throughout a geographic regional zone where other areas are free of permafrost.

**Permafrost, dry.** Perennially frozen soil or rock without ice or with an ice content lower than the pore volume, so that it does not yield excess water on thawing.

**Permafrost, ice-rich.** Perennially frozen ground that contains ice in excess of that required to fill pore spaces.

**Permafrost, marginal.** Permafrost that is very close in space, temperature, or time to thawing:

(a) Permafrost at or near the southern limit of the permafrost region, or
(b) Permafrost of which the temperature is very close to 0°C (a few tenths of a degree) anywhere in the permafrost region, or
(c) Permafrost that lasts for only a few years and then dissipates.

**Permafrost, sporadic.** Permafrost occurring in the form of scattered permafrost islands in the more discontinuous permafrost zone.

**Permafrost, submarine.** Permafrost occurring beneath the sea or ocean bottom.

**Permafrost, table.** The upper boundary of permafrost.

**Permafrost, thaw stable.** Perennially frozen soils that do not, on thawing, show loss of strength below normal, long-time thawed values or producing ground settlement.

**Permafrost, thaw unstable.** Perennially frozen soils that show, on thawing, a significant loss of strength below normal, long-time thawed values and/or significant settlement, as a direct result of the melting of the excess ice in the soil.

**Permafrost aggradation.** An increase in thickness and/or areal extent of permafrost because of natural or artificial causes as a result of climatic cooling and/or change of terrain conditions, such as vegetation succession, infilling of lake basins, or human activity.

**Permafrost degradation.** A decrease in thickness and/or areal extent of permafrost because of natural or artificial causes as a result of climatic warming and/or change of conditions, such as disturbance or removal of an insulating vegetation layer by fire or human means.

**Permafrost thickness.** The vertical distance between the permafrost table and the permafrost base.

**Pingo.** An Eskimo term for a conical, commonly more-or-less asymmetrical, mound or hill, with a circular or oval base and commonly fissured summit, occurring in the continuous and discontinuous permafrost zones, which has a core of massive ground ice covered with soil and vegetation, and which exists for at least two winters. They may be either "open" or "closed" system pingos, depending on their mode of formation.

**Polygon.** A type of patterned ground consisting of a closed, roughly equidimensional figure bounded by several sides, commonly more or less straight, but some, or all, of which may be irregularly curved. A polygon may be either "low center" or "high center," depending on whether its center is lower or higher than its margins.

**Polygon, ice wedge.** Any polygon surrounded by troughs underlain by ice wedges.

**Residual thaw layer.** The layer of thawed or unfrozen ground between the seasonally frozen ground and the permafrost table.

**Retrogressive thaw or bimodal flow slide.** A slide that consists of a steep headwall, containing ice or ice-rich sediment, which retreats in a retrogressive fashion through melting, and a debris flow formed from the mixture of thawed sediment and ice, which slides down the face of the headwall to its base.

**Rigid pavement.** A pavement structure that distributes loads to the subgrade, having as one course a portland cement concrete slab of relatively high bending resistance.

**Roadbed.** The graded portion of a highway between top and side slopes, prepared as a foundation for the pavement structure and shoulder.

**Roadbed material.** The material below the subgrade in cuts and embankments and in embankment foundations, extending to such depth as affects the support of the pavement structure.

**Seasonal frost.** Seasonal temperatures causing frost (below 0°C temperatures) that affects earth materials and keeps them frozen only during the winter.

**Seasonally frozen ground.** Ground affected by seasonal frost.

**Seasonally thawed ground.** Ground affected by seasonal thaw during the summer and seasonal frost during the winter.

**Selected material.** A suitable native material obtained from a specified source, such as a particular roadway cut or borrow area, of a suitable material having specified characteristics to be used for a specific purpose.

**Serviceability.** The ability at time of observation of a pavement to serve high-speed, high-volume automobile and truck traffic.

**Solifluction.** The process of slow, gravitational, downslope movement of saturated, nonfrozen earth material behaving apparently as a viscous mass over a surface of frozen material. Solifluction features include lobes, stripes, sheets, and terraces.

**Suprapermafrost water.** Free water in the ground above the permafrost.

**Talik.** A layer or body of unfrozen ground within the permafrost. It may be either a ''closed'' or an ''open'' talik, depending on whether it is or it is not entirely surrounded by permafrost.

**Thaw consolidation.** (1) The process of which a reduction in volume and increase in density of soil mass occurs, following thaw, in response to the escape of water under the weight of the soil itself and/or an applied load. (2) The process by which settlement due to thaw (thaw settlement) is impeded by the flow of water from the soil. Thaw consolidation is a time-dependent phenomenon that is not governed exclusively by the rate of thaw or position of the thaw front. It may proceed for many years.

**Thawing index.** The number of degree-days (the difference between the mean temperature each day and 0°C, either positive or negative) between the lowest point in the spring and the highest point the next autumn on the cumulative degree-day time curve for one thawing season. The air thawing index is determined from temperatures measured about 1.4 m above the ground surface, while that determined from temperatures measured at or immediately below a surface is known as the surface thawing index.

**Thawing settlement.** The generally differential downward movement of the ground surface resulting from escape of water on melting of excess ice in the soil and the thaw consolidation of the soil mass.

**Thermokarst (topography).** The irregular topography resulting from the process of differential thaw settlement or caving of the ground because of the melting of ground ice in thaw-unstable permafrost.

**Tundra.** A treeless, generally level to undulating region of lichens, mosses, sedges, grasses, and some low shrubs, including dwarf willows and birches, which is characteristic of both the arctic and high alpine regions outside the arctic.

**Unfrozen water content.** The ratio, expressed as a percentage, of the weight of unfrozen water to the weight of dry soil.

APPENDIX B

# Useful Conversion Factors for SI Units

| Quantity | U.S. Customary Symbol | SI Symbol | Conversion Factor |
|---|---|---|---|
| Length | ft | m | 1 ft = 0.3048 m |
| | in. | mm | 1 in. = 25.4 mm |
| | in. | m | 1 in. = 0.0254 m |
| | mile | km | 1 mile = 1.609 km |
| Area | $ft^2$ | $m^2$ | 1 $ft^2$ = 0.093 $m^2$ |
| | in.$^2$ | $mm^2$ | 1 in.$^2$ = 645.16 $mm^2$ |
| Volume | $yd^3$ | $m^3$ | 1 $yd^3$ = 0.765 $m^3$ |
| | $ft^3$ | — | 1 $ft^3$ = 0.028 $m^3$ |
| | in.$^3$ | — | 1 in.$^3$ = 28.317 L |
| | gal | L | 1 (U.S.) gal = 3.78 L |
| Velocity | ft/s | m/s | 1 ft/s = 0.305 m/s |
| | mile/hr | km/h | 1 mile/hr = 1.609 km/h |
| Acceleration | $ft/s^2$ | $m/s^2$ | 1 $ft/s^2$ = 0.305 $m/s^2$ |
| Rate of flow | $ft^3/s$ | $m^3/s$ | 1 $ft^3/s$ = 0.028 $m^3/s$ |
| | gal/hr | L/s | 1 gal/hr = 0.0013 L/s |
| Mass | lb | kg | 1 lb = 0.454 kg |
| | 1 slug (1 $lb_f/ft/s^2$) | kg | 1 slug = 14.59 kg |
| | short ton | kg | 1 short ton = 907.2kg |
| Density | $lb/ft^3$ | $kg/m^3$ | 1 $lb/ft^3$ = 16.019 $kg/m^3$ |
| | $lb/yd^3$ | $kg/m^3$ | 1 $lb/yd^3$ = 0.593 $kg/m^3$ |
| Force | $lb_f$ | N | 1 $lb_f$ = 4.448 N |
| | 1 short ton | kN | 1 $ton_f$ = 9.964 kN (short) |
| | 1 kip = 1000 $lb_f$ | 1 dyne = g $cm/s^2$ = $10^{-5}$ N | 1 $kg_f$ = 9.807 N |
| | | | 1 MN = 1000 kN |

| Quantity | U.S. Customary Symbol | SI Symbol | Conversion Factor |
|---|---|---|---|
| Pressure | lb/ft$^2$ | Pa (1 Pa = 1 N/m$^2$) | 1 lb$_f$/ft$^2$ = 47.880 Pa |
| | lb/in.$^2$ | kPa | 1 lb/in.$^2$ = 6.895 kPa |
| | ton$_f$/ft$^2$ | kPa | 1 ton$_f$/ft$^2$ = 107.252 kPa |
| | ton$_f$/in.$^2$ | MPa | 1 ton$_f$/in.$^2$ = 13.789 MPa (short) |
| Work, energy, heat | Btu | kJ (1 J = 1 Ws) | 1 Btu = 1.055 kJ |
| | ft lb$_f$ | J | 1 ft lb$_f$ = 1.356 J |
| | kWh | MJ | 1 kWh = 3.6 MJ |
| Power, heat flow rate | Btu/hr | W | 1 Btu/hr = 0.293 W |
| | hp | kW | 1 hp = 0.746 kW |
| Intensity of heat flow | Btu/ft$^2$ hr | W/m$^2$ | 1 Btu/ft$^2$ hr = 3.155 W/m$^2$ |
| Thermal conductance (heat transfer coefficient) | Btu/ft$^2$ hr °F | W/m$^2$ °C | 1 Btu/ft$^2$ hr °F = 5.678 W/m$^2$ °C |
| Thermal conductivity | Btu/ft hr °F W/m °C | — | 1 Btu/ft hr °F = 1.730 W/m °C |
| | Btu in./ft$^2$ hr °F | W/m °C | 1 Btu in./ft$^2$ hr °F = 0.144 W/m °C |
| Heat capacity | Btu/lb °F | kJ/kg °C | 1 Btu/lb °F = 4.187 kJ/kg °C |
| | Btu/ft$^3$ °F | kJ/m$^3$ °C | 1 Btu/ft$^3$ °F = 67.066 kJ/m$^3$ °C |
| Latent heat | Btu/ft$^3$ | kJ/m$^3$ | 1 Btu/ft$^3$ = 37.259 kJ/m$^3$ |

Some useful centimeter-gram-second (cgs) system of units used in heat-transfer calculations are:

| | | | |
|---|---|---|---|
| 1 cal | = 4.187 J | 1 cal/cm$^3$ | = 4187 kJ/m$^3$ |
| 1 cal/cm$^2$ s | = 41.87 kW/m$^2$ | 1 cal/g | = 4.187 kJ/kg |
| 1 cal/cm s °C | = 418.7 W/m K | 79.7 cal/g | = 144 Btu/lb |
| | = 242 Btu/ft hr °F | | = 334 J/g |
| 1 cal/cm$^3$ °C | = 4187 kJ/m$^3$ K | 1 ft$^2$/hr | = 25.806 mm$^2$/s |

# APPENDIX C

# Thermal Conductivity of Various Materials

| Material | Unit Weight | | Conductivity, $K$ | |
|---|---|---|---|---|
| | lb/ft$^3$ | kg/m$^3$ | Btu/hr ft °F | W/m K |
| Aluminum | — | — | 90–110 | 156–190 |
| Asbestos-cement board | 120 | 1922 | 0.33 | 0.57 |
| Cast iron | — | — | 29 | 50 |
| Concrete with sand and gravel mixture | 140 | 2243 | 0.75–1.0 | 1.3–1.7 |
| Concrete with lightweight aggregate | 120 | 1922 | 0.43 | 0.74 |
| Concrete asphalt | 131–138 | 2098–2211 | 0.61–0.88 | 1.05–1.52 |
| Copper | — | — | 223 | 386 |
| Ice (at 0°C or 32°F) | 57 | 913 | 1.29 | 2.23 |
| Insulating materials | | | | |
| Asbestos fibers | 9–12 | 144–192 | 0.04 | 0.07 |
| Mineral wool | 1.5–4.0 | 24–64 | 0.022 | 0.038 |
| Cellular glass board | 7.0–9.5 | 112–152 | 0.032 | 0.055 |
| Corkboard | 6–9 | 96–144 | 0.022 | 0.038 |
| Glass fiber | 9.5–11.0 | 152–176 | 0.021 | 0.036 |
| Polyurethane foam | 2.0 | 32 | 0.009–0.022 | 0.016–0.038 |
| Polystyrene—extruded, molded, or expanded | 1.5–3.0 | 24–48 | 0.019–0.022 | 0.033–0.038 |
| Straw compressed | 23 | 368 | 0.05 | 0.09 |
| Sulfur foam | 4–30 | 64–480 | 0.02–0.04 | 0.03–0.07 |
| Wood or cane fiber—interior finish (plank, pile, or lath) | 15 | 240 | 0.029 | 0.050 |

| Material | Unit Weight | | Conductivity, $K$ | |
|---|---|---|---|---|
| | lb/ft$^3$ | kg/m$^3$ | Btu/hr ft °F | W/m K |
| Insulating materials—loose fill | | | | |
| Asbestos fibers | 20–50 | 320–801 | 0.05–0.13 | 0.09–0.22 |
| Cork | 5–12 | 80–192 | 0.02–0.03 | 0.035–0.052 |
| Glass fibers | 2–12 | 32–192 | 0.02 | 0.035 |
| Mineral wool | 2–5 | 32–80 | 0.025 | 0.043 |
| Sawdust | 8–15 | 128–240 | 0.037 | 0.064 |
| Wood fiber | 2–3.5 | 32–56 | 0.025 | 0.043 |
| Rocks | 150–170 | 2403–2723 | | |
| Limestone | | | 0.75–2.9 | 1.3–5.0 |
| Sandstone | | | 1.1–2.4 | 1.8–4.2 |
| Schist | | | 0.9 | 1.6 |
| Slate | | | 2.2 | 3.8 |
| Shale | | | 0.9 | 1.5 |
| Quartzite | | | 2.6–4.1 | 4.5–7.1 |
| Granite | | | 1.0–2.3 | 1.7–4.0 |
| Snow | | | | |
| New, loose | | | 0.05 | 0.086 |
| Dense, compact | | | 0.20 | 0.340 |
| Steel | 490 | 7848 | 20–30 | 35–52 |
| Wood | | | | |
| Maple, oak, and similar hardwoods | 45 | 72 | 0.094 | 0.16 |
| Fir, pine, and similar softwoods | 32 | 51 | 0.067 | 0.12 |

# References

AASHTO, Committee on Design (1972). *AASHTO Interim Guide for the Design of Pavement Structures.*

Acum, W. E. A., and Fox, L. (1951). "Compaction of Load Stresses in a Three Layer Elastic System," *Geotechnique,* Vol. 2, pp. 293–300.

Adams, K. M. (1978). "Building and Operating Winter Roads in Canada and Alaska," Canada, Department of Indian and Northern Affairs, Environment Division, *Environmental Studies 4,* 22 pp.

Adamson, B. (1972). "Foundation with Crawl Spaces, Frost Penetration and Equivalent U-value of Floor Slab," *Frost i Jord,* No. 8, Oslo.

Aitken, G. W. (1966). "Reduction of Frost Heave by Surcharge Loading," *Proceedings of the First International Conference on Permafrost* (1963), Lafayette, Ind., National Academy of Sciences Publication 1287, pp. 319–324.

Akili, W. (1971). "Stress Strain Behaviour of Frozen Fine-Grained Soils," U.S. Highway Research Board, *Record 360,* pp. 1–9.

Alaska, State of, Department of Transportation and Public Facilities (ADOT/PF) (1982). "Guide for Flexible Pavement—Design and Evaluation," Division of Highway Design and Construction, Juneau, Alaska.

Aldrich, H. P., and Paynter, H. M. (1953). "Analytical Studies of Freezing and Thawing in Soils," U.S. Army, Corps of Engineers, Arctic Construction Frost Effects Lab, New England Division, Boston, *First Interim Report.*

Alkire, B. D., and Andersland, O. B. (1973). "The Effects of Confining Pressure on the Mechanical Properties of Sand–Ice Materials," *Journal of Glaciology,* Vol. 12, No. 66, pp. 469–481.

ALTER, A. J. (1969). "Water Supply in Cold Regions," U.S. Army, CRREL, *Monograph MIII-CS2,* Hanover, N.H., January, 91 pp.

ANDERSLAND, O. B., and ANDERSON, D. M., eds. (1978). *Geotechnical Engineering for Cold Regions,* McGraw-Hill, New York, 576 pp.

ANDERSON, D. M., and MORGENSTERN, N. R. (1973). "Physics, Chemistry, and Mechanics of Frozen Ground," North American Contribution, *Proceedings of the Second International Conference on Permafrost,* Yakutsk, USSR, National Academy of Sciences, Washington, D.C., pp. 257–288.

ANDERSON, D. M., and TICE, A. R. (1973). "The Unfrozen Interfacial Phase in Frozen Soil Water Systems," *Ecological Studies,* Vol. 4, pp. 107–125.

ARCONE, S. A., DELANEY, A. J., and SELLMAN, P. V. (1979). "Effects of Seasonal Changes and Ground Ice on Electromagnetic Surveys of Permafrost," U.S. Army, Corps of Engineers, CRREL, *Report 79–23,* October.

ARE, F. E. (1973). "Reworking of Shorelines in the Permafrost Zone," Vol. 2, *Abstract of the Second International Conference on Permafrost,* Yakutsk, USSR, Izdatelstvo "Nauka," Moscow.

ASCE (1977). "Application of Soil Dynamics in Cold Regions," Fall Convention (October 17–21, 1977), San Francisco, *Conference Preprint 3011,* 8 Papers, pp. 312.

ASHRAE (1977). "Design Heat Transfer Coefficients," American Society of Heating, *Refrigerating and Air-Conditioning Engineers Handbook,* Fundamentals Vol., Chap. 22, pp. 22–1 to 22–28.

ASSURE, A. (1963). "Discussion on Creep of Frozen Soils," North American Contribution, *Proceedings of the First International Conference on Permafrost,* Lafayette, Indiana, *National Academy of Sciences Publication 1287,* pp. 339–340.

ATKINS, R. T. (1979). "Determination of Frost Penetration by Soil Resistivity Measurements," U.S. Army, CRREL, *Special Report 79–22,* 12 pp.

AULD, R. G., ROBBINS, R. J., ROSENEGGER, L. W., and SANGSTER, R. H. B. (1978). "Pad Foundation Design and Performance of Surface Facilities in the Mackenzie Delta," *Proceedings of the Third International Conference on Permafrost,* Edmonton, Alberta, National Research Council, Vol. 1, pp. 765–771.

AVERILL, D. W., and HEINKE, G. W. (1974). "Vacuum Sewer System," *Indian and Northern Affairs Publication QS-1546-000-EE-A,* Information Canada, September.

BAKER, T. H. W., (1976). "Transportation, Preparation and Storage of Frozen Soil Samples for Laboratory Testing," *ASTM Special Technical Publication 599,* pp. 88–112.

BAKER, T. H. W. (1978). "Effect of End Conditions on the Uniaxial Compressive Strength of Frozen Sand," *Proceedings of the Third International Conference on Permafrost,* Edmonton, Canada, pp. 608–614.

BARANOV, I. J. (1959). "Geophysical Distribution of Seasonally Frozen Ground and Permafrost," *General Geocryology,* USSR Academy of Sciences, Transl. NRC TT-1121.

BARANOV, I. J., and KUDRYAVTSEV, V. A. (1966). "Permafrost in Eurasia," *Proceedings of the First International Conference on Permafrost* (1963), Lafayette, Ind., National Academy of Sciences Publication 1287, pp. 98–102.

BARNES, D. F. (1966). "Geophysical Methods for Delineating Permafrost," *Proceedings of the First International Conference on Permafrost* (1963), Lafayette, Ind., National Academy of Sciences Publication 1287, pp. 349–355.

BENDZ, J. (1977). "Permafrost Problems Beaten in Pile Driving Research," *Western Construction,* Vol. 52, No. 11, pp. 16, 18, 34, 35.

BENEDICT, J. B. (1970). "Downslope Soil Movement in a Colorado Alpine Region: Rates, Processes, and Climatic Significance," *Journal of Arctic Alpine Research,* Vol. 2, No. 3, pp. 165–226.

BERG, R. L. (1974). "Design of Civil Airfield Pavements for Seasonal Frost and Permafrost Conditions," *Report No. FAA-RD-7430,* Federal Aviation Administration, Washington, D.C., October.

BERG, R. L., GUYMON, G., and GARTNEV, K. E. (1977). "A Mathematical Model To Predict Frost Heave," *International Symposium on Frost Action in Soils,* Sweden.

BERG, R. L., and QUINN, W. F. (1977). "Use of a Light-Coloured Surface to Reduce Seasonal Thaw Penetration beneath Embankments on Permafrost," *Proceedings of the Second International Symposium on Cold Regions Engineering* (1976), Fairbanks, University of Alaska, Department of Civil Engineering, pp. 86–89.

BERG, R. L., and SMITH, N. (1976). "Observations along the Pipeline Haul Road between Livengood and the Yukon River," U.S. Army, CRREL, *Special Report 76–11,* 83 pp.

BESKOW, G. (1935). "Soil Freezing and Frost Heaving," Swedish Geological Society, Series C, No. 375, 26th Year, *Book 3;* translated by J. O. Osterberg, The Technological Institute, Northwestern University, Evanston, Ill. (1947), 145 pp.

BIRD, J. B. (1967). *The Physiography of Arctic Canada,* Johns Hopkins University Press, Baltimore, 336 pp.

BLACK, R. F. (1954). "Permafrost: A Review," *Geological Society of America Bulletin,* Vol. 65, pp. 839–856.

BOCA BASIC BUILDING CODE/1978 (1978). Building Officials and Code Administrators International, Inc., Homewood, Illinois.

BRODSKAIA, A. G. (1962). "Compressibility of Frozen Ground" (English translation), U.S. Department of Commerce, *National Technical Information Series AD715087.*

BROMS, B. B. (1965). "Design of Laterally Loaded Piles," *Journal of Soil Mechanics and Foundation,* Div. American Society of Civil Engineering, 91 (SM3), pp. 79–99.

BROWN, I. C., ed. (1967a). "Groundwater in Canada," Geological Survey of Canada, *Economic Geology Report 24,* 242 pp.

BROWN, R. J. E. (1966). "Influence of Vegetation on Permafrost," *Proceedings of the First International Conference on Permafrost* (1963), Lafayette, Ind., National Academy of Sciences Publication 1287, pp. 20–25.

BROWN, R. J. E. (1967b). "Permafrost in Canada," Canada, National Research Council, Division of Building Research, *Map NRC 9769,* and Geological Survey of Canada, *Map 1246A.*

BROWN, R. J. E. (1970). *Permafrost in Canada: Its Influence on Northern Development,* University of Toronto Press, Toronto, 246 pp.

BROWN, R. J. E. (1973). "Influence of Climatic and Terrain Factors on Ground Temperatures at Three Locations in the Permafrost Regions of Canada," *Proceedings Second International Conference on Permafrost,* Yakutsk, USSR, North American Contribution, U.S. National Academy of Sciences, pp. 27–34.

BROWN, R. J. E., and PÉWÉ, T. L. (1973). "Distribution of Permafrost in North America and Its Relationship to the Environment: A Review, 1963–1973," North American Contribution, *Proceedings of the Second International Conference on Permafrost,* Yakutsk, USSR, National Academy of Sciences, Washington, D.C., pp. 71–100.

BURMISTER, D. M. (1945). "The General Theory of Stresses and Displacements in Layered Soil Systems," *Journal of Applied Physics,* Vol. 16.

BURMISTER, D. M. (1958). "Evaluations of Pavement Systems of the WASHO Road Test by Layered Systems Methods," U.S. Highway Research Board, *Bulletin 177.*

BURNS, C. D., BRABSTON, W. N., and GRAY, R. W. (1972). "Feasibility of Using Membrane Enveloped Soil Layers as Pavement Elements of Multiple Wheel Heavy Gear Loads," U.S. Army, Corps of Engineers, Waterways Experiment Station, Vicksburg, Miss., *Misc. Paper S-72-6.*

BURT, T. P., and WILLIAMS, P. J. (1976). "Hydraulic Conductivity in Frozen Soils," Earth Science Processes, 1, pp. 349–360.

BUTKOVICH, T. R., and LANDAUER, J. I. (1959). "The Flow Law for Ice," U.S. Army Snow, Ice and Permafrost Research Establishment, *Report 56.*

CANADA ENVIRONMENT (1979). "Cold Climate Utilities Delivery Design Manual," Environmental Protection Service, Edmonton, Alberta, *Report EPS-3-WP-79-2*, 644 pp.

CANADA NATIONAL RESEARCH COUNCIL, (1980). The *Supplement to the National Code of Canada,* Associate Committee on the National Building Code, Ottawa, Ontario.

CAPPS, S. R. (1919). "The Kantishna Region, Alaska," *U.S. Geological Survey, Bulletin 687,* pp. 7–112.

CAPPS, S. R. (1940). "Geology of the Alaska Railroad," *U.S. Geological Survey, Bulletin 907,* pp. 1–197.

CARSLAW, H. S., and JAEGER, J. C. (1959). *Conduction of Heat in Solids,* 2nd ed., Oxford University Press, London.

CARSON, M. A., and KIRKBY, M. J. (1972). *Hillslope Form and Process,* Cambridge University Press, Cambridge.

CASAGRANDE, A. (1932). "Research on the Atterberg Limits of Soil," *Public Roads,* Vol. 13, No. 8, pp. 121–136.

CASAGRANDE, A. (1948). "Classification and Identification of Soils," *Transactions, ASCE,* Vol. 113, pp. 901–930.

CHAMBERLAIN, E. J. (1981). "Frost Susceptibility of Soil: Review of Index Tests," U.S. Army, CRREL, *Monograph 81–2,* 121 pp.

COMMISSION ON STANDARDIZATION OF LABORATORY AND FIELD TESTS, (1976). *Suggested Methods for Geophysical Loggings of Boreholes,* International Society for Rock Mechanics, Lisbon.

CORTE, A. E. (1969). "Geocryology and Engineering," in D. J. Varnes and G. Kiersch, eds., *Reviews in Engineering Geology,* Geological Society of America, Vol. 2, pp. 119–185.

CRORY, F. E. (1966). "Pile Foundations in Permafrost," *Proceedings of the First International Conference on Permafrost* (1963), Lafayette, Ind., National Academy of Sciences Publication 1287, pp. 467–476.

CRORY, F. E. (1973a). "Settlement Associated with the Thawing of Permafrost," North

American Contribution, *Proceedings of the Second International Conference on Permafrost,* Yakutsk, USSR, National Academy of Sciences, Washington, D.C., pp. 599–607.

CRORY, F. E. (1973b). "Installation of Driven Test Piles in Permafrost at Bethel Air Force Station, Alaska," U.S. Army, CRREL, *Technical Report 139,* 22 pp.

CRORY, F. E. (1978). "The Kotzebue Hospital: A Case Study," *Proceedings of the Conference on Applied Techniques for Cold Environments,* Anchorage, Alaska, ASCE Cold Regions Speciality Conference, May 17–19, Vol. 1, pp. 342–359.

CRORY, F. E., BERG, R. L., and BURNS, C. D. (1978). "Design Considerations for Airfield in NPRA: Applied Technique for Cold Environments," *Cold Regions Specialty Conference,* Anchorage, May 17–19, 441–458.

CRORY, F. E., ISAACS, R. M., SANGER, F. J., and SHOOK, J. F. (1982). "Designing for Frost Heave Conditions," *American Society of Civil Engineers Spring Convention,* Las Vegas, Nevada, April 26–30.

CRORY, F. E., and REED, R. E. (1965). "Measurement of Frost Heaving Forces on Piles," U.S. Army, CRREL, *Technical Report 145,* 27 pp.

DAVIS, R. M., and KITZE, F. F. (1967). "Soil Sampling and Drilling near Fairbanks, Alaska: Equipment and Procedures," U.S. Army, CRREL, *Technical Report 191,* 50 pp.

DAVISON, B. E., ROONEY, J. W., and BRUGGERS, D. E. (1978). "Design Variables Influencing Piles Driven in Permafrost," *Proceedings of the Conference on Applied Techniques for Cold Environments,* Anchorage, Alaska, ASCE Cold Regions Specialty Conference, May 17–19, Vol. 2, pp. 307–318.

DAWSON, R. N., and CRONIN, K. J. (1977). "Trends in Canadian Water and Sewer Systems Serving Northern Communities," *Proceedings of the Symposium on Utilities Delivery in Arctic Regions,* Edmonton, Alberta, Environment Canada, Environmental Protection Service, Report EPS-3-WP-77-1, pp. 1–17.

DILLON, H. B., and ANDERSLAND, O. B. (1966). "Predicting Unfrozen Water Contents in Frozen Soils," *Canadian Geotechnical Journal,* Vol. 3, No. 2, pp. 53–60.

DI PASQUALE, L., GERLEK, S., and PHUKAN, A. (1983). "Design and Construction of Pile Foundations in Yukon–Kuskokwim Delta, Alaska," *Fourth International Conference on Permafrost,* Fairbanks, July.

DOBRIN, M. D. (1960). *Introduction to Geophysical Prospecting,* 2nd ed., McGraw-Hill, New York.

EDLEFSON, N. E., and ANDERSON, A. B. B. (1943). "Thermodynamics of Soil Moisture," *Hilgardia,* Vol. 15, pp. 31–268.

EMBLETON, L., and KING, L. (1975). *Periglacial Geomorphology,* Edward Arnold, London.

ESCH, D. C. (1973). "Control of Permafrost Degradation beneath a Roadway by Subgrade Insulation," North American Contribution, *Proceedings of the Second International Conference on Permafrost,* Yakutsk, USSR, National Academy of Sciences, Washington, D.C., pp. 608–622.

ESCH, D. C. (1978). "Road Embankment Design Alternatives over Permafrost," *Proceedings of The Conference on Applied Techniques for Cold Environments,* Anchorage, Alaska, ASCE Cold Regions Specialty Conference, May 17–19, Vol. 1, pp. 159–170.

ESCH, D. C. (1981). *Thawing of Permafrost by Passive Solar Means,* Alaska Department of Transportation and Public Facilities Report.

ESCH, D. C., and LIVINGSTON, H. (1978). "Performance of a Roadway with a Peat Underlay over Permafrost," State of Alaska, *Department of Transportation and Public Facilities Report.*

ESCH, D. C., and RHODE, J. J. (1977). "Kotzebue Airport Runway Installation over Permafrost," *Proceedings of the Second International Symposium on Cold Regions Engineering* (1976), Fairbanks, University of Alaska, Department of Civil Engineering, pp. 44–61.

EVERETT, D. H. (1961). "The Thermodynamics of Frost Action in Porous Solids," *Transactions of the Faraday Society,* Vol. 57, pp. 1541–1551.

EVERETT, D. H., and HAYNES, J. M. (1965). "Capillary Properties of Some Model Pore Systems with Reference to Frost Damage," *RILEM Bulletin,* New Series, Vol. 27, pp. 31–38.

FAROUKI, O. T. (1981). "Thermal Properties of Soils," U.S. Army, CRREL, *Monograph 81–2,* 136 pp.

FEDERAL AVIATION ADMINISTRATION (1978). "Airport Pavement Design and Evaluation," *Advisory Circular AC 150/5320-6C.*

FERRIANS, O. J., JR., KACHADOORIAN, R., and GREENE, G. W. (1969). "Permafrost and Related Engineering Problems in Alaska," *U.S. Geological Survey, Professional Paper 678,* 41 pp.

FEULNER, A. J., and WILLIAMS, J. R. (1967). "Development of a Groundwater Supply at Cape Lisburne, Alaska, by Modification of the Thermal Regime of Permafrost," *U.S. Geological Survey, Professional Paper 575–B,* pp. 199–202.

FINN, W. D., and YONG, R. N. (1978). "Seismic Response of Frozen Ground," *ASCE, Journal of the Geotechnical Engineering Division,* Vol. 104, No. GT10, pp. 1225–1241.

FOSTER, R. R. (1975). "Arctic Water Supply," *Water Pollution Control,* Vol. 3, No. 3, pp. 24–28, 33.

FULWIDER, C. W., and AITKEN, G. W. (1962). "Effects of Surface Colour on Thaw Penetration beneath an Asphalt Surface in the Arctic," *Proceedings of the International Conference on the Structural Design of Asphalt Pavements,* University of Michigan, Ann Arbor, pp. 605–610.

GANDAHL, R. (1963). "Determination of the Ground Frost Line by Means of a Simple Type of Frost Depth Indicator," translated and revised by P. T. Hodgins, National Swedish Road Research Institute, Stockholm, *Report 30A.*

GARG, O. P. (1969). "Static and Dynamic Mechanical Properties of a Sandstone," M.Sc. thesis, University of Saskatchewan, Saskatoon, Canada.

GARG, O. P. (1973). "In situ Physicomechanical Properties of Permafrost Using Geophysical Technique," *Proceedings of the Second International Conference on Permafrost,* Yakutsk, USSR, National Academy of Sciences, Washington, D.C.

GERLEK, S. (1982). "The Behavior of Driven H-Piles in Frozen Silt," *Interim Report to U.S. PHS/Alaska Area Native Service,* April.

GLEN, J. W. (1952). "Experiments on the Deformation of Ice," *Journal of Glaciology,* Vol. 2, No. 12, pp. 111–114.

GLEN, J. W. (1955). "The Creep of Polycrystalline Ice," *Proceedings of the Royal Society, London,* Vol. A228, pp. 519–538.

GLEN, J. W. (1974). "The Physics of Ice," CRREL report, April.

GOLD, L. W. (1957). "A Possible Force Mechanism Associated with the Freezing of Water and Porous Materials," *Highway Research Bulletin 168, NAS-NRC,* Washington, pp 65–72.

GOLD, L. W., and LACHENBRUCH, A. H. (1973). "Thermal Conditions in Permafrost: A Review of North American Literature," North American Contribution, *Proceedings of the Second International Conference on Permafrost,* Yakutsk, USSR, National Academy of Sciences, Washington, D.C., pp. 3–25.

GONCHAROV, Y. M. (1964). "Construction of Pile Foundations in Permafrost Regions," *Problems of the North*, Vol. 10, No. 32, pp. 161–171.

GOUGHNOUR, R. R., and ANDERSLAND, O. B. (1968). "Mechanical Properties of a Sand–Ice System," *Journal of the Soil Mechanics and Foundations Division, ASCE,* Vol. 94, No. SM4, pp. 923–950.

GOUGHNOUR, R. D., and DEMAGIO, J. A. (1978). "Soil Reinforcement Methods on Highway Projects," *Symposium on Earth Reinforcement, ASCE Annual Convention*, Pittsburgh, Pa., April 27.

GRAINGE, J. W. (1969). "Arctic Heated Pipe Water and Waste Water Systems," in *Water Research,* Vol. 3, Pergamon Press, Elmsford, N.Y., pp. 47–71.

GUYMON, G. L. and LUTHIN, J. N. (1974). "A Coupled Heat and Moisture Transport Model for Arctic Soils," *Water Resources Res.,* Vol. 10, No. 5, pp. 995–1001.

GUYMON, G. L., HROMADKA, II, T. V., and BERG, R. L. (1980). "A One-Dimensional Frost Heave Model Based upon Simulation of Simultaneous Heat and Water Flux." *Cold Reg. Sci. Tech.,* 2.

GUYMON, G., BERG, R. L., JOHNSON, T. C., and HROMADKA, T. V. II. (1981). Results from a Mathematical Model of Frost Heave. *Transportation Res. Board Journal.*

HAMELIN, L. E., and COOK, F. A. (1967). *Le Périglaciaire par l'image* (Illustrated Glossary of Periglacial Phenomena), Les Presses de l'Université Laval, Sainte-Foy, Quebec, Centre d'Études Nordiques, Travaux et Documents 4, 237 pp.

HANS, R. (1960). "Seismic Refraction Soundings in Permafrost near Thule, Greenland," *International Symposium on Arctic Geology,* Calgary, Alberta, pp. 970–981.

HANSEN, B. L. (1966). "Instruments for Temperature Measurements in Permafrost," *Proceedings of the First International Conference on Permafrost* (1963), Lafayette, Ind., National Academy of Sciences Publication 1287, pp. 356–358.

HARDY, R. M., and MORRISON, H. A. (1972). "Slope Stability and Drainage Considerations for Arctic Pipelines," *Proceedings of the Canadian Northern Pipeline Research Conference,* Ottawa, Canada, National Research Council, Technical Memo. 104, pp. 249–267.

HARLAN, R. L., and NIXON, J. F. (1978). "Ground Thermal Regime," Chapter 3, *Geotechnical Engineering for Cold Regions,* McGraw-Hill Book Co., New York.

HARRISON, W. D., and OSTERKAMP, T. E. (1982). "Measurements of the Electrical Conductivity of Interstitial Water in Subsea Permafrost," *Proceedings of the Fourth Canadian Permafrost Conference* (1981), Calgary, National Research Council of Canada, Ottawa, pp. 229–237.

HARTMAN, C. W., and CARLSON, R. F. (1970). *Bibliography of Arctic Water Resources,* Institute of Water Resources, University of Alaska, Fairbanks, 344 pp.

HAYLEY, D. W. (1982). "Application of Heat Pipes to Design of Shallow Foundations on Permafrost," *Proceedings of the Fourth Canadian Permafrost Conference* (1981), Calgary, National Research Council of Canada, Ottawa, pp. 535–544.

HAYNES, F. D. (1978). "Strength and Deformation of Frozen Silt," *Proceedings of the Third International Conference on Permafrost,* Edmonton, Alberta, National Research Council, Vol. 1, pp. 655–661.

HENNION, F. B., and LOBACZ, E. F. (1973). "Corps of Engineers Technology Related to Design of Pavements in Areas of Permafrost," North American Contribution, *Proceedings of the Second International Conference on Permafrost,* Yakutsk, USSR, National Academy of Sciences, Washington, D.C., pp. 658–664.

HERJE, J. R. (1972). "Pilares and Piles in the Ground." Frost Problems, *Frost i Jord,* No. 8, Oslo.

HEUER, C. E. (1979). "The Application of Heat Pipes on the Trans-Alaska Pipeline," U.S. Army, CRREL, *Special Report 79–26,* 33 pp.

HO, D. M., HARR, M. E., and LEONARDS, G. A. (1969). "Transient Temperature Distribution in Insulated Pavements—Predictions vs. Observations," *Canada Geotechnical Journal,* Vol. 7, pp. 273–284.

HOBBS, P. V. (1974). *The Physics of Ice,* Clarendon Press, Oxford.

HOBSON, G. D., and HUNTER, J. A. (1969). "In Situ Determination of Elastic Constants in Overburden Using a Hammer Seismograph," *Geoexploration,* Vol. 7, pp. 107–112.

HOEKSTRA, P. (1965). "Conductance of Frozen Bentonite Suspensions," *Soil Science Society of America Proceedings,* Vol. 29, pp. 519–523.

HOEKSTRA, P. (1969). "The Physics and Chemistry of Frozen Soils," U.S. Highway Research Board, *Special Report 103,* pp. 78–90.

HOEKSTRA, P., and MCNEILL, D. (1973). "Electromagnetic Probing of Permafrost," North American Contribution, *Proceedings of the Second International Conference on Permafrost,* Yakutsk, USSR, National Academy of Sciences, Washington, D.C., pp. 517–526.

HOEKSTRA, P., CHAMBERLAIN, E., and FRATE, T. (1965). "Frost Heaving Pressures," U.S. Army, CRREL, *Research Report 176,* 16 pp.

HOEKSTRA, P., SELLMAN, P. V., and DELANY, A. J. (1974). "Airborne Resistivity Mapping of Permafrost near Fairbanks, Alaska," U.S. Army, CRREL, *Research Report 324,* 49 pp.

HOLTZ, R. D., and KOVACS, W. D. (1981). *An Introduction to Geotechnical Engineering,* Prentice-Hall, Inc., Englewood Cliffs, New Jersey, p. 733.

HOPKE, S. (1980). "A Model for Frost Heave Including Overburden," *Cold Regions Sci. and Tech.,* Vol. 2.

HOPKINS, D. M. (1967). "Quaternary Marine Transgressions in Alaska," in *The Bering Land Bridge* (D. M. Hopkins, ed.), Stanford University Press, Stanford, Calif. pp. 47–49.

HOPKINS, D. M., and HARTZ, R. W. (1978). "Shoreline History of Chukchi and Beaufort Seas: An Aid to Predicting Offshore Permafrost Conditions," *Annual Report RU-473,* BLM/NOAA OCSEAP, Arctic Project Office, University of Alaska, Fairbanks.

HOWARD, A. K. (1977). "Laboratory Classification of Soils—Unified Soil Classification

System," *Earth Science Training Manual 4,* U.S. Bureau of Reclamation, Denver, 56 pp.

HOWITT, F. (1971). "Permafrost Geology at Prudhoe Bay," *World Petroleum,* pp. 28.

HUGHES, O. L. (1972). "Surficial Geology and Land Classification, Mackenzie Valley Transportation Corridor," *Proceedings of the Canadian Northern Pipeline Research Conference,* Canada, National Research Council, Associate Committee on Geotechnical Research, Technical Memo. 104, pp. 17–24.

HUGHES, O. L., and TERASMAE, J. (1963). "SIPRE Ice-Corer for Obtaining Samples from Permanently Frozen Bogs," *Arctic,* Vol. 16, No. 4, pp. 270–272.

HULT, J. A. H. (1966). *Creep in Engineering Structures,* Blaisdell, Waltham, Mass.

HUNTER, J. A., JUDGE, A. S., MACAULEY H. A., GOOD, R. L., GAGNE, R. M., and BURNS, R. A. (1976). "Permafrost and Frozen Sub-Seabottom Materials in the Southern Beaufort Sea," Environment Canada, Beaufort Sea Project, *Technical Report 22,* 177 pp.

HVORSLEV, M. J., and GOODE, T. B. (1966). "Core Drilling in Frozen Ground," *Proceedings of the First International Conference on Permafrost* (1963), Lafayette, Ind., National Academy of Sciences Publication 1287, pp. 364–370.

HVORSLEV, M. J. (1949). *Subsurface Exploration and Sampling of Soils for Civil Engineering Purposes,* U.S. Army, Corps of Engineers, Waterways Experiment Station, Vicksburg, Miss., 521 pp; reprinted by the Engineering Foundation, 1962.

INGERSOLL, L. R., ZOBEL, O. J., and INGERSOLL, A. C. (1954). *Heat Conduction with Engineering, Geological and Other Applications,* University of Wisconsin Press, Madison, WI, 325 p.

ISSACS, R. M., and CODE, J. A. (1972). "Problems in Engineering Geology Related to Pipeline Construction," *Proceedings of the Canadian Northern Pipeline Research Conference,* Canada, National Research Council, Technical Memo. 104, pp. 147–179.

IVES, J. D. (1962). "Iron Mining in Permafrost: Central Labrador–Ungava," *Geographical Bulletin 17,* pp. 66–77.

IVES, J. D. (1973). "Permafrost and Its Relationship to Other Environmental Parameters in a Midlatitude, High-Altitude Setting, Front Range, Colorado Rocky Mountains," North American Contribution, *Proceedings of the Second International Conference on Permafrost*, Yakutsk, USSR, National Academy of Sciences, Washington, D.C., pp. 121–125.

JACKSON, K. A., CHALMERS, B., and UHLAMM, (1966). "Particle Sorting and Stone Migration due to Frost Heave," *Science,* Vol. 152, pp. 545–546.

JAMES, W., and SUK, R. (1977). "Least Cost Design for Water Distribution for Arctic Communities," McMaster University, Hamilton, Ontario, May.

JOHANSEN, O. (1975). "Thermal Conductivity of Soils," Ph.D. thesis, Trondheim, Norway (*CRREL Translation 637,* 1977), *ADAO44002*.

JOHNSON, T. C., COLE, D. M., and CHAMBERLAIN, E. J. (1978). "Influence of Freezing and Thawing on the Resilient Properties of Silt Soil beneath an Asphalt, Concrete Pavement," U.S. Army, CRREL, *Special Report 78–23*.

JOHNSTON, G. H. (1962). "Bench Marks in Permafrost Areas," *The Canadian Surveyor,* Vol. 16, No. 1, pp. 32–41.

JOHNSTON, G. H. (1963). "Soil Sampling in Permafrost Areas," Canada, National Research Council, Division of Building Research, *Technical Paper 155*.

JOHNSTON, G. H. (1969). "Dykes on Permafrost, Kelsey Generating Station, Manitoba," *Canadian Geotechnical Journal,* Vol. 6, No. 2, pp. 139–157.

JOHNSTON, G. H. (1973). "Ground Temperature Measurements Using Thermocouples," *Proceedings of the Seminar on the Thermal Regime and Measurements in Permafrost,* Canada, National Research Council, Associate Committee on Geotechnical Research, Technical Memo. 108, pp. 1–12.

JOHNSTON, G. H. (1981). *Permafrost: Engineering Design and Construction,* Wiley, New York, 540 pp.

JOHNSTON, G. H., and BROWN, R. J. E. (1964). "Some Observations on Permafrost Distribution at a Lake in the Mackenzie Delta, N.W.T., Canada," *Arctic,* Vol. 17, No. 3, pp. 162–175.

JOHNSTON, G. H., and LADANYI, B. (1972). "Field Tests of Grouted Rod Anchors in Permafrost," *Canadian Geotechnical Journal,* Vol. 9, No. 2, pp. 176–194.

JOHNSTON, G. H., and LADANYI, B. (1974). "Field Tests of Deep Power-Installed Screw Anchors in Permafrost," *Canadian Geotechnical Journal,* Vol. 11, No. 3, pp. 348–358.

JONES, R. H. (1980). "Development and Application of Frost Susceptibility Testing," *2nd International Symposium on Ground Freezing,* Norwegian Inst. of Tech., June 24–26.

JUDGE, A. S. (1973). "Ground Temperature Measurements Using Thermistors," *Proceedings of the Seminar on the Thermal Regime and Measurements in Permafrost,* Canada, National Research Council, Associate Committee on Geotechnical Research, Technical Memo. 108, pp. 13–22.

JUDGE, A. S. (1974), "Occurrence of Offshore Permafrost in Northern Canada," in *The Coast and Shelf of the Beaufort Sea* (J. C. Reed and J. E. Sater, eds.), Proceedings of the Symposium (January 1974), San Francisco, Arctic Institute of North America, pp. 427–437.

KAPLAR, C. W. (1968). "New Experiments to Simplify Frost Susceptibility Tests of Soils," *Highway Research Record,* Vol. 215, pp. 48–59.

KAPLAR, C. W. (1969). "Laboratory Determination of Dynamic Moduli of Frozen Soils and of Ice," U.S. Army, CRREL, *Research Report 163,* 45 p.

KAPLAR, C. W. (1971). "Some Strength Properties of Frozen Soil and Effect of Loading Rate," U.S. Army, CRREL, *Special Report 159.*

KAWASKI, K., GRUOL, V., and OSTERKAMP, T. E. (1983). "Field Evaluation Site for Ground Ice Detection," *Report No. FHWA-AK-RD-83-27,* Alaska Dept. of Transportation and Pub. Facilities.

KELLER, G. V., and FRISCHKNECHT, F. C. (1966). *Electrical Methods in Geophysical Prospecting,* Pergamon Press, Elmsford, N.Y.

KENT, D. D., FREDLUND, D. G., and WATT, W. G. (1975). "Variables Controlling Behavior of a Partly Saturated Soil," *Proceedings Conference on Soil Water Problems in Cold Regions,* Calgary, Alberta, Special Task Force, Dir. Hydrology, Am. Geophy. Union, pp. 70–88.

KERFOOT, D. E., and MACKAY, J. R. (1972). "Geomorphological Process Studies, Garry Island, N.W.T.," in *Mackenzie Delta Area,* Monograph, Brock University, St. Catharine's, Ontario, pp. 115–130.

KERSTEN, M. S. (1949). "Thermal Properties of Soils," University of Minnesota, Engineering Experiment Station, *Bulletin 28,* 227 pp.

KING, M. S., BAMFORD, T. S., and KURFURST, P. J. (1974). "Ultrasonic Velocity Measurements on Frozen Rocks and Soils," *Proceedings of the Symposium on Permafrost Geophysics,* Canada, National Research Council, Associate Committee on Geotechnical Research, Technical Memo. 113, pp. 35–42.

KINOSHITA, S. (1962). "Heave Force of Frozen Soil," *Low Temperature Sci., Kitami Tech. College Ser. A-21.*

KINOSHITA, S., and ONO, T. (1963). "Heaving Force of Frozen Ground: 1. Mainly on the Results of Field Researches," Canada, National Research Council, *Technical Translation TT-1246* (1966), 30 pp.

KITZE, F. F. (1956). "Some Experiments in Drive Sampling of Frozen Ground," U.S. Army, CRREL (ACFEL), *Misc. Paper 16,* 22 pp.

KLASSEN, H. P. (1965). "Public Utilities Problems in the Discontinuous Permafrost Areas," *Proceedings of the Canadian Regional Permafrost Conference,* Edmonton, Alberta, Canada, National Research Council, Associate Committee on Soil and Snow Mechanics, Technical Memo. 86, pp. 106–118.

KLOVE, K. and THUE, J. V. (1972). "Slab-on-Ground Foundation," *Frost i Jord,* No. 8, Oslo.

KONRAD, J. M., and MORGENSTERN, N. R. (1980). "A Mechanistic Theory of Ice Lens Formation in Fine-Grained Soils," *Canadian Geotechnical Journal,* Vol. 17, pp. 473–483.

KREITH, F. (1973). *Principles of Heat Transfer,* 3rd ed., Intext Educational Publishers, New York.

LACHENBRUCH, A. H. (1957). "Three Dimensional Heat Conduction in Permafrost beneath Heated Buildings," *U.S. Geological Survey, Bulletin 1052-B,* 19 pp.

LACHENBRUCH, A. H. (1962). "Mechanics of Thermal Contraction Cracks and Ice-Wedge Polygons in Permafrost," *Society America, Special paper 70,* 69 p.

LACHENBRUCH, A. H. (1970). "Some Estimates of the Thermal Effects of a Heated Pipeline in Permafrost," *U.S. Geological Survey, Circular 632,* 23 pp.

LACHENBRUCH, A. H., and MARSHALL, B. V. (1969). "Heat Flow in the Arctic," *Arctic,* Vol. 22, No. 3, pp. 300–311.

LACHENBRUCH, A. H., GREENE, G. W., and MARSHALL, B. V. (1966). "Permafrost and Geothermal Regimes," in *Environment of Cape Thomson Region, Alaska,* U.S. Atomic Energy Commission, pp. 149–163.

LADANYI, B. (1972). "An Engineering Theory of Creep of Frozen Soils," *Canadian Geotechnical Journal,* Vol. 9, pp. 63–80.

LADANYI, B. (1974). "Bearing Capacity of Frozen Soil," Preprints, 27th Canadian Geotechnical Conference, Edmonton, Alberta, pp. 97–107.

LADANYI, B. (1975). "Bearing Capacity of Strip Footings in Frozen Soils," *Canadian Geotechnical Journal,* Vol. 12, pp. 393–407.

LADANYI, B., and JOHNSTON, G. H. (1974). "Behaviour of Circular Footings and Plate Anchors Embedded in Permafrost," *Canadian Geotechnical Journal,* Vol. 11, No. 4, pp. 531–553.

LAMBE, T. W., and WHITMAN, R. V. (1969). *Soil Mechanics,* Wiley, New York.

LANGE, G. R. (1973). "An Investigation of Core Drilling in Perennially Frozen Gravels and Rock," U.S. Army, CRREL, *Technical Report 245,* 31 pp.

LANGE, G. R., and SMITH, T. K. (1972). "Rotary Drilling and Coring in Permafrost," in *Deep Core Drilling, Core Analysis and Borehole Thermometry at Cape Thomson, Alaska,* Part III, U.S. Army, CRREL, Technical Report 95 III, 25 pp.

LANGWAY, C. C., JR. (1967). "Stratigraphic Analysis of a Deep Ice Core from Greenland," U.S. Army, CRREL, *Research Report 77,* 132 pp.

LEWELLEN, R. I. (1973). "The Occurrence and Characteristics of Nearshore Permafrost, Northern Alaska," North American Contribution, *Proceedings of the Second International Conference on Permafrost,* Yakutsk, USSR, National Academy of Sciences, Washington, D.C., pp. 131–136.

LIGUORI, A., MAPLE, J. A., and HEUER, C. E. (1979). "The Design and Construction of the Alyeska Pipeline," *Proceedings of the Third International Conference on Permafrost* (1978), Edmonton, Alberta, National Research Council, Vol. 2, pp. 151–157.

LINELL, K. A. (1973). "Long-Term Effects of Vegetative Cover on Permafrost Stability in an Area of Discontinuous Permafrost," North American Contribution, *Proceedings of the Second International Conference on Permafrost,* Yakutsk, USSR, National Academy of Sciences, Washington, D.C., pp. 688–693.

LINELL, K. A., HENNION, F. G., and LOBACZ, E. F. (1963). "Corps of Engineers Pavement Design in Areas of Seasonal Frost," U.S. Highway Research Board *Record 33,* Washington, D.C.

LINELL, K. A., and JOHNSTON, G. H. (1973). "Engineering Design and Construction in Permafrost Regions: A Review," North American Contribution, *Proceedings of the Second International Conference on Permafrost,* Yakutsk, USSR, National Academy of Sciences, Washington, D.C., pp. 553–575.

LINELL, K. A., and KAPLAR, C. W. (1966). "Description and Classification of Frozen Soils," *Proceedings of the First International Conference on Permafrost* (1963), Lafayette, Ind., National Academy of Sciences Publication 1287, pp. 481–487.

LINELL, K. A., and LOBACZ, E. F. (1980). "Design and Construction of Foundations in Areas of Deep Seasonal Frost and Permafrost," CRREL, *Special Report 80–34,* August, 310 p.

LOBACZ, E. F., GILMAN, G. D., and HENNION, F. B. (1973). "Corps of Engineers Design of Highway Pavements in Areas of Seasonal Frost," *Proceedings of the Symposium on Frost Action on Roads,* Oslo, pp. 142–152.

LONG, E. L. (1966). "The Long Thermopile," *Proceedings of the First International Conference on Permafrost* (1963), Lafayette, Ind., National Academy of Sciences Publication 1287, pp. 487–491.

LONG, E. L (1978). "Permafrost Foundation Designs," *Proceedings of the Conference on Applied Techniques for Cold Environments,* Anchorage, Alaska, ASCE Cold Regions Specialty Conference, May 17–19, Vol. 2, pp. 973–987.

LOTSPEICH, F. B. (1971). *Environment Guidelines for Road Construction in Alaska,* Environmental Protection Agency, College (Fairbanks), Alaska.

LUNARDINI, V. J. (1978). "Theory of *n*-Factors and Correlation of Data," *Proceedings of the Third International Conference on Permafrost,* Edmonton, Alberta, National Research Council, Vol. 1, pp. 41–46.

LUNARDINI, V. J. (1981). *Heat Transfer in Cold Climate,* Van Nostrand Reinhold, New York.

LUSCHER, U., and AFIFI, S. S. (1973). ''Thaw Consolidation of Alaskan Silts and Granular Soils,'' North American Contribution, *Proceedings of the Second International Conference on Permafrost,* Yakutsk, USSR, National Academy of Sciences, Washington, D.C., pp. 325–333.

LUSCHER, U., BLACK, W. T., and NAIR, K. (1975). ''Geotechnical Aspects of Trans-Alaska Pipeline,'' *ASCE, Journal of the Transportation Engineering Division,* Vol. 101, No. TE4, pp. 669–680.

MACKAY, J. R. (1963a). ''Origin of the Pingos of the Pleistocene Mackenzie Delta Area,'' Canada, National Research Council, Associate Committee on Soil and Snow Mechanics, *Technical Memo. 76,* pp. 27–83.

MACKAY, J. R. (1963b). ''The Mackenzie Delta Area,'' N.W.T. Department of Mines Technical Survey, Geographic Branch, Ottawa, Ontario, *Memoir 8,* 202 pp.

MACKAY, J. R. (1971). ''The Origin of Massive Icy Beds in Permafrost, Western Arctic Coast Canada,'' *Canadian Journal of Earth Science,* Vol. 8, No. 4, pp. 397–422.

MACKAY, J. R. (1972). ''Offshore Permafrost and Ground Ice, Southern Beaufort Sea, Canada,'' *Canadian Journal of Earth Sciences,* Vol. 9, No. 11, pp. 1550–1561.

MACKAY, J. R. (1973). ''A Frost Tube for the Determination of Freezing in the Active Layer above Permafrost,'' *Canadian Geotechnical Journal,* Vol. 10, No. 3, pp. 392–396.

MACKAY, J. R. (1974). ''Measurement of Upward Freezing above Permafrost with a Self-Positioning Thermistor Probe,'' *Geological Survey of Canada, Paper 74–1,* Part B, pp. 250–251.

MACKAY, J. R. (1976). ''Ice Segregation at Depth in Permafrost,'' *Geological Survey of Canada, Paper 76–1A,* pp. 287–288.

MACKAY, J. R., and BLACK, R. F. (1973). ''Origin, Composition, and Structure of Perennially Frozen Ground and Ground Ice: A Review,'' North American Contribution, *Proceedings of the Second International Conference on Permafrost,* Yakutsk, USSR, National Academy of Sciences, Washington, D.C., pp. 185–192.

MACKAY, J. R., and STAGER, J. R. (1966). ''Thick Titled Beds of Segregated Ice, Mackenzie Delta Area, N.W.T.,'' *Biuletyn Peryglacjalny,* Vol. 15, pp. 39–43.

MATLOCK, H., and REESE, L. C. (1962). ''Generalized Solutions for Laterally Loaded Piles,'' Translation American Society of Civil Enginerrs, Paper 3370, Vol. 27.

MCFADDEN, T., and COLLINS, C. (1978). ''Case Study of a Water Supply for Coastal Villages Surrounded by Salt Water,'' *Proceedings of the Conference on Applied Techniques for Cold Environments,* Anchorage, Alaska, ASCE Cold Regions Specialty Conference, May 17–19, Vol. II, pp. 1029–1040.

MCPHAIL, J. F., MCMULLEN, W. B., and MURFITT, A. W. (1976). ''Yukon River to Prudhoe Bay: Lessons in Arctic Design and Construction,'' *Civil Engineering (New York),* Vol. 46, No. 2, pp. 78–82.

MCROBERTS, E. C. (1972). ''Discussion,'' *Proceedings of the Canadian Northern Pipeline Research Conference,* Ottawa, Canada, National Research Council, Technical Memo. 104, pp. 291–295.

MCROBERTS, E. C. (1973). ''Stability of Slopes in Permafrost,'' Ph.D. thesis, University of Alberta, Department of Civil Engineering, Edmonton, Alberta.

MCROBERTS, E. C. (1975). ''Some Aspects of a Simple Secondary Creep Model for Deformations in Permafrost Slopes," *Canadian Geotechnical Journal,* Vol. 12, pp. 98–105.

McRoberts, E. C., and Morgenstern, N. R. (1973). "A Study of Landslides in the Vicinity of the Mackenzie River, Mile 205 to 660," Canada Environmental–Social Program, Northern Pipelines, Task Force, Northern Oil Development.

McRoberts, E. C., and Morgenstern, N. R. (1974a). "The Stability of Thawing Slopes," *Canadian Geotechnical Journal,* Vol. 11, No. 4, pp. 447–469.

McRoberts, E. C., and Morgenstern, N. R. (1974b). "The Stability of Slopes in Frozen Soil, Mackenzie Valley, N.W.T.," *Canadian Geotechnical Journal,* Vol. 11, No. 4, pp. 554–573.

McRoberts, E. C., and Nixon, J. F. (1977). "Extensions to Thawing Slope Stability Theory," *Proceedings of the Second International Symposium on Cold Regions Engineering* (1976), Fairbanks, University of Alaska, Department of Civil Engineering, pp. 262–276.

Mellor, M. (1971). "Strength and Deformability of Rocks at Low Temperatures," U.S. Army, CRREL, *Research Report 294.*

Mellor, M., and Cole, D. M. (1982). "Deformation and Failure of Ice Under Constant Stress or Constant Strain Rate," *Cold Regions Science and Technology,* Vol. 5, pp. 201–219, Elsevier Science Publishers.

Mellor, M., and Smith, J. H. (1967). "Creep of Ice and Snow," *Physics of Ice and Snow,* ed. H. Oura, International Conference Low Temp. *Science 1,* pp. 843–856.

Mellor, M., and Testa, R. (1969). "Creep of Ice under Low Stresses," *Journal of Glaciology,* Vol. 8, No. 52, pp. 147–152.

Miller, R.D. (1966). "Phase Equilibria and Soil Freezing," *Proceedings of the First International Conference on Permafrost* (1963), Lafayette, Ind., *National Academy of Sciences Publications 1287,* pp. 193–197.

Miller, R. D. (1972). "Freezing and Heaving of Saturated and Unsaturated Soils," *Highway Research Record,* No. 393, pp. 1–11.

Miller, R. D. (1978). "Frost Heaving in Non-Colloidal Soils," *Proceedings of the Third International Conference on Permafrost,* Edmonton, Alberta, National Research Council, Vol. 1, pp. 704–714.

Molochushkin, E. N. (1973). "The Effect of Thermal Abrasion on the Temperature of Permafrost Rocks in the Coastal Zone of the Laptev Sea," Vol. 2, *Abstract of the Second International Conference on Permafrost,* Yakutsk, USSR, Izdatelstvo "Nauka," Moscow.

Morgan, R. R. (1968). "Preparation of a Worldwide VLF Effective Conductivity Map," *Report 80133-F,* Westinghouse Research Corp., Environmental Science and Technology Department, Boulder, Colo.

Morgenstern, N. R. (1981). "Geotechnical Engineering and Frontier Resource Development," *Geotechnique,* Vol. 31, No. 3, pp. 305–365.

Morgenstern, N. R., and Nixon, J. F. (1971). "One-Dimensional Consolidation of Thawing Soils," *Canadian Geotechnical Journal,* Vol. 8, No. 4, pp. 558–565.

Morgenstern, N. R., and Nixon, J. F. (1975). "An Analysis of the Performance of a Warm-Oil Pipeline in Permafrost, Inuvik, N.W.T.," *Canadian Geotechnical Journal,* Vol. 12, No. 2, pp. 199–208.

MORGENSTERN, N. R., and PRICE, E. (1965). "The Analysis of the Stability of General Slip Surfaces," *Geotechnique,* Vol. 15, No. 1, pp. 79–95.

MORGENSTERN, N. R., and Smith, L. B. (1973). "Thaw Consolidation Tests on Remoulded Clays," *Canadian Geotechnical Journal,* Vol. 10, No. 1, pp. 25–40.

MORGENSTERN, N. R., ROGGENSACK, W. D., and WEAVER, J. S. (1980). "The Behavior of Friction Piles in Ice and Ice-Rich Soils," *Canadian Geotechnical Journal,* Vol. 17, pp. 405–415.

MULLER, H. (1966). "Paleohunters in America: Origins and Diffusion," *Science,* Vol. 152, No. 3726, pp. 1191–1200.

NAKAMO, Y., and BROWN, J. (1972). "Mathematical Modelling and Validation of the Thermal Regimes in Tundra Soils," Brown, *Alaska Arctic Alp.,* Res. 4: 19–38.

NATIONAL BUILDING CODE, (1976). *Engineering and Safety Service,* New York.

NATIONAL RESEARCH COUNCIL OF CANADA. (1975). *Hydrological Atlas of Canada,* Plate 32, Permafrost.

NEUBER, H., and WOLTERS, R. (1970). "Mechanical Behavior of Frozen Soils under Triaxial Compression," *Fortschritte in der Geologie von Rheinland und Westfalen* (Krefeld, Germany), Vol. 17, pp. 499–536, Canada, National Research Council, *Translation TT-1902,* 25 pp.

NEWMARK, N. M. (1942). "Influence Charts for Computation of Stresses in Elastic Foundations," University of Illinois Engineering Experimental Station, *Circular 24,* Urbana, Ill., 19 pp.

NICHOLSON, F. H., and GRANBERG, H. B. (1973). "Permafrost and Snow-Cover Relationships, near Scheffervill, Quebec," North American Contribution, *Proceedings of the Second International Conference on Permafrost,* Yakutsk, USSR, National Academy of Sciences, Washington, D.C., pp. 151–158.

NIXON, J. F. (1973). "The Consolidation of Thawing Soil," Ph.D. thesis, University of Alberta, Department of Civil Engineering, Edmonton, Alberta.

NIXON, J. F. (1978). "Geothermal Aspects of Ventilated Pad Design," *Proceedings of the Third International Conference on Permafrost,* Edmonton, Alberta, Canada, National Research Council, Vol. 1, pp. 840–846.

NIXON, J. F., and LADANYI, B. (1978). "Thaw Consolidation," Chap. 4, *Geotechnical Engineering for Cold Regions,* McGraw-Hill Book Co., New York.

NIXON, J. F., and MCROBERTS, E. C. (1973). "A Study of Some Factors Affecting the Thawing of Frozen Soils," *Canadian Geotechnical Journal,* Vol. 10, No. 3, pp. 439–452.

NIXON, J. F., and MCROBERTS, E. C. (1976). "A Design Approach for Pile Foundations in Permafrost," *Canadian Geotechnical Journal,* Vol. 13, No. 1, pp. 40–57.

NIXON, J. F., and MORGENSTERN, N. R. (1973). "Practical Extensions to a Theory of Consolidation for Thawing Soils," North American Contribution, *Proceedings of the Second International Conference on Permafrost,* Yakutsk, USSR, National Academy of Sciences, Washington, D.C., pp. 369–376.

NIXON, J. F., and MORGENSTERN, N. R. (1974). "Thaw Consolidation Tests on Undisturbed Fine-Grained Permafrost," *Canadian Geotechnical Journal,* Vol. 11, No. 1, pp. 202–214.

NOTTINGHAM, D., and CHRISTOPHERSON, A. B. (1983). "Design Criteria for Driven Piles in Permafrost,'' *Report AK-RD-83-19,* Research Section, Division of Planning and Programming, Department of Transportation and Public Facilities, State of Alaska.

ODQUIST, F. K. G. (1966). ''Mathematical Theory of Creep and Creep Rupture,'' *Oxford Mathematical Monograph,* Clarendon Press, Oxford, England, 168 pp.

OGILVY, A. A. (1967). ''Geophysical Studies in Permafrost Regions in USSR,'' Mining Groundwater Geophysic, Geological Survey of Canada, *Economic Geology Report 26,* pp. 641–650.

O'NEILL, K., and MILLER, R. D. (1980). ''Numerical Solutions for Rigid-Ice Model of Secondary Frost Heave,'' *The Second International Symposium on Ground Freezing,* June 24–26, Norwegian Institute of Technology.

OSTERKAMP, T. E., and HARRISON, W. D. (1976a). ''Subsea Permafrost: Its Implications for Offshore Resource Development,'' *The Northern Engineer,* Vol. 8, No. 1, pp. 31–35.

OSTERKAMP, T. E., and HARRISON, W. D. (1976b). ''Subsea Permafrost at Prudhoe Bay, Alaska: Drilling Report and Data Analysis,'' *Report UAG-R-247,* Geophysical Institute, University of Alaska, Fairbanks.

OSTERKAMP, T. E., and HARRISON, W. D. (1980). ''Subsea Permafrost: Probing, Thermal Regime and Data Analysis,'' *Annual Report to the BLM/NOAA,* OCSEAP, Arctic Project Office, University of Alaska, Fairbanks.

OSTERKAMP, T. E., JURICK, R. W., GISLASON, G. A., and AKASOFU, S. I. (1980). ''Electrical Resistivity Measurements in Permafrost Terrain at the Engineer Creek Road Cut,'' Fairbanks, Alaska, *Cold Regions Science and Technology,* Vol. 3, pp. 277–286.

OUTCALT, S. I. (1980). ''A Simple Energy Balance Model of Ice Segregation,'' *Cold Regions Science and Technology,* Vol. 3, No. 2 and 3, pp. 145–152.

PARAMESWARAN, V. R. (1981). ''Adfreeze Strength of Model Piles in Ice,'' *Canadian Geotechnical Journal,* Vol. 18, No. 1, pp. 8–16.

PARKHOMENKO, E. I. (1967). *Electrical Properties of Rock* (English translation), Pergamon Press, Elmsford, N.Y.

PENNER, E. (1960). ''The Importance of Freezing Rate in Frost Action in Soils,'' *ASTM Proceedings,* Vol. 60, pp. 1151–1165.

PENNER, E. (1966). ''Pressures Developed during Unidirectional Freezing of Water-Saturated Porous Materials,'' *Proceedings of the International Conference on Low Temperature Science,* Sapporo, Japan, Vol. 1, No. 2, pp. 1401–1412.

PENNER, E. (1967). ''Heaving Pressures in Soil during Unidirectional Freezing,'' *Canadian Geotechnical Journal,* Vol. 4, No. 4, pp. 398–408.

PENNER, E. (1972). ''Influence of Freezing Rate on Frost Heaving,'' U.S. Highway Research Board, *Record 393,* pp. 56–64.

PENNER, E. (1974). ''Uplift Forces on Foundations in Frost Heaving Soils,'' *Canadian Geotechnical Journal,* Vol. 11, No. 3, pp. 323–338.

PENNER, E. (1977). ''Fundamental Aspects of Frost Action,'' *Proceedings, International Symposium on Frost Action in Soils,* University of Lulea, Sweden, Vol. 2, pp. 17–28.

PENNER, E. (1982). ''Aspects of Ice Lens Formation,'' *Proceedings of the Third International Symposium on Ground Freezing,* June 22–24, U.S. Army Corps. of Engineers, Cold Regions Research and Engineering Laboratory, Hanover, New Hampshire.

PENNER, E. and GOLD, L. W. (1971). ''Transfer of Heaving Forces by Adfreezing to

Columns and Foundation Walls in Frost-Susceptible Soils,'' *Canadian Geotechnical Journal,* Vol. 8, No. 4.

PENNER, E., JOHNSTON, G. H., and GOODRICH, L. E. (1975). ''Thermal Conductivity Laboratory Studies of Some Mackenzie Highway Soils,'' *Canadian Geotechnical Journal,* Vol. 12, No. 3, pp. 271–288.

PENNER, E. and UEDA, T. (1979). ''Effects of Temperature and Pressure on Frost Heaving,'' *Eng. Geol.,* Vol. 13, No. 1–4, pp. 29–39.

PÉWÉ, T. L. (1966). ''Ice-Wedges in Alaska: Classification, Distribution, and Climatic Significance,'' *Proceedings of the First International Conference on Permafrost* (1963), Lafayette, Ind., National Academy of Sciences Publication 1287, pp. 76–81.

PÉWÉ, T. L. (1982). ''Geologic Hazards of the Fairbanks Area, Alaska,'' *Special Report 15,* published by Division of Geological and Geophysical Surveys, State of Alaska, 109 pp.

PHUKAN, A. (1976). ''Simplified Approach to Slope Stability Analysis in Thawing Soil,'' *Proceedings of the Second International Symposium on Cold Regions Engineering,* Fairbanks, Alaska.

PHUKAN, A. (1977). ''Pile Foundation in Frozen Soils,'' *ASME Energy Technology Conference,* Houston, Tex.

PHUKAN, A. (1979a). ''Geotechnical Engineering Applications to the Chilled Gas Pipeline Design in Cold Regions,'' *Proceedings of the Specialty Conference on Pipelines in Adverse Environments,* ASCE, New Orleans, January 15–17.

PHUKAN, A. (1979b). ''Design of Foundations for Buildings in Discontinuous Permafrost,'' *U.S.-USSR Joint Seminar on Buildings in Cold Climates and on Permafrost,* Leningrad, June 25–28.

PHUKAN, A. (1980a). ''Design of Deep Foundations in Discontinuous Permafrost,'' *ASCE Spring Convention,* Technical Session on Deep Foundations, Portland, Ore., April.

PHUKAN, A. (1980b). ''Strength of Frozen Fine-Grained Soils at Warm Temperatures,'' *Second International Conference on Ground Freezing,* Trondheim, Norway, June.

PHUKAN, A. (1981a). ''The Behavior of Ice-Rich Fine-Grained Soils,'' *Permafrost Conference,* Calgary, Alberta, March.

PHUKAN, A. (1981b). ''Shallow Foundations in Continuous Permafrost,'' *ASCE Specialty Conference,* Seattle, Wash., April.

PHUKAN, A. (1981c). ''On-Shore Foundations in Arctic Conditions,'' *Arctic Technology Conference,* Ny Ålesund, Norway.

PHUKAN, A. (1981d). ''A Literature Search for Substitute Materials in Frost Protecting Layers,'' *State Project No. F 16972,* State of Alaska, Dept. of Trans. and Pub. Facilities, Div. of Plan. and Prog., Research Section, Fairbanks, 53 pp.

PHUKAN, A. (1982). ''Excavation Resistance of Artificially Frozen Soils,'' *Third Ground Freezing Conference,* Hanover, N.H., June.

PHUKAN, A., et al. (1978). ''Self-Refrigerated Gravel Pad Foundations in Frozen Soils,'' *Proceedings of the Conference on Applied Techniques for Cold Environments,* Anchorage, Alaska, ASCE Cold Regions Specialty Conference, May 17–19.

PIHLAINEN, J. A. (1959). ''Pile Construction in Permafrost,'' *Journal of the Soil Mechanics and Foundations Division, ASCE,* Vol. 85, No. SM6, Part I, pp. 75–95.

PIHLAINEN, J. A., and JOHNSTON, G. H. (1954). "Permafrost Investigations at Aklavik (Drill and Sampling) 1953," Canada, National Research Council, Division of Building Research, *Technical Paper 16,* 47 pp.

POPPE, V. (1976). "Basic Methods for Predicting Thermal Stresses and Deformations in Frozen Soils," Canada, National Research Council, *Technical Translation NRC/CNR TT-1886,* 52 pp.

PORKHAEV, G. V., TERGULYAN, Y. O., and KOELSOV, A. A. (1977). "Driving Piles in Permafrost with Hole Sinking by Steam Vibroleader," translated for *Osnovaniya, Fundamenty i Mekhanika Gruntov,* No. 3, pp. 12–14.

POULOS, H. G., and DAVIS, E. H. (1980). *Pile Foundation Analysis and Design,* John Wiley and Sons, Inc., New York, 397 pp.

PRANDTL, L. (1921). "Uber die Eindringungsfestigkeit (Harte) Plastischer Baustoffe und die Festigkeit Von Schneiden," *Zeitschrift fuer Angewandte Mathematik und Mechanik,* Vol. 1, No. 1, pp. 15–20.

PUFAHL, D. E., and MORGENSTERN, N. R. (1979). "Stabilization of Planar Landslides in Permafrost," *Canadian Geotechnical Journal,* Vol. 16, No. 4, pp. 734–747.

PUFAHL, D. E., and MORGENSTERN, N. R. (1980). "Remedial Measures for Slope Instability in Thawing Permafrost," *Second International Symposium on Ground Freezing,* Trondheim, Norway, June, pp. 1089–1101.

PUFAHL, D. E., MORGENSTERN, N. R., and ROGGENSACK, W. D. (1974). "Observations on Recent Highway Cuts in Permafrost," Canada, Department of Indian and Northern Affairs, Environmental–Social Program, Northern Pipelines, *Report 74–32,* 53 pp.

REED, R. E. (1966). "Refrigeration of a Pipe Pile by Air Circulation," U.S. Army, CRREL, *Technical Report 156,* 19 pp.

REED, S. C. (1977). "Water Supply in Arctic Regions," *Journal of the New England Water Works Association,* Vol. 84, No. 4, pp. 25–35.

REISSNER, H. (1924). "Zum Erddruckproblem," *Proceedings of the First International Congress on Applied Mechanics,* Delft, pp. 295–311.

RENDULIC, L. (1937). "A Fundamental Law of Clay Mechanics and Its Experimental Proof," (In German), Bauingenier 18, pp. 459–467.

RESEARCH INSTITUTE OF GLACIOLOGY (1975). Cryopedology and Desert Research, Academia Sinica, Lanchou, China; Canada, National Research Council, *Technical Translation NRC/CNR TT-2006,* Ottawa, 1981.

RICKARD, W., and BROWN, J. (1972). "The Performance of a Frost-Tube for the Determination of Soil Freezing and Thawing Depth," *Soil Science,* Vol. 113, No. 2, pp. 149–154.

ROBIN, G. de Q. (1972). "Polar Ice Sheets: A Review," *Polar Record,* No. 100, pp. 5–22.

ROGGENSACK, W. D. (1977). "Geotechnical Properties of Fine-Grained Permafrost Soils," Ph.D. thesis, University of Alberta, Department of Civil Engineering, Edmonton, Alberta, 423 pp.

ROGGENSACK, W. D. (1979). "Techniques for Core Drilling in Frozen Soils," *Proceedings of the Symposium on Permafrost Field Methods and Permafrost Geophysics* (October 1977), Saskatoon, Saskatchewan, Canada, National Research Council, Associate Committee on Geotechnical Research, Technical Memo. 24, pp. 14–24.

Roggensack, W. D., and Morgenstern, N. R. (1978). "Direct Shear Test on Natural Fine-Grained Permafrost Soil," *Proceedings of the Third International Conference on Permafrost,* Edmonton, Alberta, National Research Council, Vol. 1, pp. 728–735.

Rogness, D. R., and Ryan, W. L. (1977). "Vacuum Sewer System," *Indian and Northern Affairs Publication QS-1546-000-EE-A,* Information Canada, September.

Romkens, M. J. M., and Miller, R. D. (1973). "Migration of Mineral Particles in Ice with a Temperature Gradient," *Journal of Colloid Interface Sciences,* Vol. 42, pp. 103–111.

Ruedrich, R. A., and Perkins, T. K. (1973). "A Study of Factors Which Influence the Mechanical Properties of Deep Permafrost," *Paper No. SPE-4587,* 48th Annual Fall Meeting, Society of Petroleum Engineering, A.I.M.E., Las Vegas, Nevada, 15 pp.

Ryan, W. (1977). "Design Guidelines for Piping Systems," *Utilities Delivery in Arctic Regions, Environmental Protection Service, Ottawa, Ontario, EPS 3-WP-77-1.*

Ryan, W. L., and Rogness, D. R. (1977). "Pressure Sewage Collection Systems in the Arctic," *Proceedings of the Symposium on Utilities Delivery in Arctic Regions,* Edmonton, Alberta, Environment Canada, Environmental Protection Service, Report EPS-3-WP-77-1, pp. 523–552.

Saetersdal, R. (1976). "Prevention of Damage from Frost Action in Soils," *Frost i Jord,* No. 17, Oslo.

Sanger, F. J. (1968). "Ground Freezing in Construction," *Journal of the Soil Mechanics and Foundations Division, ASCE,* Vol. 94, No. SM1, pp. 131–158.

Sanger, F. J. (1969). "Foundation of Structures in Cold Regions," U.S. Army, CRREL, *Monograph MIII-C4,* 91 pp.

Savigny, K. W. (1980). "In Situ Analysis of Naturally Occurring Creep in Ice-Rich Permafrost Soil," Ph.D. thesis, University of Alberta, Edmonton.

Sayles, F. H. (1968). "Creep of Frozen Sands," U.S. Army, CRREL, *Technical Report 190.*

Sayles, F. H. (1973). "Triaxial and Creep Tests on Frozen Ottawa Sand," Northern American Contribution, *Proceedings of the Second International Conference on Permafrost,* Yakutsk, USSR, National Academy of Sciences, Washington, D.C.

Sayles, F. H., and Epanchin, N. V. (1966). "Rate of Strain Compression Tests on Frozen Ottawa Sand and Ice," U.S. Army, CRREL, *Technical Note,* Hanover, N.H.

Sayles, F. H., and Haynes, D. (1974). "Creep of Frozen Silt and Clay," U.S. Army, CRREL, *Technical Note 252,* Hanover, N.H.

Scott, R. F. (1969). "The Freezing Process and Mechanics of Frozen Ground," U.S. Army, CRREL, *Monograph II-DI.*

Seguin, M. K., and Garg, O. P. (1974). "Delineation of Frozen Rocks from the Labrador–Ungava Peninsula Using Borehole Geophysical Logging," *Proceedings of the Ninth Canadian Rock Mechanics Symposium,* Montreal, Information Canada, pp. 52–75.

Sellman, P. V., and Brown, J. (1965). "Coring of Frozen Ground, Barrow, Alaska," U.S. Army, CRREL, *Special Report 81,* 8 pp.

Shearer, J. M. (1972). "Beaufort Sea, East of Mackenzie Bay, Submarine Pingo-like Features," *Ice,* No. 38, p. 6.

Sigafoos, R. S., and Hopkins, D. M. (1952). "Soil Instability on Slopes in Regions of

Perennially Frozen Ground (Alaska)," U.S. Highway Research Board, *Special Report 2,* pp. 176–192.

SLAUGHTER, C. W., MELLOR, M., SELLMANN, P. V., BROWN, J., and BROWN, L. (1975). "Accumulating Snow to Augment the Fresh Water Supply at Barrow, Alaska," U.S. Army, CRREL, *Special Report 217,* Hanover, N.H., January.

SLUSARCHUK, W. A. and FOULGER, P. H. (1973). "Development and Calibration of a Thermal Conductivity Probe Apparatus for Use in the Field and Laboratory," Canada, *National Research Council, Division Building Research, Tech. Paper 388,* 18 pp.

SLUSARCHUK, W. A., and WATSON, G. H. (1975). "Thermal Conductivity of Some Ice-Rich Permafrost Soils," *Canadian Geotechnical Journal,* Vol. 12, No. 3, pp. 413–424.

SMITH, D. W., and GIVEN, P. W. (1977). "Evaluation of Northern Extended Aeration Sewage Treatment Plants," *Proceedings of the Second International Symposium on Cold Regions Engineering,* University of Alaska, Fairbanks, pp. 291–316.

SMITH, D. W., and JUSTICE, S. R. (1975). "Effects of Reservoir Clearing on Water Quality in the Arctic and Subarctic," *Report IWR-58,* Institute of Water Resources, University of Alaska, Fairbanks.

SMITH, D. W., and JUSTICE, S. R. (1976). "Clearing Alaskan Water Supply Impoundments Management Laboratory Study and Literature Review," *Report IWR-67,* Institute of Water Resources, University of Alaska, Fairbanks.

SMITH, M. W., and HWANG, C. T. (1973). "Thermal Disturbance Due to Channel Shifting, Mackenzie Delta, N.W.T., Canada," North American Contribution, *Proceedings of the Second International Conference on Permafrost,* Yakutsk, USSR, National Academy of Sciences, Washington, D.C., pp. 51–60.

SMITH, N. (1979). "Construction and Performance of Membrane Encapsulated Soil Layer in Alaska," U.S. Army, CRREL, *Special Report 79–16,* June.

SMITH, N., and BERG, R. (1973). "Encountering Massive Ground Ice during Road Construction in Central Alaska," North American Contribution, *Proceedings of the Second International Conference on Permafrost,* Yakutsk, USSR, National Academy of Sciences, Washington, D.C., pp. 730–735.

SPEER, T. L., WATSON, G. H., and ROWLEY, R. K. (1973). "Effects of Ground-Ice Variability and Resulting Thaw Settlements on Buried Warm-Oil Pipelines," North American Contribution, *Proceedings of the Second International Conference on Permafrost,* Yakutsk, USSR, National Academy of Sciences, Washington, D.C., pp. 746–751.

STANLEY, D. R. (1965). "Water and Sewage Problems in Discontinuous Permafrost Regions," *Proceedings of the Canadian Regional Permafrost Conference,* Edmonton, Alberta, Canada, National Research Council, Associate Committee on Soil and Snow Mechanics, Technical Memo. 86, pp. 93–105.

STEARNS, S. R. (1966). "Permafrost," U.S. Army, CRREL, Part I, Sec. A2 77 pp.

STEVENS, H. W. (1975). "The Response of Frozen Soils to Vibratory Loads," U.S. Army, CRREL, *Technical Report 265,* 103 pp.

SUTHERLAND, H. B., and GASKIN, P. N. (1973). "A Comparison of the TRRL and CRREL Tests for Frost Susceptibility of Soils," *Canadian Geotechnical Journal,* Vol. 10, No. 3, pp. 553–557.

SVDENSEN, S. D. (1972). "New Ways of Small House Foundation," *Frost i Jord,* No. 8, Oslo.

SYKES, D. J. (1971). "Effects of Fire and Fire Control on Soil and Water Relations in Northern Forests—A Preliminary Review," *Proceedings of the Symposium on Fire in the Northern Environment,* College (Fairbanks), Alaska, pp. 37–44.

TABER, S. (1916). "The Growth of Crystals under External Pressure," *American Journal of Science,* 4th Series, Vol. 41, pp. 532–556.

TABER, S. (1929). "Frost Heaving," *Journal of Geology,* Vol. 37, No. 1, pp. 428–461.

TABER, S. (1930). "The Mechanics of Frost Heaving," *Journal of Geology,* Vol. 38, pp. 303–317.

TABER, S. (1943). "Perennially Frozen Ground in Alaska: Its Origin and History," *Geological Society of America Bulletin,* Vol. 54, pp. 1433–1548.

TAYLOR, A. E., and JUDGE, A. S. (1974). "Canadian Geothermal Data Collection: Northern Wells 1955 to February 1974," Canada, Department of Energy, Mines and Resources, Earth Physics Branch, *Geothermal Series 1,* 171 pp.

TAYLOR, A. E., and JUDGE, A. S. (1975). "Canadian Goethermal Data Collection: Northern Wells 1974," Canada, Department of Energy, Mines and Resources, Earth Physics Branch, *Geothermal Series 3,* 127 pp.

TAYLOR, A. E., and JUDGE, A. S. (1976). "Canadian Geothermal Data Collection: Northern Wells 1975," Canada, Department of Energy, Mines and Resources, Earth Physics Branch, *Geothermal Series 6,* 142 pp.

TERZAGHI, K. (1943). *Theoretical Soil Mechanics,* Wiley, New York, 510 pp.

TERZAGHI, K., and PECK, R. B., (1967). *Soil Mechanics in Engineering Practice,* Wiley, New York.

TOBIASSON, W. (1973). "Performance of the Thule Hanger Soil Cooling System," North American Contribution, *Proceedings of the Second International Conference on Permafrost,* Yakutsk, USSR, National Academy of Sciences, Washington, D.C., pp. 752–758.

TOBIASSON, W. (1978). "Construction on Permafrost at Longyearbyen on Spitzbergen," *Proceedings of the Third International Conference on Permafrost,* Edmonton, Alberta, National Research Council, Vol. 1, pp. 888–890.

TRANSPORTATION RESEARCH BOARD (1974). *Roadway Design in Seasonal Frost Areas,* National Cooperative Highway Research Program, National Research Council, Washington, D.C.

TSYTOVICH, N. A. (1975). *The Mechanics of Frozen Ground,* McGraw-Hill, New York.

TSYTOVICH, N. A., KRONIK, YA. A., MARKIN, K. F., AKSENOV, V. I., and SAMUEL'SON, M. V. (1978). "Physical and Mechanical Properties of Saline Soils," USSR Contribution, *Proceedings of the Second International Conference on Permafrost* (1973), Yakutsk, USSR, National Academy of Sciences, Washington, D.C., pp. 238–247.

U.S. ARMY, (1962). "Roads, Streets, Walks and Open Storage Areas," *Flexible Pavement Design Engineering Manual, EM 1110-345-291.*

U.S. ARMY (1975). "Roadway Design in Seasonal Frost Areas," CRREL, *Report 259.*

U.S. ARMY/AIR FORCE (1965). "Arctic and Subarctic Construction: Surface Drainage Design for Airfields and Heliports in Arctic and Subarctic Regions," *Technical Manual TM5-852-7/AFM 88–19,* Chap. 17.

U.S. ARMY/AIR FORCE (1966). "Arctic and Subarctic Construction: Calculation Methods for Determination of Depths of Freeze and Thaw in Soils," *Technical Manual TM5-852-6/AFM 88–19,* Chap. 6.

U.S. ARMY, CORPS OF ENGINEERS (1965). "Soils and Geology—Pavement Design for Frost Conditions," Department of the Army, *Technical Manual TM5–818–2*.

U.S. ARMY, CORPS OF ENGINEERS, WATERWAYS EXPERIMENT STATION, (1953). "The Unified Soil Classification System," *Technical Memo. 3–357,* p. 3. Appendix B, Characteristics of Soil Groups Pertaining to Roads and Airfields.

U.S. ARMY, CORPS OF ENGINEERS, WATERWAYS EXPERIMENT STATION (1960). "The Unified Soil Classification System," *Technical Memo. 3–357,* Appendix A: Characteristics of Soil Groups Pertaining to Embankments and Foundations.

USSR (1960). "Technical Considerations in Designing Foundations in Permafrost (SN 91–60)," Canada, National Research Council, *Technical Translation TT-1033,* 64 pp.

USSR (1973). "Instructions for the Design of Bases and Foundations on Permafrost Soils Having High Ice and Salt Contents (*SN 450–72*), Gosstroi, Moscow, 25 pp. (in Russian.)

VANYAN, L. L. (1965). *Electromagnetic Depth Soundings* (English translation), Consultants Bureau, New York.

VEILLETTE, J. (1975). "Modified CRREL Ice Coring Augers," *Geological Survey of Canada, Paper 75–1,* Part A, pp. 425–426.

VESIC, A. S. (1970). "Research on Bearing Capacity of Soils," (unpublished).

VIERECK, L. A. (1973). "Ecological Effects of River Flooding and Forest Fires on Permafrost in the Taiga of Alaska," North American Contribution, *Proceedings of the Second International Conference on Permafrost,* Yakutsk, USSR, National Academy of Sciences, Washington, D.C., pp. 60–67.

VINSON, T. S. (1977). "Parameter Effects on Dynamic Properties of Frozen Soils," *ASCE Symposium on Application of Soils Dynamics in Cold Regions,* pp. 112–140.

VINSON, T. S. (1978). "Response of Frozen Ground to Dynamic Loading," in *Geotechnical Engineering for Cold Regions* (O. B. Andersland and D. M. Anderson, eds.), McGraw-Hill, New York, pp. 405–458.

VYALOV, S. S. (1959). "Rheological Properties and Bearings Capacity of Soils," U.S. Army, CRREL, *Translation 74* (1965), 237 pp.

VYALOV, S. S. (1962). "The Strength and Creep of Frozen Soils and Calculations for Ice–Soil Retaining Structures," U.S. Army, CRREL, *Translation 76* (1965), 321 pp.

VYALOV, S. S. (1973). "Interaction of Permafrost and Structures (in Russian), *Proceedings of the Third International Conference on Permafrost,* Edmonton, Canada, Vol. 2, pp. 117–136.

VYALOV, S. S., et al. (1966a). "Methods of Determining Creep, Long-Term Strength and Compressibility Characteristics of Frozen Soils," Canada, National Research Council, *Technical Translation TT-1364* (1969), 109 pp.

VYALOV, S. S. and SUSHERINE, YE. P. (1970). "Resistance of Frozen Soils to Triaxial Compression," translation by the U.S. Army Foreign Science and Technology Center, *NTIS No. AD 713981*.

VYALOV, S. S., TARGULYAN, Y. O., and VISOTSKIY, D. P. (1966b). "Interaction of Frozen Soil with Piles and Pipes during Vibratory Driving," Proceedings of the Eighth USSR Scientific Conference on Geocryology (Permafrost), Yakutsk, USSR, *Technical Translation FSTC-HT-23-944-68,* by Techtran Corp., Defense, Vol. 5, pp. 224–234.

WASHBURN, A. L. (1947). "Reconnaissance Geology of Portions of Victoria Island and Adjacent Regions," *Geological Society of America, Memo. 22*.

WASHBURN, A. L. (1956). "Classification of Patterned Ground and Review of Suggested Origins," *Geological Society of America Bulletin,* Vol. 67, pp. 823–866.

WASHBURN, A. L. (1977). *Periglacial Processes and Environments,* Edward Arnold, London, 320 pp.

WATERS, E. D. (1973). "Stabilization of Soils and Structures by Passive Heat Transfer Devices," *74th National Meeting,* March, American Institution of Chemical Engineers, New Orleans, LA, 21 pp.

WATSON, G. H., ROWLEY, R. K., and SLUSARCHUK, W. A. (1973a). "Performance of a Warm-Oil Pipeline Buried in Permafrost," North American Contribution, *Proceedings of the Second International Conference on Permafrost,* Yakutsk, USSR, U.S. National Academy of Sciences, Washington, D.C., pp. 759–766.

WATSON, G. H., SLUSARCHUK, W. A., and ROWLEY, R. K. (1973b). "Determination of Some Frozen and Thawed Properties of Permafrost Soils," *Canadian Geotechnical Journal,* Vol. 10, No. 4, pp. 592–606.

WILLIAMS, P. J. (1964). "Unfrozen Water Content of Frozen Soils and Soil Moisture Suction," *Geotechnique,* 14(3), pp. 213–246.

WILLIAMS, P. J. (1966). "Pore Pressures at a Penetrating Frost Line and Their Prediction," *Geotechnique* 16, pp. 187–208.

WILLIAMS, P. J. (1967). "Properties and Behaviour of Freezing Soils," Norwegian Geotechnical Institute, *Publication 72.*

WILLIAMS, P. J. (1977). "Thermodynamic Conditions for Ice Accumulation in Freezing Soils," *Proceedings of the International Symposium on Frost Action in Soils,* University of Lulea, Sweden, Vol. 1, pp. 42–53.

WYDER, J. E., HUNTER, J. A., and RAMPTON, V. N. (1972). "Geophysical Investigations of Surficial deposits of Tuktoyaktuk, N.W.T.," *Geol. Survey Canada, Open File 127.*

YODER, E. J., and WITCVZAK, M. W. (1975). *Principles of Pavement Design,* 2nd ed., Wiley, New York.

ZOLTAI, S. C. (1971). "Southern Limit of Permafrost Features in Peat Landforms, Manitoba and Saskatchewan," *Geological Association of Canada, Special Paper No. 9,* pp. 305–310.

# Index